T0348486

THE BIOGEOGRAPHY
OF THE AUSTRALIAN
NORTH WEST SHELF

THE BIOGEOGRAPHY OF THE AUSTRALIAN NORTH WEST SHELF

Environmental Change and Life's Response

By

BARRY WILSON
Western Australian Museum
Perth, 6000

AMSTERDAM • BOSTON • HEIDELBERG • LONDON • NEW YORK • OXFORD
PARIS • SAN DIEGO • SAN FRANCISCO • SINGAPORE • SYDNEY • TOKYO

Elsevier
30 Corporate Drive, Suite 400, Burlington, MA 01803, USA
525 B Street, Suite 1800, San Diego, CA 92101-4495, USA

First edition **2013**

Notice
No responsibility is assumed by the publisher for any injury and/or damage to persons or property as a matter of products liability, negligence or otherwise, or from any use or operation of any methods, products, instructions or ideas contained in the material herein. Because of rapid advances in the medical sciences, in particular, independent verification of diagnoses and drug dosages should be made.

Library of Congress Cataloging-in-Publication Data
Wilson, Barry Robert, 1935-
  The biogeography of the Australian north west shelf : environmental change and life's response / by Barry Wilson, Western Australian Museum, Perth. First edition.
    pages cm
  Includes bibliographical references and index.
  ISBN 978-0-12-409516-8
1. Continental shelf–Australia. 2. Biogeography–Australia. 3. Marine ecology–Australia.
4. Marine geology–Australia. 5. Australia–Environmental conditions. I. Title.
  GC85.2.A8W55 2013
  577.7'574–dc23

2013012380

British Library Cataloguing in Publication Data
A catalogue record for this book is available from the British Library

For information on all **Elsevier** publications
visit our web site at store.elsevier.com

ISBN: 978-0-12-409516-8

Working together
to grow libraries in
developing countries

www.elsevier.com • www.bookaid.org

Contents

6. Habitats and Biotic Assemblages of Intertidal Sandy and Muddy Shores

7. Benthic Shelf and Slope Habitats

8. Patterns of Life and the Processes That Produce Them

9. An Overview of the Historical Biogeography of the North West Shelf

Foreword

The Australian North West Shelf is a part of the Australian continental margin that is rich in both natural resources and marine biodiversity. This book brings a better understanding of the origins of the region's biodiversity, the habitats that support it, and the processes that sustain contemporary ecosystems in a context of ongoing environmental change. Its theme, the biogeography of the marine fauna of the region, relates directly to the primary environmental protection objectives—protection of biodiversity and the maintenance of ecosystem functions—that lie at the core of sustainable development programs underpinning the current rapid growth of industry on the North West Shelf and its coastline.

Until recently, the North West Shelf has been remote from centers of scientific research and its marine environment and biota are still poorly described. This lack of knowledge is an impediment to identification of regional conservation values and development of effective area and species management programs. It also inhibits assessment of environmental impacts of industry in the region and the implementation of environmental management programs.

Through the work of the Western Australian Museum, the Department of Environment and Conservation, and other agencies, the State Government has supported extensive biological surveys in the region. So too has the Commonwealth Government through the work of the Australian Institute of Marine Science and the Commonwealth Scientific and Industrial Research Organisation. Both Governments are currently investing heavily in marine scientific research at Ningaloo Reef and in the Kimberley, both regions with extremely high social and natural values as well as natural resources. In recent years, industry itself has undertaken environmental surveys to guide essential site selection and environmental management programs. As a result of this collaborative work, there have been significant recent advances in development of knowledge of the North West Shelf marine environment and the enormous marine biodiversity of the region has been recognized, but until now there has been no attempt to gather this information together.

This book is a valuable collation of what is presently known about the habitats and biota of the region and the historical and contemporary forces that determine their distributions. For example, the extent, geomorphic variety, and biodiversity of coral reefs in the region, varying from the slope atolls of the shelf margin to the species-rich fringing reefs in the coastal waters of the Kimberley and Pilbara coasts, are described for the first time. The geological history and contemporary oceanography of the region are reviewed, providing a basis for discussion of habitats and ecosystems present. And the levels of biodiversity of key invertebrate groups are assessed in terms of the evolutionary processes that have been responsible for the development of this region as one of the world's hotspots of marine life.

The book's author, Dr Barry Wilson, has academic training in zoology and geology and postdoctoral experience as a Research Fellow in the School of Evolutionary Biology at Harvard University. He has a lifetime of field experience and highly regarded scientific research in the region and is skilled in communicating those findings and their significance. He also is credited as being the architect of the present marine parks and reserves system in Western Australia, having contributed much to the development of marine conservation policy and led the establishment of marine reserves through his past roles as Director of Nature Conservation and inaugural Chairman of the Western Australian Marine Parks and Reserves Authority. Two of the State's first marine parks, Ningaloo and Shark Bay, which were created under Barry's term as Director of Nature Conservation, have been recognized as areas of outstanding universal significance through their inclusion in World Heritage listings.

With this background, Barry has brought a unique range of experience, skills, and insights to the production of this book. It is a timely and vital guide for wise management of the Australian North West Shelf as well as a major contribution to knowledge of Australia's natural history and heritage.

Keiran McNamara
Director General
Department of Environment and Conservation
21 March, 2013

Acknowledgments

This book was made possible by a generous grant from Chevron Australia Pty Ltd, administered by consulting firm URS Australia. It is a development of a report to the Wheatstone LNG Project in the west Pilbara. Chevron and URS staff also provided support with preparation of illustrations. In particular, I am indebted to Ceri Morgan, David Swatten, Celia Kvalsvik, Mathew Ludovico, and of the Wheatstone Central Environmental Team, and Ian Le Provost, Bob Anderson, Ian Baxter, John Davies, Damian Ogburn, Anthony Bougher, Jess Bourner, and of URS for their assistance.

The project was also supported through collaboration with the Western Australian Department of Environment and Conservation (DEC) and access to wonderful illustrative material made available by Landgate. Mark Sheridan and Michael Higgins of the DEC Marine Policy and Planning Branch prepared some of the illustrations. DEC research staff and regional staff in the Pilbara and Kimberley also provided guidance and assistance associated with the work of the Marine Parks and Reserves Authority in developing the Western Australian marine conservation reserves system.

Access to the library, library services, and natural history collections of the Western Australian Museum has been vital to this project. My colleagues among the scientific staff of the museum also have played a significant role in its conclusion. I acknowledge in particular the contributions made by Loisette Marsh who has been a source of knowledge and inspiration for very many years. Her interests in the invertebrate fauna of the North West Shelf have paralleled my own, and my ideas about biogeography of the region owe much to her input. Others who have assisted me in significant ways are Clay Bryce, Jane Fromont, Andrew Hosie, Diana Jones, Hugh Morrison, Zoe Richards, and Shirley Slack-Smith.

Recent work in the Kimberley that has led to the descriptions of the extraordinary coral reefs of that bioregion has been supported by two organizations. The Western Australian Marine Science Institution sponsored two surveys of coral reefs in the southern Kimberley and I acknowledge the interest and assistance of Steve Blake. I was also fortunate to participate in a series of field surveys in the northern Kimberley carried out by consulting firm RPS for INPEX Browse Ltd. I am grateful to INPEX and the company's Environmental Advisor Sjaak Lemmens for permission to use information and illustrations obtained during those trips and to fellow members of the RPS team who carried out the work, especially Jeremy Fitzpatrick, Mike Forde, John Huisman, and Natalie Rosser.

Participation in field surveys and environmental review studies for several other industrial development projects in the region has also contributed to this project, notably, Barrow and other west Pilbara islands (Chevron—Gorgon project), Ashburton delta and Onslow coast (Wheatstone project), Anketell Point port development (API West Pilbara Iron Ore Project), James Price Point LNG project (Woodside Energy Ltd), Scott Reef Browse development (Woodside Energy Ltd), and Irvine Island iron ore development (Pluton Resources Ltd).

I am also indebted to many scientists who have freely given advice and information: Lindsay Collins, Curtin University (coral reef geology); Ian Eliot (coastal geomorphology); Andrew Heywood, Terry Done, and Jim Underwood, AIMS (coral reef biology); Charles Sheppard, University of Warwick (coral reef biology); Peter Harris and Kristin Glen, GeoScience Australia (marine geology); Scott Condie, CSIRO; and Nick D'Adamo, Perth Regional Program, Intergovernmental Oceanographic Commission (oceanography).

Charles Sheppard (University of Warwick), Terry Done (AIMS), and Christopher Rogers (Kansas University) kindly reviewed the manuscript and made many helpful suggestions.

The photographs and other illustrations used in this book have come from many sources and these are acknowledged in the relevant captions. The generosity of these contributors is greatly appreciated.

Introduction

There are "hotspots" of biodiversity on our planet and the Australian North West Shelf is one of them. This continental shelf is a wide ramp, almost 2500 km long, forming the north-western margin of Australia. It has a long and varied geological history and a huge diversity of marine habitats that have led to the development over time of a marine fauna that is rich in species and distinctive in many ways. This chapter gives a summary of the marine environment and species-rich biota of this remote coast of north-western Australia and seeks to explain the distribution patterns of its species.

While our wonderful living planet is probably not unique in the universe, there is no doubt that, with its extraordinary abundance and diversity of life and complex ecosystems, it is a rare thing. There is broad recognition in our generation of our place as a derivative and participant in the life processes of our world. There is need for better understanding of the origins and diversity of life and the way the multitude of species organize themselves into integrated, functional, productive ecosystems that make our world livable. However, the spread of the kinds of plants and animals around the world is not uniform. There are places, mainly those that are ice-bound, near-waterless, or lightless, where conditions are not favorable for life and the diversity of life forms is low. There are other places, mostly moist, warm and sunlit, and with complex geomorphology and diverse microhabitats, where life thrives, the variety of species is large and the ecosystems they construct are beautiful, intricate, and so complex to be seemingly beyond comprehension. These, including Australia's North West Shelf, are the planet's hotspots of biodiversity.

The North West Shelf is defined by natural physiographic boundaries and comprises mainly carbonate sediments. It is one of the largest structures of its kind in the modern world and its distinctive character is of long standing through geological time. The history of the marine fauna of this huge structure is one of the untold stories of life on earth. During the past 50 million years, the shelf has seen the replacement of its original temperate marine fauna with a tropical one of quite different evolutionary origins. It has seen periods of mass extinction followed by evolutionary diversification and recovery and periods, like the present, when the region has supported an extremely rich marine fauna. During its periods of recovery and species richness, it has been a center of marine evolution, as well as a "sink" episodically receiving tropical species from the great Indo-West Pacific region, and a "source" of species that have established new evolutionary lines in the temperate waters of higher latitudes on the south-western shores of Western Australia.

Understanding the range of species diversity that inhabits our world, and the factors that create and maintain it, is a key to sustaining the habitats that support us. The way to begin such a study in any given region is to assemble what is known about the kinds of plants and animals that are present, the ways they arrange themselves in functional groups, how they are distributed ecologically and in space, and the environmental forces that determine these things. This is the branch of natural science known as biogeography that is the central theme of this chapter.

There are several aspects of biogeography that are vital. One is to understand how the biota of a region relates to that of adjacent regions, especially in regard to exchange of larval recruits and breeding adults (connectivity). Another is to understand that the patterns of distribution we see today are not constants but are highly dynamic and changeable. What we see is a snapshot in an ongoing process of evolutionary change. Very long ago, with little experience of the world but blessed with profound insights about how it operates, Greek philosopher Heraclitus (c. 2500 BP) wrote that "nothing endures but change." This principle is the very essence of biogeography. The physical processes that control contemporary ecological and spatial distributions of plants and animals, especially geology and climate, are ever-changing and the evolutionary processes of life itself are the means by which life forms adapt to a changing environment. The patterns of life we see today are an outcome of everything that has happened before.

With a different human-oriented timescale, historians put this matter neatly with the phrase "there can be no future without a past" and of course there is no present without a past either. While contemporary biological distribution patterns can be described as they are, they cannot be understood without knowledge of the history of environmental change that caused them to be this way. And without that understanding, we cannot judge what may happen next and how we might arrange our own affairs accordingly.

The main drivers of the development of the modern North West Shelf and its diverse biota have been the sequence of tectonic and sedimentary events of the last 60 million years (the Tertiary) and especially the history of dramatic climate and sea level change during the last 2 million years (the Quaternary). Tectonic events established the position of the shelf in relation to the major climatic and biogeographic regions of the world.

They also set up those basic structures of this part of Australia's continental margin that enable and limit the connectivity of its biota with that of other regions. Within that framework, the modern marine biota has evolved in response to sedimentary and climatic events that happened in the Tertiary and Quaternary periods and those that are ongoing. Geological history and the history of climatic change provide essential context for biogeographical studies in our region.

In the present context, while being one of the world's hotspots of marine biodiversity, until now the North West Shelf has been remote from centers of scientific research and its marine flora and fauna are inadequately studied. Nevertheless, there is a rapidly growing body of information and it is timely for us to assemble and review what we know about the life forms, life assemblages, and ecological processes of this biologically rich and diverse marine bioregion. This is the purpose of this chapter in regard to marine habitats and the distribution of marine plants and certain invertebrate animals that play key ecological roles on the North West Shelf. Following the Introduction, it is arranged with three main subject areas—a brief description of the physical environment (Chapter 2), an account of the main kinds of marine habitats and invertebrate associations (Chapters 3–7) and a discussion of the biogeographic processes that operate in the region and their outcomes (Chapters 8 and 9).

It is not possible for one author to deal effectively with the details of subject matter of such a wide scope. In this study, a broad outline of the physical characteristics of the region and its habitats and biota is given and the analysis of biogeographic patterns and evolutionary history focuses on corals and especially molluscs, taxa that are within the experience of this author. Like the evolution of species and faunas, knowledge grows and develops over time. The information and ideas presented here are intended as a framework, imperfect of course, that may encourage others to explore the matters discussed. The aim is a better understanding of the processes that determine and sustain the biodiversity and ecology of this resource-rich and extremely interesting part of the Australian coastline.

1.1 THE GEOGRAPHIC SCOPE OF THIS STUDY

Krummel[1] introduced the term North-West Australian Shelf for the north-western continental margin of Australia. The whole region is most commonly referred to now as the North West Shelf (Figure 1.1) and is defined as the continental shelf and marginal terraces and platforms (to depths of 2000 m) of the north-western coast from North West Cape to Melville Island, a distance of about 2400 km.[2,3] In this study, it is noted that the north-western continental shelf margin actually continues on across the Arafura Sea and terminates on the southern side of the so-called Bird's Head Peninsula of Irian Jaya.

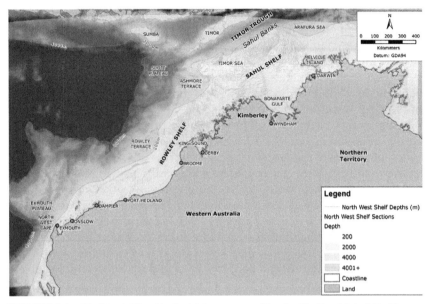

FIGURE 1.1   The North West Shelf with its two parts (Rowley Shelf and Sahul Shelf), marginal terraces (Rowley Terrace and Ashmore Terrace), and marginal plateaux (Exmouth Plateau and Scott Plateau). *Modified after Ref. 3.*

In recent years, it has become common practice in the Australian petroleum and gas industry to restrict the term North West Shelf to the offshore Carnarvon Basin, distinguishing that resource-rich area from the Browse and Bonaparte Basins further north. This report does not limit the concept of the North West Shelf in that way but follows the area designation based on geological and geomorphological criteria.

The North West Shelf was subdivided by Fairbridge[4] into two parts, the south-western Rowley Shelf and north-eastern Sahul Shelf, a division that is maintained here. The Sahul Shelf is that part east of the longitude of Ashmore Island (123°E) that lies within the Arafura and Timor Seas, bordering the Timor Trough and the waters of Indonesia. The Rowley Shelf is that part, west of Ashmore Island, facing international waters of the northeastern Indian Ocean and includes the marginal Exmouth and Scott Plateaux.

The North West Shelf, as so defined, does not correspond with the North-west Marine Region subject to a planning study by Commonwealth agencies. That region is an administrative unit comprising waters under Commonwealth jurisdiction (i.e., from the State 3-mile territorial baseline to the 200 nm EEZ boundary) and extends from off Kalbarri on the West Coast of Western Australia to the Northern Territory border. It does not extend to the northern limit of the North West Shelf but does extend beyond its southern limit to include the Dirk Hartog Shelf. Nevertheless, this biogeographic review of the North West Shelf has direct relevance to the North-west Marine Region and draws heavily on material published for that bioregional planning study.[5]

1.2 THE INTENT AND UTILITY OF BIOGEOGRAPHY

The objectives and usefulness of historical biogeography complement those of taxonomy—two aspects of biology that are often studied by the same practitioners. Both disciplines are descriptive initially, seeking to recognize and describe common patterns—of similarities among an array of organisms in the former and geographical distribution in the latter—and they provide a classificatory framework that supports the development of knowledge in other branches of marine science. Both may lead to explanatory and predictive interpretations with significant scientific utility.

In today's circumstance of increasing human impact on the marine environment, along with changing climate and accompanying sea level and oceanographic changes, a biogeographic classification of Australian waters, based on causal factors with a predictive capacity, becomes a crucial tool in environmental management. In Australia, marine biogeography has found recent application in the field of resource and environmental management and conservation. The idea is that designation of sections of the marine environment as distinct biogeographic units allows representative parts of them to be selected as special management regimes, for example, for conservation reserves. The importance of this is clear from recent efforts internationally and in Australia to identify "bioregions" that facilitate the selection of representative marine conservation reserves.[6–10]

The utility of a biogeographic classification is diminished if key terminology is poorly defined and if classification units are inconsistently applied, which is presently the case in Australian biogeographic studies. A three-level biogeographic area classification is adopted in this report, using terminology that is consistent with the broadly accepted, contemporary view.

The three levels are as follows:

Level 1. Realm. The largest biogeographic spatial unit—internally coherent at high taxonomic levels as a result of shared and unique evolutionary history; with high levels of endemism at generic and higher rank.

 The term Region has been used at this level by many marine biogeographers (following Ekman[11]) but Realm is preferred here for the sake of conformity with common practice in terrestrial biogeography. Region is used as a generic term for biogeographic

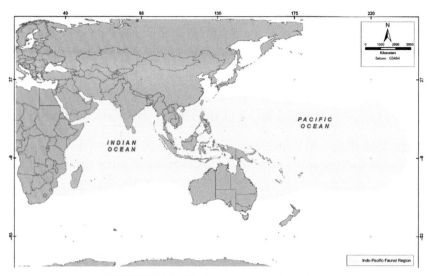

FIGURE 1.2   The Indo-West Pacific Realm.

categories of any rank. The marine shelf and biota of northern Australia, including the North West Shelf, are regarded as a subunit of the vast Indo-West Pacific Realm (Figure 1.2).

Level 2. Province. Places where species endemism has developed to the extent that, over time, the fauna has taken on a definite provincial character as a result of restricted gene flow and regional speciation. (Definition derived from Briggs.[12])

Identification of geographic (or oceanographic) features (biogeographic barriers) that restrict gene flow and promote speciation is an important step in designation of biogeographic provinces. Biogeographers generally recognize two criteria that may be used to delineate between provinces, level of endemism and faunistic similarity measured in terms of species and higher taxa. Ideally, both these criteria should be quantified, but the level of taxonomic knowledge does not often allow this to be assessed objectivity. There is no general agreement on the degree of faunistic similarity that is required or on how it should be measured.

In this biogeographic study of the North West Shelf, a concept of "biogeographic province" is applied that is based on the following criteria:

- an area clearly delimited by geomorphic characteristics that are explainable in terms of its geological and climatic history
- an area that is bounded by "geographic barriers" that are thought to apply to at least some marine organisms, i.e., as an impediment to dispersal and a driver of vicariant speciation

- an area that exhibits 10% or more species endemism
- an area that exhibits a significant (undefined) level of faunistic dissimilarity to that of the neighboring areas under consideration.

Level 3. Bioregion. An area distinguishable from other areas within a province by distinctive ecosystems, biotic assemblages, and habitats.

> The presence of endemics is not essential to the diagnosis of a bioregion, although they may be present. Emphasis is on habitat and ecological factors rather than on speciation history. Faunistic dissimilarities between the bioregions of a province often may be explained by habitat differences alone.
>
> Biogeographic units of this level (or levels) are variously named. Three terms that have been used, each with a slightly different emphasis, are biotic province,[13,14] bioregion,[15] and ecoregion.[16] The term bioregion has come into common usage, as a subcategory of province and sometimes as a replacement of it, in the context of conservation and natural resource management planning. In Australia, the Integrated Marine and Coastal Regionalisation of Australia (IMCRA) bioregion definitions[10] are based on faunistic, habitat, and physical criteria. They are a useful planning tool as a guide for selecting representative areas for reservation and other management decisions. In this study, the IMCRA bioregions are treated as a third tier of biogeographic area, subordinate to province.
>
> Later in this report (Chapter 4), the term "coral reef bioregion" is used for areas where there are coral reefs that have distinctive biogeomorphic characteristics that derive from different geological histories and oceanographic conditions. They equate roughly with the IMCRA mesoscale bioregions for this part of the Australian coastline.

An aim of this biogeographic study is to consider whether or not the North West Shelf meets the criteria for designation as a distinctive province within the Northern Australian Subrealm and to review evidence for subdivision at the level of bioregion.

Recent designations of major planning areas on the Australian coast refer to the major units as "Regions," for example, the North-west Marine Region that includes most of the North West Shelf.[5] These areas are based on administrative, jurisdictional boundaries, and only approximately on biogeographic criteria. The term Region applied to them should not be used in the sense of traditional biogeography. The areas are not equivalents of the Regions (Realms) or Provinces currently recognized in the biogeographic literature.

1.3 BIOGEOGRAPHIC CLASSIFICATION OF THE NORTHERN AUSTRALIAN COAST

1.3.1 The Indo-West Pacific Realm

In 1856, Forbes[17] designated a marine "Indo-Pacific Province" extending from the east coast of Africa to the eastern-most islands of Polynesia, including the northern coast of Australia, describing it as the "realm of reef-building corals" (Figure 1.2). Subsequently, marine biogeographers have adopted that concept though usually amending the term to Indo-West Pacific, recognizing that the coastal faunas of the eastern Pacific (western coast of the tropical Americas) are distinctively different.[11,12,18] This vast biogeographic realm is the most species-rich marine area on earth.

In Australia, marine biogeographers have long acknowledged that the tropical northern marine fauna is part of the Indo-West Pacific biogeographic realm, usually ranking it as a subrealm or subregion, and that it is quite different to that of the temperate southern coast which is regarded as a distinct biogeographic realm.[11,12,19–22] There is a very high level of species endemism on the southern coast of Australia and many taxonomic differences at subgeneric and generic level between it and the northern coast, resulting from the different evolutionary histories of the two biogeographic regions.[21] While the tropical marine fauna of the Indo-West Pacific, including northern Australia, had its origins in the pan-tropical, early Tertiary Sea of Tethys, the temperate fauna of the south had very different origins in high latitudes of the Southern Hemisphere. The tropical northern and temperate southern marine faunas overlap on the east and west coasts (Figure 1.3). This circumstance came about as a result of the repositioning of the Australian continent astride the Tropic of Capricorn in the Miocene.[21]

1.3.2 Australian Coastal Marine Biogeographic Provinces

Australian coastal marine biogeographic provinces, both north and south, have been delineated by many authors, not always in the same way.[23–30]

Several authors have recognized the presence of many endemic species in the tropical marine fauna of the north-west coast, not seen on the tropical eastern Australian coast or elsewhere in the Indo-West Pacific realm. Hedley[24] designated the region from Shark Bay to the Gulf of Carpentaria as a distinct biogeographic area that he named the Damperian Province (Figure 1.4). Whitley[26] and Clark[27] followed this arrangement. However, not all later authors were convinced that separate provincial status was justified. In a study of the bivalve family Cardiidae, Wilson and

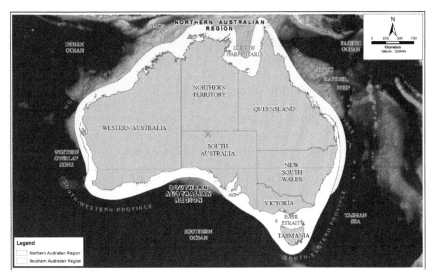

FIGURE 1.3   The primary Australian biogeographic regions. *After Ref. 20.*

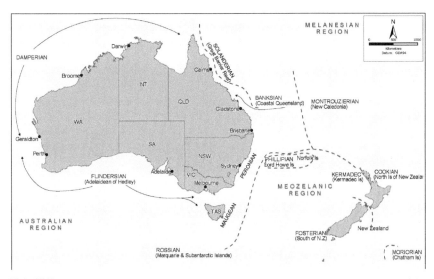

FIGURE 1.4   The biogeographic provinces of tropical northern Australia proposed by Whitley.[26]

Stevenson[31] found "little support for recognition of this region [Damperian] as a distinct faunal province" even though they listed four endemic species (15%) of a total of 27 cardiid species in the region. On the basis of gastropod distributions, Wells[32,33] considered that "the waters of tropical Australia should be considered as a single Tropical Australian Province." This matter is further considered later in this report.

1.3.3 The IMCRA Bioregions

IMCRA Version 4.0[10] introduced a third-tier classification of Australian waters involving both benthic and pelagic bioregions. The pelagic bioregions are beyond the scope of this report. The benthic bioregions included two categories referred to as

- "provincial bioregions that reflect biogeographic patterns in distributions of bottom-dwelling fish" and
- "mesoscale regions on the continental shelf" (Figure 1.5).

The provincial bioregions, based on demersal fish data,[34] include two types defined as "provinces" and "transitions" (biotones). These data and the analysis of them are an important advance in knowledge of the little-known bottom fauna on the outer continental shelf and slope around Australia. Although provisional, they indicate that the distribution patterns displayed by the deep benthic fauna may be quite different to those of inner shelf and coastal species, a conclusion that should be no surprise. The environmental history of the outer shelf and slope, below the lowest Quaternary sea level, subject in places to shelf-edge subsidence, and populated by benthic species with different reproductive and dispersal strategies to those of the inner shelf, has resulted in different geographic distribution patterns.

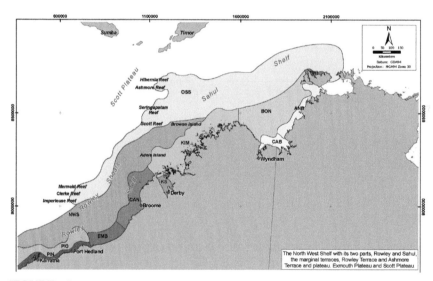

FIGURE 1.5   The IMCRA coastal mesoscale bioregions of tropical north-western Australia. TWI, Tiwi; ANB, Anson-Beagle; BON, Bonaparte Gulf; CAB, Cambridge-Bonaparte; OSS, Oceanic Shoals; KIM, Kimberley; KS, King Sound; CAN, Canning; NWS, North West Shelf; EMB, Eighty Mile Beach; PIO, Pilbara Offshore; PIN, Pilbara Nearshore. *Redrawn from Ref. 8.*

Much more information is needed before an account of the historical biogeography of the outer shelf and slope component of the North West Shelf fauna can be attempted. Partly for this reason, partly because the two IMCRA benthic bioregion categories are contradictory (they overlap on the middle and outer shelf) the IMCRA "provincial bioregions" will not be considered further in this study. However, the IMCRA "mesoscale regions," as a third level of biogeographic area, provide a useful classification of the inner shelf and coastal habitats of the North West Shelf and will be applied. They include 10 coastal bioregions between North West Cape and Melville Island based primarily on geomorphological criteria.

References

1. Krummel O. *Handbuch der Ozeanographie*, 2 vols, 1897, Stuttgart.
2. Purcell PG, Purcell RR, editors. *The North West Shelf of Australia*. Petroleum Exploration Society of Australia: Perth; 1988.
3. Bradshaw MT, Yeates AN, Beynon RM, Brakel AT, Langford RP, Totterdell JM, et al. Palaeographic evolution of the North West Shelf. In: Purcell PG, Purcell RR, editors. *The North West Shelf of Australia*. Petroleum Exploration Society of Australia: Perth; 1988. p. 29–54.
4. Fairbridge RW. The Sahul shelf, northern Australia: its structure and geological relationships. *J R Soc West Aust* 1953;**37**:1–34.
5. Department of the Environment, Heritage, Water and the Arts . *The north-west marine bioregional plan. Bioregional profile*. Canberra: Australian Government; 2008.
6. Department of Conservation and Land Management. *A representative marine reserve system for Western Australia. Report of the Marine Parks and Reserves Selection Working Group.* Department of Conservation and Land Management, Perth; 1994.
7. Muldoon J, editor. *Towards a Marine Regionalisation for Australia*. Proceedings of a workshop held in Sydney, New South Wales, 4–6 March 1994. Great Barrier Reef Marine Park Authority; 1995, 233 pp.
8. Thackway R, Cresswell ID. *Interim Marine and Coastal Regionalisation for Australia: an ecosystem-based classification for marine and coastal environments. Version 3.3.* Canberra: Environment Australia, Commonwealth Department of the Environment; 1998.
9. ANZECC Task Force on Marine Protected Areas. *Guidelines for establishing the National Representative System of Marine Protected Areas.* Australian and New Zealand Environment Council, Task Force on Marine Protected Areas. Environment Australia, Canberra; 1998.
10. Commonwealth of Australia . *A guide to the Integrated Marine and Coastal Regionalisation of Australia. Version 4.0.* Canberra: Department of the Environment and Heritage; 2006.
11. Ekman S. *Zoogeography of the sea*. London: Sidgwick and Jackson; 1953, 417 pp.
12. Briggs JC. *Marine zoogeography*. New York: McGraw-Hill; 1974, 475 pp.
13. Dice LR. *The biotic provinces of North America*. Ann Arbor: University of Michigan Press; 1943, 78 pp.
14. Matvejev SD. *Biogeografija Jugoslavije—osnovni principi*. Belgrad: Bioloski Institut NR Srbije; 1961.
15. Fisk P. Bioregions and Technologies. A new planning tool for stable state economic development. Unpublished report, Centre for Maximum Potential Building Systems, Inc. West Arizona State University; 1983.

16. Spalding MD, Fox HE, Allen GR, Davidson N, Ferdana ZA, Finlayson M, et al. Marine ecoregions of the world: a bioregionalisation of coastal and shelf areas. *Bioscience* 2007;**57**(7):573–83.
17. Forbes E. *The physical atlas of natural phenomena;* 1856.
18. Abbott RT. The genus *Strombus* in the Indo-Pacific. *Indo-Pacific Mollusca* 1960;**1**(2):33–146.
19. Doderlein L. Ondopazifische Eurylae. *Abh Bayr Akad Wiss Math Nat Abt* 1927;31, Munchen [not seen].
20. Wilson BR, Gillett K. *Australian shells.* Sydney: Read; 1971, 168 pp.
21. Wilson BR, Allen GA. Major components and distribution of marine fauna. In: Dyne GR, Walton DW, editors. *The fauna of Australia. General articles.* Canberra: Australian Government Publishing Service; 1987. p. 43–68, [chapter 3].
22. Ponder WF, Wells FE. Distribution and relationships of marine and estuarine fauna. In: Beesley PL, Ross GJB, Wells A, editors. *Mollusca: the southern synthesis. Fauna of Australia,* vol. 5. Melbourne: CSIRO Publishing; 1998. p. 77–80, Part B viii, p. 565–1234.
23. Hedley C. The effect of the Bassian isthmus upon the existing marine fauna; a study in ancient geography. *Proc Linn Soc New South Wales* 1904;**28**:876–83.
24. Hedley C. Zoogeography. In: Jose AW, Carter HJ, editors. *Australian encyclopedia 2,* 2 vols. Sydney: Angas & Robertson; 1926. p. 743–4.
25. Cotton BC. Fissurellidae from "Flindersian" Region, Southern Australia. *Rec South Aust Museum* 1930;**4**:219–22.
26. Whitley G. Marine zoogeographical regions of Australia. *Aust Nat* 1932;**8**:166–7.
27. Clark HL. The echinoderm fauna of Australia. *Publ Carnegie Inst* 1946;**566**:1–567.
28. Bennett I, Pope EC. Intertidal zonation of the exposed rocky shores of Victoria, together with a rearrangement of the biographical provinces of temperate Australian shores. *Aust J Mar Freshw Res* 1953;**4**:105–59.
29. Knox GA. The biogeography and intertidal ecology of the Australian coasts. *Oceanograph Mar Biol Annu Rev* 1963;**1**:341–404.
30. Poore GCB. Biogeography and diversity of Australia's marine biota. In: Zann LP, Kailola P, editors. *The marine environment technical annexe 1, State of the Marine Environment Report for Australia.* Townsville: Great Barrier Reef Marine Park Authority; 1995. p. 75–84.
31. Wilson BR, Stevenson SE. Cardiidae (Mollusca, Bivalvia) of Western Australia. *Western Australian Museum Special Publication* No. 9; 1977. 114 pp.
32. Wells FE. Comparative distributions of the mudwhelks *Terebralia sulcata* and *T. palustris* in a mangrove swamp in northwestern Australia. *Malacol Rev* 1980;**13**:1–5.
33. Wells FE. Zoogeographic affinities of prosobranch gastropods on offshore coral reefs in northwestern Australia. *Veliger* 1986;**29**:191–9.
34. Last P, Lyric V, Yearsley G, Gledhill D, Gomon M, Rees T, et al. *Validation of the national demersal fish datasets for the regionalisation of the Australian continental slope and outer shelf (> 40 m depth).* Australia: Department of Environment and Heritage and CSIRO Marine Research; 2005.

The Contemporary Physical Environment

2.1 GEOLOGY, ORIGINS, AND HISTORY OF THE NORTHWEST CONTINENTAL MARGIN

The major controls on paleographic development of the North West Shelf region have been climatic sea level change and tectonism.[1] The contemporary outcomes of these forces, the sedimentary history of the region and its changing oceanography, have been fundamental to the historical biogeography of the North West Shelf and its marine biota. The origins, geology, and sediments of the North West Shelf have been reviewed by Carrigy and Fairbridge,[2] Jones,[3,4] Stagg and Exon,[5] Bradshaw et al.,[1] Exon and Colwell,[6] James et al.,[7] and Baker et al.[8]

The North West Shelf forms Australia's northwestern continental margin. It evolved as a set of overlapping rim basins from a Mesozoic rifted-arch system that developed in high southern latitudes along the margin of the Tethyan Ocean and Gondwanaland.[9] Rifting in the Early Cretaceous ruptured Gondwanaland creating the Indian and Australasian land masses, separated by a rift valley that became a major sedimentary depo-center (antecedent of the modern Perth and southern Carnarvon Basins). With subsequent sea floor spreading, the Indian and Australasian land masses drifted apart, the eastern side of the rift valley forming the western margin of the new Australian continent. In this way, the northwestern and western continental margins (North West Shelf and Rottnest-Dirk Hartog Shelves, respectively) developed as a continuous, curved, open-sided sedimentary "super basin" referred to as the Westralian Superbasin.[10]

In the mid-Tertiary, there was no clear division between a northwestern continental shelf and a western one (Figure 2.1), but in the Late Tertiary, uplift of the Cape Range Peninsula at the northwest "corner" created a narrow restriction partially separating the North West Shelf from the continental shelf of the West Coast.

13

FIGURE 2.1   The approximate position of the northwestern shoreline in the Miocene, showing the wide continental shelf (pale blue), a cluster of outer shelf bioherms (antecedents of the modern shelf-edge atolls and platform reefs) and the absence of a shelf constriction where the Cape Range Peninsula is now located. *After Ref. 1*

The present separation of Australia and New Guinea by the shallow shelf waters of the Arafura Sea and the Gulf of Carpentaria has not been a permanent feature in recent geological history. The northwestern side of the Australasian continental margin and outer shelf benthic habitats are continuous between northern Australia and West Irian. During Quaternary periods of low sea level, the coastline was also continuous and New Guinea and Australia were connected by a wide "land bridge." The Indonesian islands of Aru were part of the Australasian land mass, and relict populations of marsupials and other typical Australian and New Guinea animals may be found there. We may think of the North West Shelf in those times as extending all the way from North West Cape to the south coast of West Irian, the northern end of it in close proximity to the antecedent of the modern East Indies Triangle with its immensely high biodiversity (Section 9.1) and without any direct connectivity with the east coast of Australia.

2.1.1 The Main Geological Elements of the North West Shelf and Adjacent Shores

Coastal geomorphology, marine habitats, and the distributions of marine organisms are hugely influenced by the very different geological regimes along the modern North West Shelf coastline (Figure 2.2). The main onshore geological elements of the North West Shelf region are

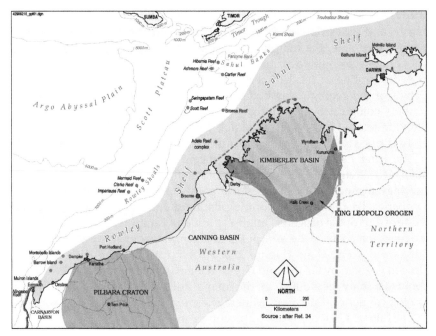

FIGURE 2.2   General geological features of the North West Shelf region. *Drawn by Environmental GIS.*

two Pre-Cambrian blocks, three sedimentary basins that have onshore and offshore parts extending across the width of the shelf to the continental margin, and, off the Kimberley coast, an offshore sedimentary basin that does not extend to the coast.

2.1.1.1 Pre-Cambrian Blocks

In the north of Western Australia, the Kimberley Block comprises two elements, the Kimberley Basin and the King Leopold-Halls Creek Orogen (Figure 2.2).

The Kimberley Basin is formed mainly of Proterozoic rocks of the Kimberley Group comprising undeformed but metamorphosed sandstones and siltstones, basic volcanics, and intrusive basalts. There are also areas of shallow Quaternary alluvial and colluvial sediments and, in the Bonaparte Archipelago and on the coast of the adjacent mainland, Cainozoic laterite deposits. The contemporary terrestrial surface of the basin is a plateau, primarily a very ancient subaerial erosion surface developed on flat-lying rocks.[11–13] Its deeply dissected northwestern side is submerged, and the margin of the basin lies 50 km or more seaward of the present coastline where it is overlapped by sediments of the Browse basin. On this part of the Kimberley coast, the seabed of the inner shelf is an inundated terrestrial land surface with highly irregular bathymetry and superficial marine and terrestrial sediments over the Proterozoic basement.[14] On the northeastern side, the margin of the Kimberley Basin is also submerged but not dissected and the submerged area is much narrower.

A description of the Kimberley Basin is given by Griffin and Grey.[15] Details of the geology may be found in the maps and explanatory notes of the Geological Series, Bureau of Mineral Resources, sheets SD 52-5 (Londonderry), SD 52-9 (Drysdale), SD 51-12 (Montague Sound), SD 51-16 (Prince Regent), SD 51-15 (Camden Sound), and SE 51-3 (Yampi).

The King Leopold-Halls Creek Orogen is a complex band of highly deformed Archaeozoic and Proterozoic metamorphic and igneous rocks curving around the southwestern and southeastern margins of the Kimberley Basin and incorporating its marginal sedimentary rocks. The western end of the orogen is intensely folded (Yampi Fold Belt) and forms the rocky ria coast of the Yampi Peninsula and the adjacent islands of the Buccaneer Archipelago (Figure 2.3). A description of the King Leopold Orogen is given by Griffin and Grey.[16] Details of the geology of the Yampi Peninsula may be found in the maps and explanatory notes of the Geological Series, Bureau of Mineral Resources, sheet SE 51-3 (Yampi).[17]

The Pilbara Craton (Figure 2.2) occupies an egg-shaped area of the north-west corner of the continent comprising Pre-Cambrian granite-greenstone,

FIGURE 2.3   An aerial view looking south across Talbot Bay, an almost landlocked embayment on the northern side of the Yampi Peninsula, illustrating the ria landscape of the Kimberley Bioregion. The geology is intensely folded and faulted and the landscape inundated by the Holocene Post-Last Glacial Transgression. In the foreground are two peninsulas of Molema Island with a vast mud bank between them. At the top left corner, a portion of South Turtle Reef is visible. Note also the three small patch reefs in the central channel. These reefs are rich in coral species in spite of the extreme muddy conditions. *Photo: Shakti Chakravarty, Airborne Research Australia.*

sedimentary, and volcanic rocks aged from 2.5 to 3.6 Ga. Much of its northern margin is overlain by younger sedimentary rocks of the Carnarvon Basin, but igneous and volcanic rocks of the craton are exposed along the shores of the coast in the vicinity of the Dampier Archipelago. In this area, the complex Pre-Cambrian geology results in landscape of high relief and Quaternary inundation has produced a ria-like coastline.[18] Descriptions of the geology are given by Tyler and Griffin.[17]

2.1.1.2 Sedimentary Basins

The contemporary North West Shelf comprises a series of four sedimentary basins representing the northeastern components of the Westralian Superbasin.[19] From north to south, these are Bonaparte, Browse, Offshore Canning, and North Carnarvon Basins (Figure 2.4). Onshore, the basins either simply lap onto Pre-Cambrian elements or abut or merge with cross-trending Paleozoic basins of the continental land mass.[1] Carbonate sediments became dominant in the Late Cretaceous, with a prograding carbonate wedge on the outer shelf and slope. Today the North West Shelf is one of the world's major carbonate ramps, supporting an abundant and species-rich benthic and demersal fauna.

The Bonaparte Basin occupies the Sahul Shelf. It contains up to 18 km of sedimentary and volcanic rocks ranging in age from Cambrian to

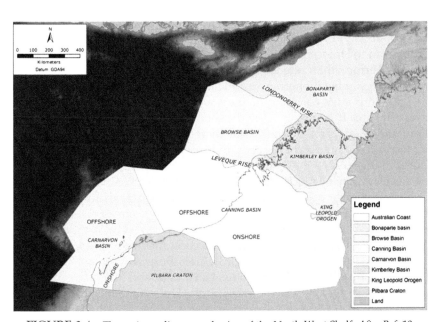

FIGURE 2.4 The major sedimentary basins of the North West Shelf. *After Ref. 19.*

Quaternary. Except for a very small area of coastal land at the bottom of Joseph Bonaparte Gulf, this basin does not have an onshore component, but it overlaps the submerged northeastern margin of the Kimberley Block some distance from the shore. In the Late Tertiary, collision of the Australian and Eurasian plates dramatically modified the physiography of this part of the continental shelf resulting in the formation of the Timor Trough and ongoing subsidence of its margin. Bordering the Timor Trough, there are two topographic highs, the Ashmore Platform on which Ashmore, Cartier, and Hibernia Reefs are situated and the Sahul Platform that bears a series of large biogenic reefs, banks, and shoals. The Bonaparte Basin holds rich hydrocarbon resources, and there is an extensive geological literature.[20-23]

Further south the Browse Basin contains thicknesses of more than 11 km of sedimentary and volcanic rocks ranging in age from Early Permian to Quaternary. This basin is entirely offshore. Its nearshore margin onlaps the Proterozoic basement of the Kimberley Block but does not reach the shore. The Browse Basin is structurally complex and subject to ongoing subsidence, like the Bonaparte Basin to the north but perhaps not at the same rate. Downwarping of the continental margin is responsible for formation of the Ashmore Terrace, a marginal plateau beyond the present shelf edge. Scott Reef and Seringapatam Reef are growing bioherms on anticlinal trends on that marginal terrace, apparently keeping pace with subsidence. The Mesozoic sediments of the Browse Basin are also rich in hydrocarbon reservoirs and geological exploration is ongoing.

The Canning Basin occupies the middle part of the North West Shelf with about one-third of its area offshore.[1,4,24,25] It contains sediments from Ordovician to Holocene age and a subsiding marginal shelf known as the Rowley Terrace that extends to depths of around 600 m. The Rowley Shoals are growing bioherms on anticlinal trends of this marginal terrace. The contemporary coastline of the Canning Basin spans a distance of about 700 km. It is mainly a sandy shore of low relief with Quaternary dune and mud flat sediments, but there are also low rocky headlands of Mesozoic and Quaternary rocks in the north and Quaternary limestone shores in the south. The northern boundary of the Canning Basin crosses the coast at Cape Leveque at the tip of the Dampier Peninsula and marks an abrupt change of coastal geology from Proterozoic metasedimentary, metamorphic and igneous rocks to Cretaceous-Cainozoic sedimentary rocks, with a corresponding abrupt change of coastal geomorphology. Details of the onshore geology along the coast may be found in the maps and explanatory notes of the Geological Series, Bureau of Mineral Resources (Sheets SE 51 for Pender Bay, Broome, La Grange, and Mandora).

The Carnarvon Basin is the most southerly sedimentary basin of the North West Shelf with a southern part that extends onto the Dirk Hartog Shelf of the West Coast.[4,8,26] This basin has both onshore and offshore components and a depth of about 15 km of sediment. In the east, along the 100 km or so between Cape Lambert and Cape Preston, the offshore Carnarvon Basin onlaps the submerged margin of the Proterozoic Pilbara Craton and does not reach the shore. The western part of the basin overlies the coast between Cape Preston and North West Cape, forming mainly sandy shores with low relief, long beaches, extensive tidal flats and mangroves and major deltas. There are also stretches of Quaternary coastal limestone and Late Pleistocene coral limestone benches along the shore. As a result of Holocene inundation, there are many limestone islands, shoals and fringing and patch reefs. Because of the rich hydrocarbon resources of the Carnarvon Basin, there is a very extensive and ever-growing geological literature.

2.1.2 Cainozoic Phases of Deposition

Deposition of carbonate sediments dominated the North West Shelf in the Late Cretaceous and that circumstance continues. Through the Cainozoic, there have been four major phases of marine deposition resulting from sea level change and transgression across the North West Shelf.[27]

These phases occurred in the

- Paleocene to Early Eocene
- Upper Middle to Late Eocene
- Late Oligocene to Middle Miocene
- Pliocene to the Present.

During each of these major sedimentary phases, there were also minor episodic phases of transgression and regression. The four major transgressions across the North West Shelf, and their minor secondary transgressive episodes, are believed to be partly the result of shelf downwarping and subsidence, but they also corresponded with global warming events. They roughly correspond with the cyclic Tertiary climatic changes that shaped the evolution of the unique terrestrial flora and fauna that characterizes the modern Australian continent.[28]

The first two of the major transgressive, sedimentation phases occurred when this section of the shelf margin lay within high latitudes prior to the northward drift of the continent. The benthic marine fauna of the region at that time was warm temperate with a large cosmopolitan component. But the new position of the Australian continent in the Miocene, straddling the Tropic of Capricorn, brought northern Australia, including the North West Shelf, into the tropical zone, where it has remained. That tectonic event created the primary marine biogeographic situation we see today, with a northern tropical coastal and shelf fauna and a southern temperate one, the two overlapping on the West Coast (Section 1.3; Figure 1.3).

It may be supposed that transgressive events brought on by sea level rise are accompanied by the following conditions that are conducive to rapid evolution of benthic marine species:

- higher sea temperatures
- expansion of the world's tropical zone
- wide continental shelves and vast expansion of benthic shelf habitat.

Conversely, regressive events may correlate with

- lower sea temperatures
- contraction of the world's tropic zone
- narrow continental shelves and contraction of benthic shelf habitat.

It is no coincidence that the Late Oligocene-Middle Miocene transgression around the equatorial zone of world seas was accompanied by major evolutionary radiation of marine fauna. All of the families of tropical marine invertebrates were established by that time and many of them underwent rapid speciation and radiation as a result of a massive increase in available shelf habitat and the geographic expansion of the tropical zone. For example, the gastropod family Cypraeidae (cowries) radiated extensively in the Indo-West Pacific region during the Miocene with a significant increase in both genera and species.[29]

During the Late Oligocene-Middle Miocene transgression, the North West Shelf was as wide or wider than it is today and was virtually continuous with the Dirk Hartog Shelf and the Rottnest Shelf to the south (Figure 2.1). Marine Miocene limestone deposits containing diverse benthic fossil faunas are exposed on Barrow Island, Cape Range, and other onshore localities of the Carnarvon Basin. It is likely that the marine shelf fauna of the northwest coast at that time was at least as species rich as it is today.

Little is known of the marine fauna of northwestern Australia during the Late Miocene-Early Pliocene regression. Deep Pliocene strata have been encountered in drillings on the North West Shelf, but there are no exposures of Pliocene rocks on the islands or mainland of either the northwest or west coasts. If this fall of sea level correlated with lower global temperature and resulted in latitudinal contraction of the tropical zone, significant extinction of coastal marine fauna may have occurred, especially at the southern end of the North West Shelf.

Nor is there a helpful Late Pliocene fossil record in northwestern Australia. An exception may be the Jurabi Member on the western side of the Cape Range Peninsula that may be Late Pliocene[30] and represents the beginnings of a Late Pliocene-Quaternary transgression. There are fossiliferous Pliocene marine deposits at depth on the Swan Coastal Plain, but insufficient information is available of their faunistic composition to judge their biogeographic composition. Nevertheless, it is clear that the marine tropical zone and fauna extended to beyond latitude 35° on the Rottnest Shelf and east along the South Coast during the Late Pliocene. There are species-rich shell beds of that age with many tropical shallow benthic species on the Roe Plain on the shores of the Great Australian Bight and in South Australia.[31]

The Pleistocene eustatic sea level changes and accompanying transgressive-regressive cycles on the North West Shelf are even less well known. A Pleistocene record of sea level change on some of the shelf-edge atolls is now emerging.[32] There are Pleistocene high stand deposits containing fossil molluscs and other organisms on the coastal plains of the Pilbara and Canning coasts, but they remain unstudied. Apparently, there are none on the Kimberley coast, perhaps as a result of regional subsidence in that region (see Section 2.1.5). Nevertheless, there is now a fairly accurate global sea level curve (Figure 2.12), and it is possible to estimate the approximate positions of the shoreline during the Pleistocene low stand periods, especially the last one.

Clearly, during the regressions of the Late Miocene-Early Pliocene and the Pleistocene eustatic low stands, the North West Shelf was very much narrower than it is today. The cyclic transgressive and regressive events, drastically expanding and contracting the area of shelf habitat and the tropical zone, must have had a profound influence on the evolution of the shelf benthic marine fauna of the northwest and west coasts.

2.1.3 Morphology and Bathymetry of the North West Shelf

The International Hydrographic Organization (IHO) defines a continental shelf as "a zone adjacent to a continent extending from the low water line to a depth at which there is usually a marked increase of slope toward the oceanic depths." Such an increase usually occurs in the vicinity of the 200-m isobath. A continental slope is defined by the IHO as "the slope seaward from the shelf to the upper edge of a continental rise or the point where there is a general reduction of slope." Marginal to continental shelf, on or beyond the continental slope, there may be plateaux and terraces that are more or less level surfaces with one or more steep sides. These and other deep-sea geomorphic features of the Australian continental margin have been described by Heap and Harris.[33]

For most of its length, the outer margin of the North West Shelf is a "roll over zone," and there is no marked increase of slope. Consequently, the shelf edge is taken arbitrarily to be the 200-m isobath.[1] However, there are two areas where an abrupt change of slope occurs. The first is located northwest of the Montebello Islands where there is a distinct edge at a depth around 180 m and a very steep slope below that to the 1000-m isobath (Figure 2.5). The second is along the northern margins of the Sahul

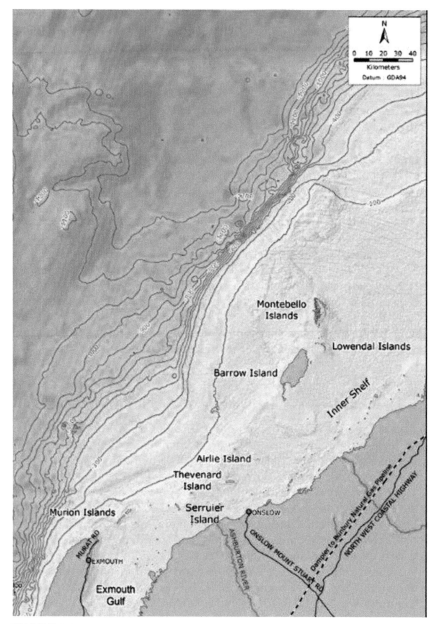

FIGURE 2.5 The steep continental margin west of the Montebello and Barrow Islands at the western end of the North West Shelf. *Drawn by URS Australia.*

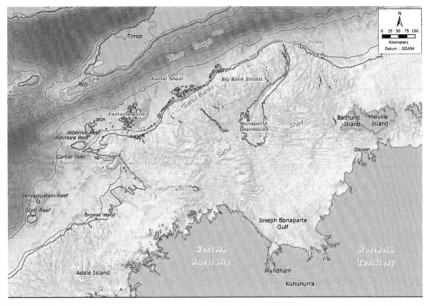

FIGURE 2.6   The steep continental margin of the Sahul Shelf adjacent to the Timor Trough in the Timor Sea. *Drawn by URS Australia.*

Shelf where it borders the Timor Trough and along the southwest face of Johnson and Woodbine Banks southeast of Ashmore Reef (Figure 2.6).

Within the offshore petroleum and gas industry, it has become common practice to define the North West Shelf as the whole of the continental margin including the continental shelf plus its continental slope and marginal terraces and plateaux to depths of 2000 m.[34] Clearly, this concept relates to the regional hydrocarbon resource and not to the shelf as a geomorphic structure. Nevertheless, although the principal concerns of this study are the habitats and biota of the shelf itself, there is utility in applying a broader concept of the North West Shelf that includes the continental slope and deep-sea terraces, most notably the subsided Ashmore and Rowley Terraces from which arise a series of the oceanic reefs that are a major biogeomorphic feature of the region.

2.1.3.1 *Divisions of the Shelf*

The North West Shelf has been subdivided into depth zones in several different ways. The terms inner, middle, and outer shelf are often used loosely to categorize shelf habitats and biota. Attempts to delineate formal depth zones have produced variable results. There are obvious changes in benthic and demersal fauna across the width of the shelf, but so far, there have been few attempts to document either the habitats or species associations. Primary causal factors in biotic distributions across the shelf are depth and characteristics of seabed substrate. We might expect benthic habitat-species associations to form mosaic patterns rather than be arranged in clear parallel depth zones. Nevertheless, there is a relation between substrate and depth, and a classification of the shelf into depth zones is a helpful first step toward understanding the distribution patterns of benthic habitats and biota. The objective must be to devise a classification of the shelf that is not entirely arbitrary but relates to the geological, oceanographic, and historical factors that have determined the contemporary distribution of substrate types.

For planning purposes, the report of the Marine Parks and Reserves Selection Working Group[35] referred to "nearshore" and "offshore" environmental types in the western part of the North West Shelf. The boundary between them was set arbitrarily as the 10-m isobath, that being the approximate seaward limit of the nearshore zone characterized by more turbid water and a seabed rock platform covered by thin sand veneer and macroalgae.[36] Subsequently, this became the basis for designation of the IMCRA Pilbara (nearshore) and Pilbara (offshore) meso-scale bioregions (Commonwealth of Australia 2004). However, the nearshore and offshore zones were not intended to equate with Inner and Outer Shelf zones.

The North West Shelf Joint Environmental Planning Study report on the Pilbara coast[37] designated a nearshore biome called the "coastal zone," with two subbiomes at 0-10 and 10-20 m. The offshore part of the shelf was subdivided into three biomes, determined by statistical analysis of research trawl data on fish species associations. The three offshore biomes so determined were Inner Shelf (20-70 m), Mid-Shelf (70-120 m), and Outer Shelf (120-200 m). These coastal, inner shelf and mid-shelf biomes do not equate with geomorphic or hydrodynamic criteria, and it is yet to be shown whether they apply to benthic invertebrate communities as well as fish associations.

The North-west Marine Region Planning Study report[8] recognized inner, middle, and outer shelf/slope zones but bounded them at 0-30, 30-200, and >200 m, respectively. The source of these divisions and the criteria on which they were based were not specified.

On the basis of hydrodynamics and bathymetry, Dix et al.[38] and James et al.[7] recognized three zones, an Inner-ramp at 0-50 m, a Mid-ramp at 50-120 m, and an Outer-ramp at >120 m (Figure 2.7). These may be taken as synonyms of Inner Shelf, Middle Shelf, and Outer Shelf. These boundaries are based on hydrodynamic, geological, and eustatic criteria. The 50-m boundary is the limit of fair-weather wave influence and is a natural inner shelf hydrodynamic boundary.[7] Within this inner shelf zone, there is often a fraction of terrestrial siliciclastic sediment. In the Kimberley, the 50-m isobath marks the approximate submerged margin of the Pre-Cambrian Kimberley Block which is a helpful coincidence.

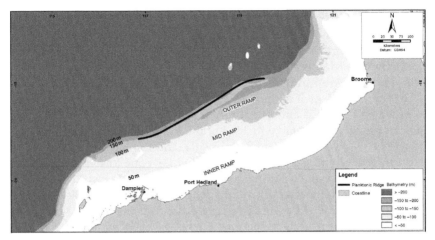

FIGURE 2.7   Divisions of the North West Shelf into Inner Shelf, Middle Shelf, and Outer Shelf sections, based on hydrodynamic and geological characteristics and eustatic history. *After Ref. 7; redrawn by URS Australia.*

The 115-m isobath is about the limit of the storm wave base and thereby a natural middle shelf hydrodynamic boundary. On the North West Shelf, it is also the approximate limit of coarse biogenic sand and gravel with a significant fragmental fraction. Beyond 120 m, in the outer shelf/slope zone, the sea floor sediments are primarily of pelagic origin. Along the length of the North West Shelf, this isobath corresponds approximately with the top of a distinct escarpment or cliff with its seaward toe at about 120-125 m. This feature is believed to be the shoreline during the Last Glacial Maximum (LGM). Above this isobath, the shelf was subaerially exposed at least once in the Quaternary. In other words, the inner and outer shelf zones were terrestrial habitat during the lowest sea level of the LGM.

A natural seaward boundary of the outer shelf would normally be the shelf edge, where such exists. Northwest of the Montebello Islands and along the margin of the Sahul Shelf, an outer shelf margin is obvious at 180 m. Elsewhere on the North West Shelf, as noted earlier, an outer margin is impossible to determine on topographic criteria because there is no clear edge to the shelf itself, but for the purposes of species distribution studies, the outer limit of the outer shelf may remain undefined.

Accordingly, in this study, the shelf is divided in a manner defined by hydrodynamic and geological characteristics and eustatic history, with the following three zones:

- Inner Shelf 0-50 m
- Middle Shelf 50-120 m
- Outer Shelf >120 m (local circumstances used to determine an outer margin, if necessary).

2.1.3.2 Geomorphic Features of the Continental Shelf

The surface of most of the Rowley Shelf slopes gradually from the shore to the shelf margin,[4] although the inner shelves of the West Pilbara and especially the Kimberley have complex topography. Topographically, the sea floor of the Sahul Shelf is much more varied.

The continental shelf of the Canning Basin off the coasts of the Kimberley and East Pilbara is almost entirely featureless. Exceptions are Baleine Bank and shoals associated with the Lacapede Islands close to the shore on the western side of the Dampier Peninsula, and Lynher Bank and shoals on the middle shelf associated with the Cape Leveque Rise that marks the boundary between the Canning and Browse Basins.

Complex seabed topography of the inner shelf in the Kimberley Bioregion represents the submerged terrestrial landscapes of the Kimberley Plateau's dissected western margin. There are high, submerged ridges and hills and Late Pleistocene valleys with paleochannels representing ancient drainage systems (Figure 2.8). Deeply incised canyons cut by major rivers into the terrestrial margin of the Kimberley Plateau continue across the middle shelf. There are also contemporary tidal scour channels and accretive mud banks, shoals, and reefs. Browse Reef and the Heywood and Echuca Shoals are thought to be ancient bioherms arising from the contemporary outer shelf close to the shelf margin.

On the inner shelf of the West Pilbara (Carnarvon Basin), there are expanses of limestone pavement on the seabed, and many shoals, banks, reefs, and islands. Most of these topographic high areas are remnants of

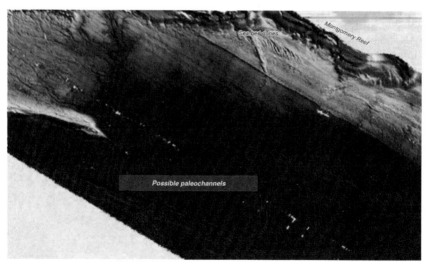

FIGURE 2.8   A sidebeam image of Late Pleistocene paleochannels on the sea floor south of Montgomery Reef, Kimberley Bioregion. *Image courtesy AIMS.*

the Late Pleistocene coastal plain inundated by the LGM transgression, or Holocene accretive sedimentary structures. An important feature is the sinuous bank over an anticlinal trend on which the Montebello-Lowendal-Barrow Island group and associated shoals are situated (Figures 2.5 and 4.32). The outer shelf of the West Pilbara is featureless except for Rankin Bank and Glomar Shoal that are high points just inside the 100-m isobath and would have been islands during the LGM and possibly coral reefs during intermediate stages of the eustatic sea level cycles.

Most of the inner and middle Sahul Shelf north of Bonaparte Gulf is also a drowned terrestrial landscape with a complex topography (Figure 4.7). It has a major central depression, known as the Bonaparte Depression, that has an area of $37,170 \, km^2$ and maximum depth of around 150 m.[8,39] Today, the Bonaparte Depression is a deep marine basin ($\sim$150 m) ringed by incised paleo-drainage channels. In the southwest, there is a deep subsea canyon called the Penguin Deep that was also once an incised river valley with a major deltaic fan at its mouth. Today, these subsea canyons channel tidal water on and off the shelf and slope of the Timor Sea.

During the LGM, when the Timor Sea was reduced to a narrow seaway, the Bonaparte Depression was an estuarine gulf, probably with extensive mangrove habitats, surrounded by an aerially exposed, grassy-plain.[22,40,41] The gulf opened to the sea along a sunken graben structure known as the Malita Shelf Valley. Several large rivers of the Australian continent and many smaller rivers off the surrounding plains drained into the gulf.[42]

The northwest-facing side of the Sahul Shelf, along the margin of the Java Trough, bears a series of submerged banks (bioherms) known as the Sahul Banks, including clusters arising from the continental slope and a series along the 200-m isobath of the shelf edge. The boundary of the Bonaparte and Browse Basins is a southwest-facing ridge called the Londonderry Rise that lies along the shelf edge. There is another series of submerged banks along the southern margin of the Londonderry Rise southeast of Ashmore Reef, including Johnson and Woodbine Banks just inside the 200-m isobath. Cartier Reef and Barracouta Shoal lie on the continental slope below that isobath.

2.1.3.3 Geomorphic Features of the Continental Slope

The continental margin of the North West Shelf has been subjected to downwarping. Several marginal plateaux and terraces have been the result. Along most of the length of the Rowley Shelf, at depths from 200 to 2000 m, there is a subhorizontal zone on the continental slope referred to as the Rowley Terrace (southern part) and Ashmore Terrace (northern part). Beyond this is the Exmouth Plateau in the south and Scott Plateau in the north, at depths between 2000 and 4000 m.

2.1.3.4 Shelf Terraces

A striking feature of the North West Shelf, throughout its length from North West Cape to the Arafura Sea, is the occurrence of semihorizontal surfaces, separated by escarpments or short steep slopes. They have been described by many authors.[2-4,22,40,43-47] These seabed topographic features have been described as "cuestas" (Fairbridge[44]), "submerged shorelines" (James et al.[7]), and "terraces, notches, and scarps" (Jongsma[47]). Although there are discontinuities and variations in level, some of these terraces may be traced and correlated over long distances.

The terraces have varied and complex origins. Fairbridge[44] and Carrigy and Fairbridge[2] presented three possible explanations of Sahul Shelf terraces involving subsidence with tilting, eustatic sea level change or faulting (Figure 2.10). They concluded that none of these alternatives alone explain their origins completely and that these processes have operated in various combinations.

The deepest and oldest of the shelf terraces lie beyond the shelf edge, e.g., on the marginal Rowley and Ashmore Terraces at depths of 200-400 m and are thought to be Late Tertiary structures, partly due to subsidence of the continental margin and unrelated to Quaternary eustatic sea level change.[3] They appear to be erosional incisions into a Miocene surface. In many places, they are covered by subsequent sediments but remain clearly evident in seismic profiles. Some inner shelf terraces have tectonic origins, e.g., terraces of the inner shelf in the Kimberley region that are bounded by deep faults.[48] However, most of the terraces on the North West Shelf are thought to be Quaternary strandlines, either with or without the additional effects of subsidence.

The most widespread and persistent of the submerged North West Shelf strandline terraces can be traced, with only minor breaks, along almost the entire shelf margin from North West Cape to the Arafura Sea. In the west, on the Rowley Shelf, this terrace lies at around 120-125 m,[3,7] while in the Timor Sea, it lies at 102-143 m[45] and, in the Arafura Sea, it lies at around 125 m.[47] Fairbridge[44] and Glenn[22] identified it on the Sahul Shelf as a 10-15-m high scarp at around the 100-112-m isobath. The seaward escarpment of this terrace is interpreted as the low stand shoreline of the LGM (Late Pleistocene Marine Isotope Stage (MIS) 2).[7] It rises abruptly, sometimes cliffed, with a height of up to 30 m, and has a face of cemented grainstone. The topographically irregular top is commonly rocky with a prolific benthic epifauna.

On the seabed of the middle and inner shelf in both the Bonaparte and Browse Basins, there are series of widespread shelf terraces at various depths. The origins and mode of formation of these structures are thought to be related to the combined effects of eustatic sea level change and subsidence.[22] In the southwest, on the Rowley Shelf, any submerged Pleistocene terraces above the LGM strandline appear to have been obliterated.

At the western end of the North West Shelf, there are also terraces on coastal land that record Quaternary high sea level and tectonic uplift events. On the western side of the Cape Range Peninsula, high above present sea level, there are high terraces, thought to be of early Pleistocene and perhaps Pliocene age that are evidence of local uplift.[30,49] On the eastern side of Exmouth Gulf, on Urala Station in the vicinity of the Ashburton delta, there are also at least two early or middle Pleistocene high benches with estuarine and shallow embayment fossil faunas that are undescribed. There are high limestone benches at several localities along the coast of the Canning Bioregion (Figure 8.10), attributed to the Bossut Formation and thought to be of early or middle Pleistocene age.[50,51]

On shores of the Pilbara coast, at the western end of the Rowley Shelf, there is commonly a high terrace or benches several meters above mean sea level, comprising either consolidated dunal limestone or coral limestone, the latter being the remains of Late Pleistocene coral reefs (Figure 8.8). These benches are dated as Late Pleistocene (MIS Stage 5e), that is, the Last Interglacial period of high sea level. Similar benches of coastal limestone (but not coralline) occur along the shores of the North West Shelf as far north as North Head on the west coast of the Dampier Peninsula (Figure 8.9). However, there are no comparable Late Pleistocene, high sea level terraces on the Kimberley coast, or elsewhere on the shores of the Sahul Shelf—evidence of Late Quaternary regional subsidence in that region.

2.1.4 Sediments of Benthic Shelf Habitats

2.1.4.1 Sediment Types

Surficial sediments of the North West Shelf have been described by Carrigy and Fairbridge[2] and Baker et al.[8] Detailed studies on the sediments of the Sahul Shelf have been published by Van Andel and Veevers[45] and Glenn.[22] Jones,[4] Brown,[52] and Dix et al.[38] have described the sediments and sedimentary processes on the Rowley Shelf. James et al.[7] described eight principal sediment facies on the Rowley Shelf, and the following notes are derived mainly from that source (Figure 2.9).

Overall, the North West Shelf of today is a low depositional environment. Contemporary sedimentation rates are low. Surficial sediments are predominantly carbonates, the carbonate component exceeding 90% in the majority of samples.[4] The sea floor is strongly affected by cyclonic storms, long-period swells, and large internal tides, resulting in accumulation of mainly coarse-grained sediments.[7] Medium, coarse, and very coarse-grained calcarenites and lag gravels, consisting mainly of relict organic materials, are spread widely over the shelf.[3] Regions of fine-grained sediments are restricted to the nearshore zone, local sediment traps, and the outer shelf margin and slope.

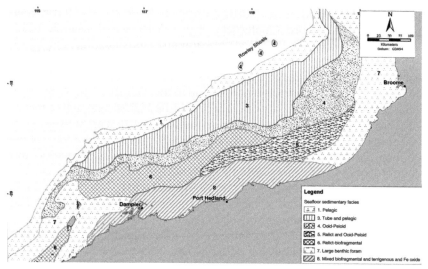

FIGURE 2.9   Substrate facies of the Rowley Shelf. *After Ref. 7; redrawn by URS Australia.*

On the inner shelf, terrigenous sediment with relatively low carbonate content occurs mainly in two coastal zones, the Kimberley and Pilbara associated with areas of moderate rainfall and river discharge. On the Kimberley coast and in Bonaparte Gulf, there are deep mud deposits in gulfs and landlocked bays, especially in association with estuaries and mangrove habitats. Nearshore sediments along the Pilbara coast between Dampier and the delta of the De Grey River are mixed biofragments and terrigenous quartz and iron-impregnated carbonate grains (facies 8). Elsewhere, inner shelf sediments are predominantly foraminiferal sand and gravel (facies 7).

On the mid-shelf of the Rowley Shelf, the sediments are generally medium to coarse-grained ooid/peloid sands with a fraction of coarser biogenic skeletal material (facies 4 and 5) or coarse biogenic sand and gravel with a minor fraction of ooids and peloids (facies 6). For the most part, the mid-shelf sediments are palimpsest, with diverse particle types of different ages, variably mixed by contemporary biological and hydro-dynamic processes. They display complexities that are largely explainable in the context of modern and late Quaternary oceanography. The ooid and peloid sands are believed to have been formed in a shallow evaporative environment during the first phase of the post-LGM but now "stranded" at a depth and in a lowered salinity zone that is unsuitable for continued sediment formation of this kind. During the sea level "stillstand" of the past 6000 years or so, this part of the continental shelf has been covered by greater than 100 m of water and subjected to the forces of episodic storm waves and periodic internal waves that have reworked the sediment, even piling it into huge sand waves that are spaced hundreds of meters apart, with rock pavement exposed in the swales between.

Mid-shelf sediments of the Sahul Shelf are also predominantly carbonate sand and gravel, but there is mud in holes and valleys.[8] The Bonaparte Depression on the Sahul Shelf is a mid-shelf sediment trap with a bed of silty clay sediment, an outcome of its alternating Quaternary history as marine basin and estuarine gulf, although there are coarse-grained sediments in the southern part closer to the coast.

On the outer Rowley Shelf, beyond the 120-m strandline, to depths of 200 m or more, sediments include a relictual element with calcareous tubes, ooids/peloids, and biofragments (facies 3). The tubes are thought to be those of infaunal crustaceans while the biofragments include parts of the shallow water bivalves *Anadara* and *Pinna*. James *et al.*[7] interpreted these materials as representing a former benthic community of the inner shelf in the Late Pleistocene to Holocene period. These relictual elements are reworked and mixed and overlain today by modern pelagic sediment. Beyond the 200-m isobath, on the continental slope, pelagic sediments dominate, becoming progressively muddier with increasing depth.

A noteworthy feature of the modern outer shelf along 300 km or so between longitudes 116°30′E and 119°30′E is the presence of an elongate mound of pelagic sediment[3,4,7] (Figure 2.7). There is another, smaller structure of this kind further north between 122°E and 123°E. These are contemporary constructional features. They are composed of fine pelagic carbonate sand and mud with aragonite needles and the remains of pteropods, pelagic foraminifera, sponge spicules, and some benthic foraminifera and echinoids (facies 2). Radiocarbon dating of material from these deposits has yielded dates of 0.35 and 9.2 ka, indicating contemporary age. These deposits are interpreted as a result of high surface productivity along the shelf edge, associated with upwelling and nutrient enhancement in the zone of near-surface shear between the Holloway Current and Indian Ocean water. It is, thereby, a contemporary outer shelf deposition process that began with the advent of the Holloway Current in the post-LGM period.

Another feature of the North West Shelf is the presence in many areas of limestone pavement, exposed on the seabed or covered beneath a veneer of carbonate sediment. On the nearshore part of the inner shelf, especially in the west, the pavement is apparently of Pleistocene age and was probably lithified during aerial exposure during periods of low sea level. However, limestone pavement also exists on the outer shelf to depths of at least 282 m.[4] That deep limestone is commonly brown or gray, platy, or nodular rock consisting of skeletal material poorly cemented in a micritic matrix. Sometimes it is oolitic and pelletal or chalky white limestone. It is thought to be a thin layer of Pleistocene origin. Jones[3] discussed the issue of formation of limestone seabed pavement in these submerged conditions. On the inner shelf, exposed limestone pavement is generally covered with growth of macroalgae. In deeper water, it may bear an epifaunal cover of sponges, bryozoans, and alcyonarians. Such benthic epifaunal communities are significant contrasts to the infaunal invertebrate communities of areas covered with carbonate sands.

2.1.4.2 Summary of Rowley Shelf Sediments, After James et al.[7]

Outer-ramp Facies

(1) The sea floor on the continental slope and marginal platforms below 210 m are blanketed by pelagic sediment, mostly planktonic foraminifer-pteropod carbonate sand or mud, becoming progressively muddier below 500 m.

(2) Over a distance of about 300 km (between longitudes 16.5° and 120°), there is a conspicuous ridge around 45 m high and 20 km wide that runs along the margin of the shelf, at a depth of 130-175 m, beyond the escarpment formed along the LGM shore. It is composed of fine-grained sediment composed of pelagic carbonate sand and mud.

(3) A zone of calcareous tubes, biofragments, ooids, peloids, and pelagic mud along the shelf margin at depths of 110-210 m. The modern component of the sediment includes numerous pteropods and benthic detritivorous foraminifera.

Mid-ramp Facies

(4) A band of ooid/peloid sand lies along most of the length of the Rowley Shelf at depths of 70-130 m. The sand includes a coarser fraction of mollusc, echinoderm, bryozoan, zooxanthellate coral, foraminifera, and rhodolith remains. Video images of the seabed in this zone reveal a meagre community of benthic molluscs and other invertebrates. Radiocarbon dating revealed that the ooids were formed over a narrow time period between 15.4 and 12.7 ka, corresponding with the beginning of the rapid sea level rise during the first phase of the post-LGM.

(5) A patch of relictual fragments and ooid/peloid sand, at depths of 40-70 m, similar to facies 4 except that the proportion of relict and biofragments is greater than that of the ooid/peloids. Living benthic biota is sparse, principally bivalves, azoothanthellate corals, clypeaster echinoids, and bryozoans, with patches of pavement bearing prolific sponge growth.

(6) Sand and gravel comprising relict intraclast and biofragments at depths of 40-70 m. The sand is mostly benthic foraminifera and mollusc particles and less than 10% ooids and peloids. The coarser material is mostly mollusc and bryozoan fragments.

Inner-ramp Facies

(7) Foraminifera sand and coarse sand—inshore at depths less than 60 m. Comprising poorly sorted sand with variable proportions of biofragments, locally with patches of terrestrial muds and quartz and feldspar grains, and an abundant and diverse benthic fauna and flora. Results of a systematic dredging study of the benthic habitats on this type off the Canning coast were published by Fry *et al.*[53]

(8) Mixed biofragmental-terrigenous sand and gravel—inshore at depths less than 40 m, in the central Dampier region. Sediments of this zone have a high proportion of carbonate particles impregnated with iron oxide and up to 20% sand-sized quartz and crystalline rock fragments. There is usually an abundant and diverse modern biota a systematic dredging study of the benthic fauna in this habitat type in the Dampier Archipelago.[54]

2.1.5 Tectonism and Shelf Subsidence

Gondwanaland began to break up in the Late Jurassic, and at that time, the continental margin of the northwestern region began to subside. Carbonate sediments became dominant in the Late Cretaceous, most of the sedimentation deposited over preexisting block-faulted topography. A regional northwest tilt of the North West Shelf began in the Cretaceous and continued throughout the Cainozoic. This tectonic activity resulted in the formation of several major marginal terraces, e.g., the Rowley and Ashmore Terraces and the Exmouth and Scott Plateaux (Figure 1.1).

Following northward drift through the early Tertiary and collision of the Australian and Eurasian continental plates in the Miocene, tectonic activity was reactivated along the North West Shelf.[1] In the Carnarvon Basin, Miocene to recent faulting disrupted the regional subsidence pattern. There was also some local uplift in the Pliocene resulting in the development anticlines that form the backbone of the emergent Montebello-Lowendal-Barrow chain of islands and the Cape Range, Rough Range, and Giralia Range on the mainland. Elevation of Cape Range created the Cape Range Peninsula and the uplift of terraces along the western coast of the peninsula, at least two of which are Pleistocene fossil coral reefs (antecedents of the contemporary Ningaloo Reef).

Further north in the Offshore Canning Basin, there are major northeast trending anticlinal structures that had their origins in the Miocene. One of these is located on the deep marginal Rowley Terrace[4] and provided a topographic high upon which a series of bioherms developed (the Rowley Shoals) their growth keeping pace with ongoing marginal subsidence. Similar anticlinal structures occur on the Ashmore Terrace of the outer

Browse Basin, and they too provided topographic highs on a subsiding marginal terrace for the development of bioherms (Scott and Seringapatam Reefs).[5]

There is also evidence of subsidence along the inner shelf of the Kimberley coast. Gregory[55] first described the ria-like morphology of the Kimberley Block, and since then it has been understood that this area must have undergone strong and recent subsidence. The terracing of the inner shelf off the Kimberley coast required significant Late Cenozoic subsidence.[2,43,44,46] Teichert and Fairbridge[43] described and discussed the coral reefs of the shelf edge and inner Sahul and Rowley Shelves and suggested that tilted regional subsidence was a fundamental factor in their development. They concluded that eustatic sea level change and subsidence have been interacting causal factors in the development of the Kimberley reef structures we see today (Figure 2.10). Ongoing subsidence of the inner shelf of the Kimberley coast would explain the absence of coastal Pleistocene fossil reefs visible above contemporary sea level.

In the Pliocene, there was rapid downwarping and flexing of the crust along the Australian northwestern continental margin,[56] especially along the shelf edge of the Sahul Shelf and strong downwarping within the Timor Trough associated with the Australian and Eurasian plate collision.[57] These tectonic events resulted in reactivation of faulting activity

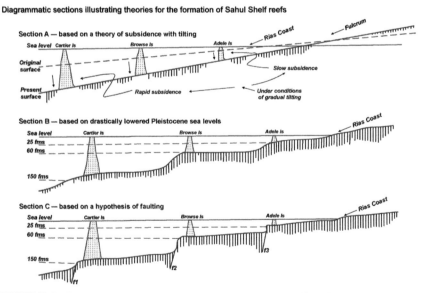

FIGURE 2.10   Diagrammatic representation of the Sahul Shelf profile indicating titled subsidence. *From Ref. 44; redrawn by URS Australia.*

and extensive leakage of hydrocarbons that may have been a factor in the development of the many carbonate reefs during that time.[58]

Models of continental margins predict a subsidence rate of about 1 cm/ka or less for the edge of a rifted continent following a rift event.[59] There is evidence that the rate of subsidence of the Sahul Shelf is significantly greater than this. Subsidence rates on the flanks of the Bonaparte Basin have been estimated at around 10-15 cm/ka for the last several million years.[60] Higher rates of subsidence in the order of 15 cm/ka, that is, a fall of around 150 m, during the Quaternary have been suggested.[22] Subsidence rates at South and North Scott Reef on the Rowley Shelf margin have averaged 0.45 and 0.29 m/ka, respectively.[32]

A pronounced tilt of the whole Australian continent, along a NW/SE axis, with ongoing subsidence of the northern side, has been proposed[61] (Figure 2.11) and the magnitude of the downward motion of the northern Australian margin estimated to be 250-300 m since the mid-Miocene (averaging 3 cm/ka). This is an average figure over a longer period and consistent with an increased subsidence rate during the Pliocene downwarping along the outer Sahul Shelf.

2.1.5.1 Section Summary

Tectonic activities since the Miocene have had profound consequences for the distribution of the modern benthic shelf and shore marine biota on the North West Shelf, perhaps the most significant being the following:

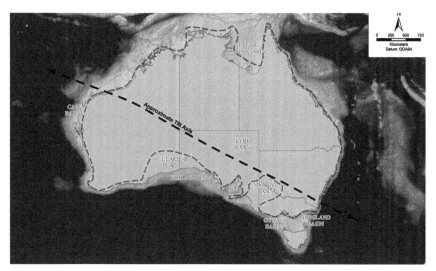

FIGURE 2.11   A "tilting continent"—an hypothesis that the Australian continent is tilted along NW-SE axis. *After Ref.* [61]*, figure 3; redrawn by URS Australia.*

(1) The northward Tertiary drift of Australia (and the other southern continents) brought it partly into the world's tropic zone, straddling the Tropic of Capricorn, creating the basic division of Australian coastal waters into a northern tropical biogeographic region and a southern warm temperate biogeographic region.

(2) The initial Miocene impact of the Australasian and Eurasian plates established a temporary shallow sea connection[62,63] that may have allowed the passage of marine animals lacking pelagic larvae from the Tertiary pan-tropical Tethyan biogeographic region onto the Australian continental margin.[64]

(3) Tertiary folding, faulting, and fracturing created topographic highs on the seabed that, coupled with ongoing subsidence of the shelf margin and possibly hydrocarbon leakage, facilitated the development of coral reefs.

(4) The emergence of the Cape Range anticline in the Pliocene established a partial biogeographic barrier between the North West Shelf and the Dirk Hartog-Rottnest Shelf of the west coast that has resulted in vicariant speciation and a degree of endemism on either side (see Section 9.5.2.1).

(5) Quaternary subsidence of the Kimberley margin facilitated the development of species-rich fringing coral reefs along the inner shelf of that region following the post-LGM transgression.

2.1.6 Eustatic Sea Level Change

2.1.6.1 Global Quaternary Sea Level Change

Global climate over the past 3 million years has oscillated between glacial and interglacial conditions with transfer of immense amounts of water between the earth's ice sheets and the oceans and corresponding eustatic oscillations of sea level.[65] The trigger for these climate and sea level variations is believed to relate to cyclic changes in the Earth's orbit and solar radiation (the Milankovitch cycles) and the insolation of the world's atmosphere and oceans, but there are complex feedback effects that complicate the outcomes. Through the Pleistocene period, there were 17 major sea level cycles occurring at around 100,000 year intervals with maximum amplitudes of 120-140 m, but there have also been smaller scale variations superimposed over the major glacial and interglacial cycles. Using a variety of techniques, the stages of the sea level oscillations have been documented and numbered (MISs).

The eustatic rises and falls of sea level through the Pleistocene resulted in drastic transgressions and regressions of the sea across continental shelves that had profound effects on the spatial distributions of both marine and terrestrial plants and animals. The last of the major eustatic cycles, with an amplitude of 125 m and pulses and irregularities, was the dominant controlling factor in the establishment of the modern benthic habitats of the North West Shelf, especially on the middle and inner shelf where habitats alternated between marine and terrestrial.

The Chappel and Shackelton graph of eustatic sea level change during the past 150,000 years of the Quaternary[65] (Figure 2.12) is a global approximation, although it is known that the timing of the major highs and lows, as well as the exact levels achieved, is subject to regional variation, partly as a result of isostatic and tectonic variability and partly because there is a lack of synchronicity between the northern and southern hemispheres.[66,67]

The Last Interglacial was a period of global warming that peaked at about 125,000 years BP (Figure 2.12). It was marked by sea level at a little above that of today and is referred to as substage MIS-5e. There followed a series of oscillations between warm interstadials (designated as MIS stages 5d-a, 4, and 3) and increasingly cold stadials. The last glacial period (Last Glacial Maximum [LGM] designated as MIS-2) began about 30,000 years BP and lasted 11,000 years. At its nadir 19,000 BP, global sea level fell to about 125 m below that of today. Then as global climate warmed sea level rose to its present level 6000 years before the present.[68,69]

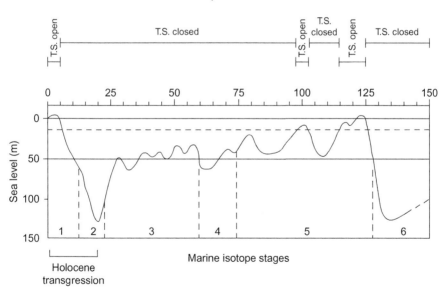

FIGURE 2.12 The Late Quaternary sea level curve and marine isotope stages over the last 150,000 years (after Ref. 65). The bars across the top indicate periods when the Torres Strait (T.S.) was open and closed and the horizontal dotted line represents the −12 m depth of the Torres Sill and the height of sea level needed before the Torres Strait was opened. The solid horizontal line represents the −50 m depth of the Arafura Sill and the height of sea level need to inundate the Gulf of Carpentaria.

The rate of post-LGM sea level rise was not uniform, varying from 5 to 40 mm per year. There were three major pulses of meltwater resulting from events of rapid decay of the polar ices sheets, causing sudden increases in the rate of sea level rise.[70] These are known as the Melt Water Phases (MPW), and they occurred at 19,000, 14,000, and 11,500 years BP, each phase lasting several hundred years with rise rates of about 30, 40, and 25 mm per year, respectively (Figure 2.13). They are of fundamental importance in the present context because they represent periods when the growth of many coral reefs was unable to keep pace with sea level rise and reef communities "drowned." The intervening periods of slower sea level rise were periods of coral reef growth.[71] It is likely that other benthic shelf and coastal habitats, especially mangals, were affected in similar ways by varied rates of sea level rise with stop-go phases of growth as they adjusted their positions in relation to regression of coastline and, in the case of the benthic shelf habitats, changes in substrate.

2.1.6.2 Sea Level Change on the North West Shelf

Human occupation of northern Australia began long before the Last Glacial period. It is salutary to remind ourselves that subsequent changes of climate must have had profound economic and social consequences for the

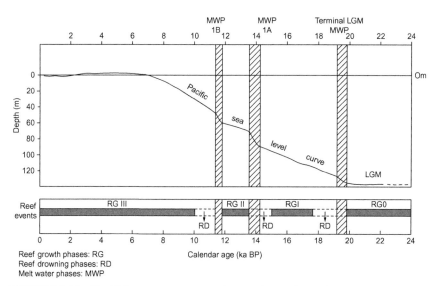

FIGURE 2.13 The Holocene transgression showing the Pacific sea level curve and its relationship to three meltwater phases (MWP) and alternating phases of coral reef growth (RG0, I, II, and III) and reef drowning (RD). The same sequence of events would have also affected mangroves and other shore habitats and even benthic shelf habitats. *Modified from Ref. 71, figure 11.*

early Australians. The rapid post-LGM sea level rise resulted in loss of very large areas of terrestrial territory, a band nearly 300 km wide in the case of the coastal land in the Canning Basin and most of the area is now occupied by the Gulf of Carpentaria. For the marine benthic shelf organisms that are the present subject, the reverse was true. For them, the LGM was a period when the continental shelf was very narrow, and benthic shelf habitat was severely restricted while the interglacial and present high sea level periods were times of an exceptionally wide shelf and vastly increased area of benthic habitat. However, it is important to keep in mind that both the LGM and the last interglacial were brief periods of extreme lows and highs. For the greater part of Quaternary time, sea level has averaged around −50 m below the present and the continental shelf has been of intermediate width.

Figure 2.5 shows the approximate position of the shoreline at the western end of the North West Shelf during the LGM when, globally, sea level was around −125 m, and there was a very wide coastal plain northeast of the Cape Range Peninsula. The present middle and inner shelf islands of the West Pilbara were then hills on the coastal plain. Figure 2.6 shows the comparable position of the shoreline of the Sahul Shelf at the same time. The LGM shoreline is identifiable today as a 10-30 m high escarpment or steep ramp along the outer part of the contemporary mid-shelf over much of the length of the North West Shelf. Evidence that this was the approximate position of the LGM shoreline is found in the following observations:

- Palynological data suggest that the emergent part of the Sahul Shelf at 18,000 years BP was covered by open grassland and eucalypt woodland in a dry, arid climate, with mangroves in coastal estuaries.[72]
- Calcareous nodules from areas surrounding the Bonaparte Depression are interpreted as having had terrestrial origins.[42]
- The presence of fossil brackish-water bivalves from the floor of that now marine embayment.[42]
- The presence of sand, consisting primarily of calcareous ooids and peloids, on the mid-ramp of the Rowley Shelf in the depth range 70-130 m.[7] The ooids were dated at 15.4-12.7 ka, i.e., formed during the early phases of the post-LGM transgression. They were interpreted as having been formed in shallow (less than 5 m) tidal-dominated coastal settings with elevated salinity, now stranded on the mid-shelf in an environment where they could not have been created.
- The discovery of remains of shallow water molluscs including species of *Anadara* and *Pinna* (species that now inhabit soft substrata of the nearshore, inner shelf) seaward of the −125 m escarpment, in the −110 to −210 m depth range.[7]

There is some ambivalence about the depth of the LGM terrace and a need to substantiate this evidence and establish more precisely the position of the shore at that time (Section 2.1.5). While James *et al.*[7] describe it as a cliff or escarpment located along the −125-m isobath of the Rowley Shelf, Jones[4] gives −120 m. In regard with the Sahul Shelf, Fairbridge[44] refers to "the 10-15 m high scarp that commonly occurs in the vicinity of the 55-60 fathom (100-110 m) isobath," while Van Andel and Veevers[45] refer to "a small scarp 10 to 15 m high" marking a change of slope between −110 and −130 m on the Sahul Shelf. Glenn[22] describes an "outer shelf terrace" at −112 m, taking that measure at the foot of the escarpment. This variability may be accounted for, in part at least, by whether the depth of the terrace is taken at the foot of the escarpment or the top and by varied rates of shelf subsidence. Nevertheless, it seems that a conspicuous escarpment marking the low stand LGM shoreline may be recognized along most of the North West Shelf from the Sahul Banks to at least the Montebello Islands.

Dix et al.[38] noted a series of paleo-embayments that existed along the coast of the Rowley Shelf during the LGM period of low sea level. At that time, the continental shelf of the region was very narrow, occupying the space now classed as the outer shelf and a little of the upper slope. That represents a significant loss of marine benthic shelf habitat during the Last Glacial period that was recovered in the Holocene by the post-LGM transgression over a period of only about 12,000 years. Such a gross change in the area of available benthic habitat must have had severe ecological, demographic, and biogeographic consequences for communities of benthic organisms.

However, the story is more complicated than that. The transgression brought on dramatic impacts on sea floor sedimentation of oceanographic conditions and other aspects of climate change, as well as sea level change.[7,38] To begin with, the southerly flow of warm, low salinity, tropical water, now referred to in that region as the Holloway Current (Section 2.2.1), was weak or did not operate during the LGM period. It is thought to have recommenced only around 12,000 years ago, halfway through the post-LGM transgression.[7,73–75] Most of the shallow carbonate deposition that is such a feature of the mid-shelf ceased about 10,000 years ago with the resumption of the Holloway current. Not only did the Late Quaternary sea level drastically change the available area of benthic shelf habitat, but associated oceanographic aspects of climate change altered the nature of the sea floor sediment over large areas of the shelf.

Another aspect of fundamental importance to the shelf biota is the rate of the post-LGM transgression. From the time it began at around 19,000 years BP until it reached its present level 6000 years BP, sea level rose 125 m, an average rate of nearly 1 m/100 years. On the North West Shelf at its widest point (Canning Basin near Cape Keraudren), it translates to an annual advance of the shoreline of about 13 m. (Stoddart[76] estimated that in "the East Indies the sea transgressed across flat-lying land at rates of up to 10-15 m/day.") However, as noted earlier, rate of sea level rise, and thereby transgression across the sloping shelf, was very variable. Clearly, benthic shelf habitat during the time of the transgression was highly unstable, and it is reasonable to question whether the 6000 years of subsequent stillstand (more or less) was enough time for community stability to have been achieved.

The benthic habitats that existed along the narrow LGM continental shelf, the rate and progress of the post-LGM transgression, and the processes by which new communities were established in the new marine benthic and shore habitats are of fundamental concern to the biogeographic history of the North West Shelf. It is clear that benthic habitats of the shelf were in a state of dynamic revision throughout the Holocene.

2.1.6.3 Benthic Shelf and Shore Habitats During the LGM

How might the primary benthic shelf and shore habitat types that exist today have been represented along the LGM shore of the North West Shelf? And how did the reconstruction of benthic shelf habitats proceed as the Holocene transgression progressed? Inner shelf and most of middle shelf benthic habitats were completely destroyed through the 11,000 years of the LGM. Even the shelf-edge atolls and platform reefs along the shelf margin would have been severely disrupted requiring reconstruction of their coral and other assemblages.

Rocky shore habitat may not have been as well developed during the Last Glacial as it is today. The shoreline must have been far beyond the margins of the Kimberley Block and the Pilbara Craton with their Proterozoic rocks and dissected land surfaces. There would have been no ria-coastlines like those of the modern Kimberley Bioregion and Dampier Archipelago, and it is very unlikely that there were outcrops of igneous and metamorphic rocks along the coast. There may have been coastal outcrops of Tertiary sedimentary rocks, and it is possible that there were Quaternary limestone shores, particularly along the escarpment now submerged 125 m or so below the surface. However, it seems likely that the shore was predominantly sandy, like the present coasts of the Canning, Eighty Mile Beach, and most of the Pilbara Bioregions.

The major rivers that enter the sea along the Pilbara and Kimberley coasts today would certainly have extended to the LGM shore. Their paleochannels are recognizable in the modern seabed bathymetry, especially on the mid and outer parts of the Sahul Shelf. They would have had estuaries and mangrove habitats, although the climate then may have been a little more arid than it is now and river outflow may have been less. It is reasonable to assume that rocky shore, beach, and mangal communities followed the sea landward as the post-LGM transgression progressed, although the process could not have been a steady one because of the unsteady rate of sea level rise. The response of coral reefs may have been different. Montaggioni[71] has discussed the conditions necessary for coral reef development. The first of these concerns the presence of preexisting substrate. In this regard, the Sahul Shelf and the Rowley Shelf are very different.

There are today many shoals and topographic high areas on the seabed along the Sahul Shelf margin that are too deep for modern coral reef growth but whose surfaces would have been in the coral growth zone during the LGM. As the transgression progressed, many of the Sahul Banks appear to have become submerged because of rapid subsidence as well as eustatic sea level rise. However, the mid and inner parts of the Sahul Shelf are topographically complex and may have presented suitable rocky substrata for reef growth. The Sahul Shelf may well have been an important refuge for corals and coral reef communities throughout the Holocene.

There may have been rocky substrate on the Rowley Shelf along the LGM shoreline providing suitable for coral reef development, but if that were the case, none survived sea level rise (although the atolls on the continental slope terraces were rejuvenated). Furthermore, the middle shelf is a virtually featureless soft sediment plain. Not until sea level had risen to around the present −50 m level would extensive rocky substrate have been encountered. That occurred around 10,000 years BP, following the third major meltwater phase (Figure 2.13). The modern reefs of the Kimberley and Pilbara coasts were established and grew during the last phase (RGIII) of coral reef growth. It follows that coral reefs may not have been extensive on the Rowley Shelf over the 20,000-year period between the beginning of the Last Glacial and the middle Holocene, except perhaps at the Rowley Shelf slope atolls.

The history of reef growth at the North West Shelf slope atolls is also problematic. The fate of atoll reefs during the LGM and earlier Pleistocene low stand periods has been much discussed. Stoddart[76] outlined the problem. During the LGM, most modern atolls were left as limestone islands with steep or cliffed sides and crater-like central valleys. Coral growth may have retreated down the sides around the perimeter of the islands but, at best, could only have been narrow fringing structures adhering to the perimeter walls, populated by coral species adapted to high energy, exposed situations. Many coral species and communities that live in lagoons are not capable of survival in such habitats. Reef communities of lagoons less than 130 m deep would have been obliterated. The proposal that coral reef communities simply followed the sea level down and back up again with the post-LGM transgression is too simplistic. Long-distance connectivity with distant reefs would have been essential for reestablishment of coral reef communities of the original kind.

Scott Reef offers a fine example of this problem. The deep (40-60 m) coral communities of South Scott Reef lagoon (Figure 4.8) include many foliose and delicate branching species that do not live in shallow backreef habitats or high energy habitats of the perimeter walls. The reestablishment of that community must have relied on recruitment from deep lagoon or deep bank communities within the region or from more distant reefs further north on resumption of the Indonesian Throughflow (ITF) and Holloway Currents as the post-LGM transgression developed. Underwood et al.[77] have shown that the contemporary coral communities of Scott Reef are essentially self-populating in the short-term ecological sense, but long-distance larval recruitment must occur even as rare events for reef recovery to be effective at geological timescales. This matter will be discussed later in the section dealing with connectivity (Section 8.4).

In the case of the Rowley shelf-edge atolls, there is a further complication. In the absence of a strong Holloway Current, the LGM sea conditions of the narrow shelf were probably nutrient rich (affected by upwelling— Section 2.2.5) so that reef organisms adapted to oligotrophic oceanic water may not have fared very well. There is likely to have been a change in the species composition of reef communities from oceanic to continental assemblages, even among those that survived sea level change.

There are other aspects of Holocene environmental change that may have affected coral reef growth on the North West Shelf, sea temperature being an important one. There remains uncertainty about how coral growth responded to LGM sea surface temperatures.[71] The geographic distribution of coral reefs during that period is poorly known because most of the evidence that may have been preserved is at present submerged. Such evidence does exist in both the Atlantic and Indo-Pacific oceans but none is yet known for the Australian North West Shelf. At Ningaloo Reef, close to the margin of the tropical zone, temperature conditions during the LGM may not have been optimal for coral growth, but it is not known whether coral reef communities adapted to lower temperatures and persisted there during the LGM. However, unfavorable temperature may not have been an issue at the northern end of the North West Shelf. A study of foraminifera from deep-sea cores[73] found no evidence of lowered temperatures in the Timor Sea during the LGM from which we could assume that ocean temperature in the northern part of the North West Shelf has remained within the range for coral growth through the Holocene.

2.2 THE OCEANOGRAPHY OF THE NORTH WEST SHELF

2.2.1 Water Characteristics

A report to the Department of the Environment and Water Resources[78] has summarized the physical properties of the water overlying the North West Shelf.

Broadly speaking, surface sea temperature (SST) of coastal waters in the Kimberley averages 28.5 °C and in the Pilbara 27.3 °C. Offshore, near the shelf margin, SST is similar averaging 28.5 °C in the north and 28.6 °C off the Pilbara. There is some seasonal variation. Near the shelf margin, around the Ichthys gas field of the Browse Basin, SST is consistently around 30 °C in March and 26-27 °C in July.[79] In the nearshore coastal waters of the Dampier Archipelago, the annual sea temperature range is approximately 20-30 °C, although in shallow embayments the range is greater, from 18 °C in winter to 34 °C in summer.[80] At a Pilbara mid-shelf location, there is seasonal variation in surface temperatures ranging from 35 °C in March-April to 24 °C in July.[81]

Offshore waters are stratified with a distinct thermocline at depths of around 30-50 m in summer and 70-120 m in winter.[78] The shelf break is an area of intense boundary mixing where mixed water intrudes onto the continental shelf. At a Pilbara mid-shelf station with a depth of 120 m, intrusion (from the continental slope) of cold-bottom water around 20 °C in summer has been observed,[81] interpreted as evidence of weak upwelling associated with a reversal of shelf-margin currents toward the northeast. Bottom temperatures as low as 12 °C were found below the thermocline on the continental slope and outer shelf in the Browse Basin.[79]

Lying in the path of the Indonesian Through Flow, salinity of water along the shelf margin is in the 34-35 ppt range. Salinity of coastal waters is generally higher but spatially and seasonally variable. Figures between 34.8 and 35.2 ppt are normal, but near the tropical monsoonal Kimberley coast and close to the major rivers of the Pilbara coast, temporary dilution by river discharge is a seasonal feature.

Nutrient concentrations are generally low in surface waters of the North West Shelf,[82] especially near the shelf margin where measures of N (μM) and P (μM) are usually around 0.05/12.8 and 0.11/0.85, respectively.[78] Nutrient concentrations below the thermocline are significantly higher and seasonal nutrient enrichment of benthic habitats on the outer and middle shelf may be expected as a result of upwelling. This effect is enhanced in the vicinity of topographic high features on the sea floor of the continental slope and outer shelf and greatly influences growth of algae, corals, and other fauna of the shelf-margin reefs and submerged banks.

Nutrient concentrations in nearshore coastal waters of the North West Shelf are relatively high. Measures of N (μM) and P (μM) from the nearshore waters of the Kimberley are usually around 0.21/16.6 and 0.19/1.15, respectively, and of the Pilbara coast, they are 0.14/11.65 and 0.14/0.86.[78] Nearshore nutrients are derived from benthic primary production in the photic zone of the inner shelf and inputs from the adjacent terrestrial environment, mainly via river discharge.

Turbidity is an important controller of ecosystems and species distributions. It is usually measured as Light Attenuation Coefficients or concentration of Total Suspended Solids (TSS). Light attenuation associated with increased concentration of suspended particles affects species composition of benthic communities, especially in regard with phototrophic invertebrates that depend on light for part of their nutrition. Conversely, concentration of suspended organic particles may have positive effects on heterotrophic benthic invertebrates that filter-feed directly on suspended organic particles. However, sedimentation that comes with high concentrations of suspended particles may also have negative effects on sessile invertebrates by smothering and interference with larval settlement.

Offshore waters of the shelf margin are generally clear with low concentrations of suspended particles and high light penetrations. Coastal waters, on the other hand, as recipients of fine terrestrial sediment and with increased nutrient concentrations, tend to be more turbid with higher concentrations of suspended particles. In the shallows of the inner shelf, turbidity is greatly influenced by disturbance and resuspension of fine sediments on the sea floor by tidal flows and waves (especially storm waves during cyclones). Resuspension of seafloor sediments during periods of spring tide is a major factor in the high turbidity of coastal waters of both the Kimberley and the Pilbara where macrotidal conditions prevail. Due principally to seasonal weather (wave) conditions, the turbidity of nearshore waters tends to be greater in summer than in winter.

Forde[83] described spatial and seasonal variations of turbidity in Mermaid Sound of the Dampier Archipelago. As expected, there was a gradient in TSS with highest concentrations occurring near the shore at the head of the sound (surface water figures up to 2.64 mg/L in March and up to 7.54 mg/L in early December). In open sea conditions at Nelson Rocks near the entrance to the sound where the water is clearer, the mean of a sample taken in March was 0.62 mg/L.

It is often said that the coastal waters of the Kimberley are turbid and seemingly an unsuitable environment for coral reef development.[84] However, the information now available indicates that turbidity in the Kimberley Bioregion is highly variable, ranging from 0.8 to 13 mg/L but is most commonly less than 2 mg/L.[85] These figures are consistent with the extreme macrotidal conditions and fine muddy seafloor sediments in the coastal waters of the Kimberley. The highest TSS concentrations recorded have been from samples taken from "water boils" during periods of spring tide. The lowest concentrations occur during periods of neap tide. Turbidity is moderate for much of the time at most nearshore localities but becomes very high, in places, during periods of extreme tidal flow. It may also be enhanced and prolonged during and after cyclonic storms and by the complexity of the coastline and sea floor topography.

These results suggest that, for benthic invertebrates such as corals, the damaging effects of sedimentation associated with high turbidity are likely to be transient rather than chronic, at least around the offshore islands of the Kimberley archipelagos. And further, with such variable coastal topography and tidal currents, the impacts of sedimentation on Kimberley benthic fauna are likely to be very variable at local spatial scales, enhancing the patchiness of communities like coral reefs.

2.2.2 Regional Circulation Patterns in the Eastern Indian Ocean

Circulation patterns on and adjacent to the North West Shelf are thought to be an outcome of complex interaction between tides, wind, and regional forcing by large-scale regional ocean currents at the shelf break.[86,87]

Chapter 2 of *The north-west marine bioregional plan, bioregional profile*,[88] summarizes ocean circulation patterns in the Indo-Australian Basin, that is, the area of the northeastern Indian Ocean between the northwest coast of Australia and the Indonesian islands of Sumatra, Java, and the Lesser Sunda Islands. There have been many oceanographic studies in the region that provide details and further references.[81,89–93] Nevertheless, there remains much to be learned about the sources and pathways of water affecting the North West Shelf. The primary source appears to be a flow of warm, low salinity, oligotrophic tropical water from the western Pacific, known as the Indonesian Throughflow (ITF), also called the Indo-Pacific Through-flow. Driven by a gradient in ocean level from the western Pacific to the Indian Ocean, it enters the Indo-Australian Basin through passages between the islands of eastern Indonesia (referred to as the "ITF gateways"). Once in the Indo-Australian Basin, ITF water flows in two directions (Figure 2.14).

South of Java, ITF surface water joins the South Java Current to form the westward flowing South Equatorial Current, but a portion of this recurves and flows by circuitous routes southward, bringing mixed Pacific and central Indian Ocean water to the western end of the North West Shelf at about latitude 20°S.[92] A branch of the Eastern Gyral Current also delivers central Indian Ocean water to that area.

ITF water also flows directly across the Timor Sea onto the Sahul Shelf margin and joins a seasonal (around autumn) southwesterly flow known as the Holloway Current.[93] The Holloway Current flows along the North West Shelf margin, driven by a steric height gradient, augmented by easterly winds during the autumn-winter monsoon season.[91,94] Water from this region has been identified as the major source of the very important eastern boundary Leeuwin Current off the West Coast.[95] However, recent authors have proposed that the source of the Leeuwin Current is complex with contributions of tropical water from both the seasonal Holloway

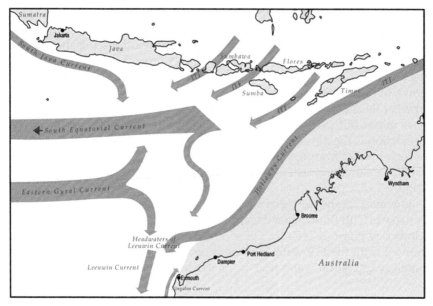

FIGURE 2.14   Generalized current system in the eastern Indian Ocean—constructed as a summary of figures by Refs. 92,93 *Drawn by Environmental GIS.*

Current and more indirect flows via the branches of the South Equatorial and Eastern Gyral Currents.[92,93] By this account, the headwaters of the Leeuwin Current lie at the western end of the North West Shelf (~20°S, 116°E) fed with both tropical Pacific and central Indian Ocean water (Figure 2.14). On its way through the Indonesian archipelago, the ITF passes through the East Indies Triangle (Section 8.1) with its immensely species-rich benthic shelf and reef habitats including the coral platform reefs and shelf atolls of the Molluca, Banda, and Flores Seas. When it operates, it is a means of immigration of planktotrophic marine species from that greatest of all centers of marine biodiversity to the North West Shelf and it is a fundamental biogeographic driver in the region. However, there is potential for pelagic connectivity also with the northeastern Indian Ocean via the South Java and Eastern Gyral Currents.

2.2.3 Coastal Currents on the Inner and Middle Shelf

Coastal currents on the inner- and middle-ramps of the North West Shelf are generally along-shore, wind-driven Ekman transport flows with seasonal reversal associated with the reversal of winter southeasterly trade winds and the summer monsoon.[87] However, those general trends are locally overridden by the effects of astronomic semidiurnal tides. Spring tide amplitudes range from 4 m on the West Pilbara coast to over

11 m in the southern Kimberley Bioregion. In the archipelagos of the West Pilbara and on the bathymetrically complex inner shelf of the Kimberley Bioregion, local currents are driven by tidal and wind forcing and may be strong and multidirectional and reverse direction on a twice daily basis. In the Buccaneer Archipelago of the southern Kimberley Bioregion, ebbs and flows of spring tides commonly induce very high velocities with local small-scale shears and dramatic whirlpools.

A minor but possibly biogeographically important feature of coastal circulation is the development of a wind-driven, northward, inshore flow of cooler water from the West Coast known as the Ningaloo Current that extends onto the western end of the North West Shelf in summer[96] (Figure 2.15).

In winter, at the northern end of the North West Shelf, water from the Bonaparte Gulf flows around Cape Londonderry and southwest along the Kimberley coast, driven by the southeast monsoon.[22,97] Whether this is derived from the ITF water delivered onto the Sahul Shelf via the Timor Gateway or coastal water that originates in the Arafura Sea is unknown.

2.2.4 Seasonal Variations

On the North West Shelf, the Holloway Current is a rather narrow surface current (~100 km wide and less than 300 m deep) that flows most strongly and persistently along the shelf break during the southeast trade-wind season from March to July. In spring and summer months, with the onset of the northwest monsoon and the reversal of the regional winds of northwestern Australia from southeasterly to westerly, the Holloway Current weakens and frequently reverses its direction of flow.

As a result of turbulence, and eddying and internal waves, there is an area of intense boundary mixing along the shelf margin.[78] In autumn and winter when the Holloway Current flows most strongly, there is mixing of oligotrophic ITF oceanic water and shelf coastal water by means of eddies associated with the large-scale, regional, oceanic circulation and possibly cyclonic storms. But in summer months, when the Holloway Current is weak or absent and likely to reverse direction, the water column along the shelf margin may stratify, allowing complex tidal currents to dominate water circulation patterns of the outer and middle shelf. Internal waves generated by barotropic tides along the shelf break may cause bottom turbulence and raise cooler, nutrient-rich water from the deep sea higher in the water column where it could intersect with the upper slope and outer and middle shelf with profound impacts on benthic and pelagic trophic systems.

On the shelf slope, the surface layer of warm, oligotrophic ITF water overlays cooler, more nutrient-rich waters of the deep sea. The boundary

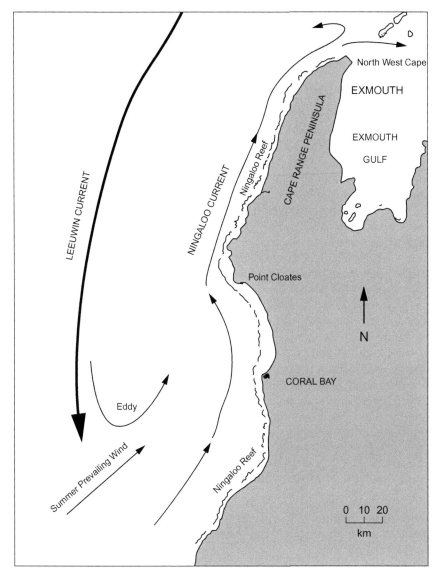

FIGURE 2.15 The Ningaloo Current—a cool, coastal, wind-driven current flowing near-shore, principally from September until mid-April. The Ningaloo Current is counter to the southerly Leeuwin Current and may flow around North West Cape onto the western end of the North West Shelf. *Modified from Ref.* [96]

between these two water bodies is marked by a sharp thermocline that is a barrier to convective mixing of nutrients. When the Holloway Current flows strongly in the autumn-winter months, upwelling would tend to be suppressed along the shelf break. However, during the Northwest

Monsoon period, when the influence of the Holloway Current is thought to be subdued or lacking, there is episodic upwelling of nutrient-rich deep-sea water onto the outer shelf.

2.2.5 Variations in Geological Time

Intensity of flow of the ITF varies interannually. The differential gradient in sea levels between the Pacific and Indian Oceans is greatest during La Niña years when the ITF flows most strongly. In El Niño years with lesser Pacific-Indian Ocean sea level gradients, the ITF and Leeuwin Current flow more weakly. An affect of the ENSO cycle on the Holloway Current might also be expected but is yet to be confirmed.[94]

On a longer, geological timescale, flow of the ITF and the Holloway Current are intermittent, developing strongly in periods of global warming and high sea level like the present but weakening or perhaps ceasing altogether during glacial periods of low sea level.[73–75,98] On evidence from seabed sediments, James *et al.*[7] have shown that during the LGM and the early stages of the post-LGM transgression, the shelf-margin current flow was arrested and the North West Shelf was a site of carbonate sedimentation with strong offshore winds that probably enhanced upwelling. Along the North West Shelf margin, the Holloway Current resumed flowing strongly about 12,000 years ago, inhibiting upwelling along the shelf margin and establishing the oceanographic conditions and the biotic connectivity services it provides at the present time. From this, it seems likely that during glacial periods coastal waters along the break of the narrow continental shelf may have been nutrient enriched, unlike the oligotrophic conditions of the present.

2.2.6 Section Summary and Conclusions

Several conclusions may be drawn that have relevance to the biogeography of the region. Three primary biogeographic units may be recognized that are interactive but have distinctly different origins and ecosystem functions.

(1) First, there is the shallow, oligotrophic surface water mass that forms the sterically and wind-driven Holloway Current flowing seasonally southwest along the North West Shelf margin.

The Holloway Current, when it flows strongly, may have two significant impacts on the shelf benthic biota—suppressing upwelling along the shelf margin and carrying planktic biota, including planktotrophic larvae of benthic shelf and reef animals that originated, at least historically, from the species-rich shelf areas of the East Indies Triangle.

Where the surface water mass of the Holloway Current floods around pinnacles and banks of hard substrate along the North West Shelf margin, it would deliver the larvae of oceanic reef organisms and establish rich coral reef ecosystems adapted to oceanic oligotrophic conditions. Along its inner margins, where mixing with outer shelf water occurs, it would deliver larvae of benthic shelf organisms. However, the connectivity that the Holloway Current provides is seasonal (March to July), apparently variable interannually (strongest in La Niña years) and intermittent at geological timescales (operating most effectively in interglacial periods of high sea level). It is important to note that connectivity by this means applies only to marine species with pelagic larvae.

The distances between the coral reefs of the Banda and Flores Seas and the Sahul banks are of the same order as those between the Ashmore complex, Seringapatam, Scott Reef, and the Rowley Shoals. At present, all these reefs lie directly in the path of the ITF and Holloway Currents, and perhaps, they should be regarded as units of the same biogeographic system. The biogeomorphic (and political) divisions represented by the two continental shelves and the Java Trough that separates them into Indonesian and Australian components have little relevance to the planktotrophic reef and benthic shelf ecosystems served by this ocean current system.

The fate of the shelf-edge coral reef communities during the glacial periods is conjectural. Although sea temperatures at those times are likely to have remained well within ranges tolerable to corals,[73] coral reefs might not have supported the same community types as those of the present day. The response of shallow coral communities to lowered sea level may have been to follow it down (and up again later), although the fringing reef habitats created around the sides of the atolls would have been very different to the back reef and lagoon habitats prevailing during high sea level periods. Also of importance to the shelf-margin coral communities may have been a change from oligotrophic to nutrient-rich conditions. It follows that the shelf-edge reef community assemblages and species composition may have changed during the oscillating Quaternary periods of climate and sea level. Species requiring oligotrophic, oceanic conditions may have become (temporally) regionally extinct. If it is the case that the Holloway Current switches on and off over geological time with episodes of climate change, perhaps we should regard the present-day species-rich shelf-edge reefs as transient communities that come and go or at least require reconstruction with renewed interglacial oligotrophic and high sea level conditions.

(2) The North West Shelf itself provides complex matrices of benthic shelf, reef, and shore habitats. The areas and configuration of these diverse habitats, and the assemblages of benthic organisms they support, are constantly changing in response to climate change episodes (Section 2.1.5), but taken as a whole, this system is "permanent" (though evolving) and may be thought of as the default biogeographic unit of the region. It is persistent through geological time whether the Holloway Current flows or not. Nevertheless, the benthic, reef, and shore faunas of the North West Shelf are enriched and refurbished by intermittent connectivity with the East Indies Triangle delivered by the Holloway Current during global warming periods like the present.

Inner shelf habitats are serviced primarily by limited nutrient delivery from terrestrial run-off and *in situ* primary production. Middle shelf and outer shelf habitats are serviced by nutrient delivery through upwelling and planktonic primary production along the shelf margin during the summer months when the Holloway Current is weak or absent and, historically, perhaps more so during periods of global cooling when it may not flow at all. In contemporary times, where the plankton-rich layer below the thermocline intercepts the seabed, benthic habitats support species-rich filter-feeding epibenthic invertebrate communities. In glacial periods, with extended duration of upwelling, this community type, best developed on hard substrates, may have been a major feature of a narrow North West Shelf benthic fauna.

The modern, species-rich, inner shelf fringing reefs of the Kimberley and West Pilbara flourish in nutrient-rich coastal waters. In the Kimberley, it is likely that *in situ* primary production is high and that it is augmented by fluvial run-off from the high rainfall hinterland. In the West Pilbara, *in situ* primary production probably is also high, but there is limited episodic fluvial run-off from an arid hinterland. However, upwelling along the relatively narrow continental margin at the western end of the North West Shelf probably ensures a seasonal supply of nutrients across most of the shelf. This may be a factor in the high biodiversity of the West Pilbara benthic shelf, reef, and shore faunas (Section 9.2). Unlike the oceanic shelf-edge coral reefs, the continental coral assemblages of the inner shelf fringing and patch reefs do not suffer a drastic change from oligotrophic to nutrient-rich conditions during periods when the Holloway Current is weak or does not flow. They are already adapted to nutrient-rich conditions. While changing sea levels during the climate change episodes of the Quaternary would have required spatial rearrangements, changing nutrient conditions may not have been of concern to them.

(3) The bathyal, deep-sea benthic ecosystems of the continental slope and marginal terraces of the North West Shelf have little connectivity with the benthic fauna of the shelf and little evolutionary affinity with it. However, the boundary between the benthic slope and shelf faunas probably moves up and down in response to sea level change and, over geological time, limited exchange of species is at least possible.

Though very poorly studied to date, it seems that the bathyal benthic fauna of the slopes and terraces is probably rich in biomass and species (Section 8.2). This ecosystem is supported by the rain of organic material from the rich planktonic communities in the water mass below the thermocline and the oligotrophic surface layer. We may assume that it flourishes in upwelling conditions and that the absence of the oligotrophic surface water and extended duration of upwelling in the glacial phases was an advantage.

References

Geology and Geomorphology

1. Bradshaw MT, Yeates AN, Beynon RM, Brakel AT, Langford RP, Totterdell JM, et al. Palaeographic evolution of the North West Shelf. In: Purcell PG, Purcell RR, editors. *The North West Shelf of Australia*. Perth: Petroleum Exploration Society of Australia; 1988. p. 29–54.
2. Carrigy MA, Fairbridge RW. Recent sedimentation, physiography and structure of the continental shelves of Western Australia. *J R Soc West Aust* 1954;**37**:65–95.
3. Jones HA. Late Cenozoic sedimentary forms on the northwest Australian continental shelf. *Mar Geol* 1971;**10**(4):M20–6.
4. Jones HA. Marine geology of the Northwest Australian Continental Shelf. *Bur Miner Resour Geol Geophys Bull* 1973;**136**:1–102.
5. Stagg HMJ, Exon NF. Geology of the Scott Plateau and Rowley Terrace, off northwestern Australia. *BMR Bull* 1981;**213**:1–47.
6. Exon NF, Colwell JB. Geological history of the outer North West Shelf of Australia: a synthesis. *AGSO J Aust Geol Geophys* 1994;**15**(l):177–88.
7. James NP, Bone Y, Kyser TK, Dix GR, Collins LB. The importance of changing oceanography in controlling late Quaternary carbonate sedimentation on a high-energy, tropical, oceanic ramp: north-western Australia. *Sedimentology* 2004;**51**(6):1179–205.
8. Baker C, Potter A, Tran M, Heap AD. Sedimentology and geomorphology of the North-west Marine Region of Australia. *Geoscience Australia Record 2008/07*, Canberra; 2008. 220 pp.
9. Veevers JJ. Morphotectonics of Australia's northwest margin. In: Purcell PG, Purcell RR, editors. *The North West Shelf of Australia*. Perth: Petroleum Exploration Society of Australia; 1988. p. 19–27.
10. Yeates AN, Bradshaw MT, Dickins JM, Brakel AT, Exon NF, Langford RP, Muljolland SM, Totterdell JM, Yeung M. The Westralian Superbasin, and Australian link with Tethys. In: McKenzie KG, editor. *Shallow Tethys 2,* Wagga Wagga Proceedings. A A Balkema Rotterdam; Boston MA. 1987. p. 199–213.
11. Jutson JT. The physiography (geomorphology) of Western Australia. *Bull Geol Surv West Aust* 1934;**61**:240.
12. Wright RL. Geomorphology of the West Kimberley area. *CSIRO land research*, vol. 9. Canberra: CSIRO; 1964. p. 103–118.
13. Young RW. Structure heritage and planation in the evolution of land forms of the East Kimberley. *Aust J Earth Sci* 1992;**9**:141–51.
14. Brooke BP. Geomorphology, part 4. In: Wells FE, Hanley JR, Walker DI, editors. *Survey of the marine biota of the southern Kimberley islands*. Perth: Western Australian Museum; 1995. p. 67–80 [unpublished report No. UR286].
15. Griffin TJ, Grey K. Kimberley Basin. In: *Geology and Mineral Resources of Western Australia*. Western Australia Geological Survey; 1990a. p. 293–304 [Memoir 3].

16. Griffin TJ, Grey K. King Leopold and Halls Creek Orogens. In: *Geology and Mineral Resources of Western Australia*. Western Australia Geological Survey; 1990b. p. 232–55 [Memoir 3].
17. Tyler IM, Griffin TJ. *Explanatory notes on the Yampi 1:250 000 Geological Sheet, Western Australia*. 2nd ed. Geological Survey of Western Australia; 1993.
18. Semeniuk V. The Pilbara coast: a riverine coastal plain in a tropical arid setting, northwestern Australia. *Sediment Geol* 1993;**83**:235–56.
19. Trendall AF, Cockbain AE. Basins Introduction. In: *Geology and Mineral Resources of Western Australia*. Western Australia Geological Survey; 1990. p. 289–93 [Memoir 3].
20. Gunn PJ. Bonaparte Basin: evolution and structural framework. In: Purcell PG, Purcell RR, editors. *The North West Shelf of Australia*. Perth: Petroleum Exploration Society of Australia; 1988. p. 275–85.
21. Mory AJ. Bonaparte Basin. In: *Geology and mineral resources of Western Australia*, Geological Survey of Western Australia; 1990. p. 380–415 [Memoir 3].
22. Glenn K. *Sedimentary processes during the Late Quaternary across the Kimberley Shelf, Northwest Australia*. Thesis, University of Adelaide; 2004.
23. Glenn K. Water properties of Ashmore Reef. In: Russell BC, Larson CJ, Glasby RC, Willan RC, Martin J, editors. *Understanding the cultural and natural heritage values and management challenges of the Ashmore Region. Records of the museums and art galleries of the Northern Territory*; 2005. p. 9–12 [Supplement No. 1].
24. Middleton MF. Canning Basin. In: *Geology and mineral resources of Western Australia*, Geological Survey of Western Australia; 1990. p. 425–453 [Memoir 3].
25. Horstman EL, Purcell PG. The offshore Canning Basin. In: Purcell PG, Purcell RR, editors. *The North West Shelf of Australia*. Perth: Petroleum Exploration Society of Australia; 1988. p. 253–7.
26. Hocking RM. Carnarvon Basin. In: *Geology and mineral resources of Western Australia*, Geological Survey of Western Australia; 1990. p. 457 495 [Memoir 3].
27. Quilty PG. Cenozoic sedimentation cycles in Western Australia. *Geology* 1977;**5**:336–40.
28. Frakes LA, McGowran B, Bowler JM. Evolution of Australian environments. In: Dyne GR, Walton DW, editors. *Fauna of Australia*, vol. 1A. Canberra: Australian Government Publishing Service; 1987. p. 1–16.
29. Kay EA. Evolutionary radiations in the Cypraeidae. In: Taylor J, editor. *Origin and evolutionary radiations in the mollusca*. Oxford, UK: Oxford University Press; 1996. p. 211–20.
30. Wyroll KH, Kendrick GW, Long JA. The geomorphology and Late Cenozoic geomorphological evolution of the Cape Range—Exmouth Gulf region. In: Humphreys WF, editor. *The biogeography of Cape Range. Western Australia. Records of the Western Australian Museum*; 1993. p. 1–23 [Supplement No. 45].
31. Ludbrook NH. Quaternary molluscs of the western part of the Eucla Basin. *Bull Geol Surv West Aust* 1978;**125**:1–286.
32. Collins LB, Testa V, Zhao J, Qu D. Holocene growth history and evolution of the Scott Reef carbonate platform and coral reef. *J R Soc West Aust* 2011;**94**(2):239–50.
33. Heap AD, Harris PT. Geomorphology of the Australian margin and adjacent seafloor. *Aust J Earth Sci* 2008;**55**:555–85.
34. Purcell PG, Purcell RR, editors. *The North West Shelf of Australia*. Perth: Petroleum Exploration Society of Australia; 1988.
35. Department of Conservation and Land Management . *A representative marine reserve system for Western Australia, report of the marine parks and reserves selection working group*. Perth: Department of Conservation and Land Management; 1994.
36. Dix GR. High energy, inner shelf carbonate facies along a tide-dominated non-rimmed margin, northwestern Australia. *Mar Geol* 1989;**89**(3–4):347–62.
37. Lyne V, Fuller M, Last P, Butler A, Martin A, Scott R. *Ecosystem characterisation of Australia's North West Shelf*. Technical Report No. 12. North West Shelf Joint Environmental Management Study; 2006.

38. Dix GR, James NP, Kyser TK, Bone Y, Collins LB. Genesis and dispersal of carbonate mud relative to late Quaternary sea-level change along a distally-steepened carbonate ramp (Northwestern Shelf, Western Australia). *J Sediment Res* 2005;**75**(4):665–78.

39. Lees BG. Recent terrigenous sedimentation in Joseph Bonaparte Gulf, Northwestern Australia. *Mar Geol* 1992;**103**(1–3):199–213.

40. Lavering IH. Quaternary and modern environments of the Van Dieman Rise, Timor Sea and potential effects of additional petroleum exploration activity. *BMR J Aust Geol Geophys* 1993;**13**(4):281–92.

41. Hantoro WS. Quaternary sea level variations in the Pacific-Indian Ocean gateways: response and impact. *Quat Int* 1997;**37**:73–80.

42. van Andel TH, Heath GR, Moore TC, McGeary DFR. Late Quaternary history, climate, and oceanography of the Timor Sea, northwestern Australia. *Am J Sci* 1967;**265**(9):737–58.

43. Teichert C, Fairbridge RW. Some coral reefs of the Sahul shelf. *Geogr Rev* 1948;**38**(2):222–49.

44. Fairbridge RW. The Sahul Shelf, northern Australia: its structure and geological relationships. *J R Soc West Aust* 1953;**37**:1–34.

45. van Andel TH, Veevers JJ. Morphology and sediments of the Timor Sea. *Bur Miner Res Bull* 1967;**3**:.

46. Jongsma D. Eustatic sea level changes in the Arafura Sea. *Nature* 1970;**228**:150–1511.

47. Jongsma D. Marine geology of the Arafura Sea. *Bur Miner Resour Geol Geophys Bull* 1974;**157**:1–73, plates 1–6, map.

48. Elliot RML. Browse Basin. In: *Geology and mineral resources of Western Australia*, Geological Survey of Western Australia; 1990. p. 535–561 [Memoir 3].

49. Van der Graff WJE, Denman PD, Hocking RM. Emerged Pleistocene marine terraces on the Cape Range, Western Australia. *Annual report of the Geological Survey Branch of the Mines Department for the year 1975*; 1976. p. 62–69.

50. Brunnschweiler RO. The geology of the Dampier Peninsula, Western Australia. *Bureau of Mineral Resources*, Report 13; 1957.

51. Semeniuk V. Holocene sedimentation, stratigraphy, biostratigraphy, and history of the Canning Coast, north-western Australia. *J R Soc West Aust* 2008;**91**(1):53–148.

52. Brown RG. Holocene sediments and environments, Exmouth Gulf, Western Australia. In: Purcell PG, Purcell RR, editors. *The North West Shelf of Australia*. Perth: Petroleum Exploration Society of Australia; 1988. p. 85–93.

53. Fry G, Heyward A, Wassenberg T, Colquhoun J, Pitcher R, Smith G, et al. *Benthic habitat surveys of potential LNG hub locations in the Kimberley region*. A joint CSIRO and AIMS report for the Western Australian Marine Science Institution; 2008. 131 pp.

54. Taylor JD, Glover EA. Diversity and distribution of subtidal benthic molluscs from the Dampier Archipelago, Western Australia; results of the 1999 dredge survey (DA2/99). *Records of the Western Australian Museum*; 2004. p. 247–291 [Supplement No. 66].

55. Gregory JW. *The nature and origin of Fiords*. London: John Murray; 1913, 542 pp.

56. O'Brien GW, Glenn KC. Natural hydrocarbon seepage, sub-seafloor geology and eustatic sea-level variations as key determiners of the nature and distribution of carbonate build-ups and other benthic habitats in the Timor Sea. In: Russell BC, Larson CJ, Glasby RC, Willan RC, Martin J, editors. *Understanding the cultural and natural heritage values and management challenges of the Ashmore Region. Records of the museums and art galleries of the Northern Territory*; 2005. p. 31–42 [Supplement No. 1].

57. O'Brien GW, Lisk M, Duddy IR, Hamilton J, Woods EP, Cowley R. Plate convergence, foreland development and fault reactivation: primary controls on brine migration, thermal histories and trap breach in the Timor Sea. *Mar Pet Geol* 1999;**16**(6):533–60.

58. O'Brien GW, Woods EP. Hydrocarbon-related diagenetic zones (HRDZs) in the Vulcan Sub-basin, Timor Sea: recognition and exploration implications. *Aust Pet Prod Explor Assoc J* 1995;**35**:220–52.

59. McKenzie D. Some remarks on the development of sedimentary basins. *Earth Planet Sci Lett* 1978;**40**(1):25–32.
60. Kennard JM, Deighton I, Edwards DS, Colwell JB, O'Brien GW, Boreham CJ. Thermal history modelling and transient heat pulses: new insights into hydrocarbon expulsion and 'hot flushes' in the Vulcan Sub-Basin, Timor Sea. *APPEA J* 1999;**39**(1):177–207.
61. Sandiford M. The tilting continent: a new constraint on the dynamic topographic field from Australia. *Earth Planet Sci Lett* 2007;**261**:152–63.
62. Hall R. Cenozoic tectonics of SE Asia and Australasia, In: Howes JVC, Noble RA, editors. Proceedings of the international conference on petroleum systems of SE Asia and Australia. Indonesian Petroleum Association; 1997.
63. Hall R. The plate tectonics of Cenozoic SE Asia and the distribution of land and sea. In: Hall R, Holloway D, editors. *Biogeography and geological evolution of SE Asia*. Leiden: Backhuys Publishers; 1998. p. 99–131.
64. Wilson BR, Clarkson P. *Australia's spectacular cowries. A review and field study of two endemic genera, Zoila and Umbilia*. San Diego: Odyssey Publishing; 2004 396 pp.

Sea Level Change

65. Chappel J, Shackelton NJ. Oxygen isotopes and sea-level. *Nature* 1986;**324**:137–40.
66. Lambeck K, Nakada M. Late Pleistocene and Holocene sea-level change along the Australian coast. *Palaeontology* 1990;**89**:143–76.
67. Lambeck K, Esat TM, Potter E-K. Links between climate and sea levels for the past three million years. *Nature* 2002;**419**:199–206.
68. Yokoyama Y, Esat TM, Lambeck K. Coupled climate and sea level changes deduced from Huon Peninsula coral terraces of the last ice age. *Earth Planet Sci* 2001;**193**:579–87.
69. Peltier WR. On eustatic sea level history: the last glacial maximum to Holocene. *Quat Sci Rev* 2002;**21**:377–96.
70. Fairbanks R. A 17,000-year glacio-eustatic sea level record: influence of glacial melting rates on the Younger Dryas event and deep-ocean circulation. *Nature* 1989;**342**:637–42.
71. Montaggioni LF. History of Indo-Pacific coral reef systems since the last glaciation: development patterns and controlling factors. *Earth Sci Rev* 2005;**71**:1–75.
72. van der Kaars WA. Palynology of eastern Indonesian marine piston-cores: a Late Quaternary vegetational and climatic record for Australasia. *Palaeogeogr Palaeoclimatol Palaeoecol* 1991;**85**(3–4):239–302.
73. Wells PE, Wells GM. Large-scale reorganisation of ocean currents offshore Western Australia during the Late Quaternary. *Mar Micropaleontol* 1994;**24**:157–86.
74. Matinez JI, De Deckker P, Chivas AR. New estimates for salinity changes in the western Pacific Warm Pool during the last glacial maximum: oxygen-isotope evidence. *Mar Micropaleontol* 1997;**32**:311–40.
75. Gingele LX, de Deckker P, Hillenbrand CD. Late quaternary fluctuations of the Leeuwin current and palaeoclimates on the adjacent land masses: clay mineral evidence. *Aust J Earth Sci* 2001;**48**:867–74.
76. Stoddart DR. Continuity and crisis in the reef community. *Micronesica* 1974;**12**(1):1–9.
77. Underwood JN, Smith LD, Van Oppen MJH, Gilmour JP. Multiple scales of genetic connectivity in a brooding coral on isolated reefs following catastrophic bleaching. *Mol Ecol* 2007;**16**:771–84.

Oceanography

78. Brewer D, Lyne V, Skewes T, Rothslisberg, P. *Trophic systems of the north west marine region*. Report to the Department of the Environment and Water Resources. Cleveland: CSIRO; 2007. 157 pp.

79. INPEX. *Biological and ecological studies of the Bonaparte Archipelago and Browse Basin. Marine sediment and water quality.* Inpex, Perth, Western Australia.

80. Jones DS, editor. Marine biodiversity of the Dampier Archipelago, Western Australia 1998–2002. *Records of the Western Australian Museum*; 2004 [Supplement No. 66].

81. Holloway PE, Nye HC. Leeuwin current and wind distributions on the southern part of the Australian North West Shelf between January 1982 and July 1983. *Aust J Mar Freshw Res* 1985;**36**:123–37.

82. Condie SA, Dunne JR. Seasonal characteristics of the surface mixed layer in the Australasian region: implications for primary productivity and biogeography. *Aust J Mar Freshw Res* 2006;**57**:569–90.

83. Forde MJ. *Technical report on suspended matter in Mermaid Sound, Dampier Archipelago.* Department of Conservation and Environment; 1985 [Bulletin 215].

84. Bassett-Smith PW. On the formation of the coral-reefs on the N.W. coast of Australia. *Proc Zool Soc Lond* 1899;157–9. London, UK.

85. McAlpine KW, Masini RJ, Sim CB, Daly T. Background concentrations of selected metals and Total Suspended Solids in the coastal waters of the Kimberley Region. *Marine Technical Report Series MTR 6*. Perth, Western Australia: Office of the Environmental Protection Authority; 2012.

86. Brink KH, Bahr F, Shearman RK. Alongshore currents and mesoscale variability near the shelf-edge off northwestern Australia. *J Geophys Res* 2007;**112**:C05013, 19 pp.

87. Condie SA, Andrewartha JR. Circulation and connectivity on the Australian North West Shelf. *Cont Shelf Res* 2008;**28**:1724–39.

88. Department of the Environment, Heritage, Water and the Arts . *The north-west marine bioregional plan, bioregional profile.* Canberra: Australian Government; 2008.

89. Cresswell G, Frische A, Peterson J, Quadfasel D. Circulation in the Timor Sea. *J Geophys Res* 1993;**98**:369–79.

90. Meyers G, Bailey RJ, Worby AP. Geostrophic transport of Indonesian throughflow. *Deep Sea Res* 1995;**42**(7):1163–74.

91. Kronberg M. *Ocean circulation over the North West Shelf of Australia—does it impact on the Leeuwin Current?* PhD Thesis, University of Copenhagen, Denmark; 2004.

92. Domingues CM, Maltrud ME, Wijffels S, Church JA, Tomczak M. Simulated Lagrangian pathways between the Leeuwin Current System and the upper-ocean circulation of the southeast Indian Ocean. *Deep Sea Res II* 2007;**54**:797–817.

93. D'Adamo N, Fandry C, Buchan S, Domingues C. Northern sources of the Leeuwin Current and the "Holloway Current" on the North West Shelf. *J R Soc West Aust* 2009;**92**:53–66.

94. Godfrey JS, Ridgeway KR. Seasonal behaviour and possible mechanisms of the Leeuwin Current, Western Australia. *Trop Ocean Atmos Newsletter* 1984;**26**:16–7.

95. Nof D, Pichevin T, Sprintall J. "Teddies" and the origin of the Leeuwin Current. *J Phys Oceanogr* 2002;**32**(9):2571–88.

96. Taylor JG, Pearce AF. Ningaloo Reef currents: implications for coral spawn dispersal, zooplankton and whale shark abundance. *J R Soc West Aust* 1999;**82**:57–65.

97. Holliday D, Beckley LE, Weller E, Sutton AL. Natural variability of macro-plankton and larval fishes off the Kimberley, north-western Australia: preliminary findings. *J R Soc West Aust* 2011;**94**(2):181–94.

98. Okada H, Wells P. Late Quaternary nannofossil indicators of climate change in two deep-sea cores associated with the Leeuwin Current off Western Australia. *Palaeogeogr Palaeoclimatol Palaeontol* 1997;**131**(3–4):413–32.

3

Intertidal Rocky Shores

Intertidal rocky shores provide diverse habitats and generally support species-rich biota. Many rocky shore species that inhabit the upper-littoral and lower-littoral zones are narrowly confined to those zones and comprise distinctive intertidal assemblages. On the other hand, most rocky shore mid-littoral invertebrates are also found on rocky substrata in the adjacent shallow sub-tidal zone.

The most critical factors that determine the ecological characteristics of rocky shores are their level (height) within the intertidal zone, their degree of exposure to wave action, the magnitude of tidal range, and the "rugosity" of the surface, i.e., the complexity of the surface in terms of loose stones, ledges, pools, crevices, and gutters. There is great variation in the invertebrate assemblages of rocky shores along the coast and islands of the North West Shelf as a result of varied ocean conditions and the varied geomorphic forms and shore profiles that different rock types produce. Of the latter, rock platforms—near horizontal surfaces in the intertidal zone—provide the most diverse habitats and the most varied biota.

This section considers rock types, rock platforms and their geomorphic features, and the characteristic species assemblages that inhabit rocky shores of the North West Shelf.

3.1 ROCK TYPES

The rock types along the coast of the North West Shelf are varied, forming shores of many different profiles, surfaces, textures, rugosities, and other habitat characteristics that affect the lives of the invertebrates and fishes that inhabit them. While there is no evidence to suggest that rock type is a major factor in the geographical distribution of rocky shore species (with the exception of certain taxa that bore only in carbonate rocks), it certainly affects their local ecological distribution and relative abundance.

There are two sections of the North Shelf coast where Proterozoic rocky shores are dominant—the northwest Kimberley (Kimberley and

Bonaparte Bioregions) and the Dampier Archipelago where stretches of rocky shore are interrupted by short beaches and mangals. Quaternary limestone shores are common on the Pilbara coast but rare in the Kimberley Bioregion except for superficial biogenic encrustations over the surfaces of rock platforms.

Shores of the Canning Bioregion are characterized by long stretches of sandy beach, sometimes with mangal habitats, and rocky shores are represented by only occasional rocky headlands of Quaternary or Mesozoic rocks. The shore of the Eighty Mile Beach Bioregion, between Cape Jaubert and Cape Keraudren, is a stretch of beach around 230 km long without rocky shore habitat and without offshore islands. The importance of that coastline as a biogeographic barrier to obligate rocky shore species is discussed later.

3.1.1 Rocky Shores of Precambrian Igneous and Metamorphic Rocks

The Kimberley Bioregion from Cape Londonderry to Cape Leveque is a macrotidal, ria-coast. The shores are dominantly rocky (between short beaches and mangal habitats) comprising mainly Proterozoic igneous rocks and metamorphosed bedded sandstone. In the Bonaparte Bioregion, east of Cape Londonderry along the western side of Joseph Bonaparte Gulf, the coastline is also predominantly of Proterozoic sandstones but it is not a ria-coast and has a low profile of cliffs and sloping rocky shores.

On shores of these kinds, the intertidal zone is generally steep and narrow with a highly irregular profile, prolific growth of macroalgae, and a rich cryptic intertidal invertebrate fauna. The Proterozoic sandstones (mainly King Leopold and Warton Sandstones) are flat-bedded and strongly jointed and generally feature angular ledges and blocks (Figures 3.1 and 3.2). They often form terraced intertidal rock platforms with wide mid-littoral and lower-littoral zones. In the north Kimberley, there are also basalt shores that may have smooth, sloping ramps (Figure 3.3) or fields of rounded boulders or cobbles (Figure 3.4). In the northern Bonaparte Archipelago, many islands have a cap of Tertiary laterite that produce scree slopes and rocky shores of irregularly shaped laterite boulders (Figure 3.5).

Rocky shore habitats of Archaeozoic igneous and metamorphic rocks occur on the shores of the peninsulas and islands of the central Pilbara coast, where the Pilbara Craton outcrops along the coast between Cape Keraudren and Cape Preston and in the Dampier Archipelago. Sections of this coast are also drowned and ria-like and characteristic shore profiles are of massive basalt headlands and boulders tumbling into the sublittoral zone (Figure 3.6). Rock platforms are not developed on the Archaeozoic rocky shores of the Dampier Archipelago (but do occur on Pleistocene limestone shores that form the perimeter of the archipelago—see below).

FIGURE 3.1 West Montalivet Island, Bonaparte Archipelago, north Kimberley. Flat-bedded and strongly jointed Warton Sandstones form rock platforms and angular blocks and ledges on a seaward shore. *Photo: Natalie Rosser.*

FIGURE 3.2 Flat-bedded Warton Sandstone at high tide in Sampson Inlet, Camden Sound, North Kimberley. During the Last Glacial Maximum, this long, winding, narrow inlet was a river canyon incised into the western margin of the Kimberley Plateau but it is now flooded by the sea. *Photo: Barry Wilson.*

FIGURE 3.3 A smooth basalt (Carson Volcanics pillow larva) shore at North Maret Island. Note the overlying laterite cliff and fallen boulders. The basalt surface represents the unconformity with the laterite cap. It outcrops at about the intertidal zone in many places on the eastern shore of North Maret Island. *Photo: Barry Wilson.*

FIGURE 3.4 A shore of basalt cobbles of Carson Volcanics in St. George Basin, Kimberley Bioregion. Note the flat-topped Mt. Trafalgar mesa (King Leopold Sandstone) in the background. *Photo: Barry Wilson.*

3.1.2 Rocky Shores of Mesozoic Sedimentary Rocks

In the northern part of the Canning coast, from Cape Leveque to Cape Missiessy, many rocky headlands are of early Cretaceous sandstones that vary greatly in lithology. They include hard, sometimes silicified, fine to coarse

FIGURE 3.5 Brunei Bay, North Maret Island. Tumbled laterite boulders from the Tertiary laterite cap often form rocky shores in the Bonaparte Archipelago. *Photo: Barry Wilson.*

FIGURE 3.6 Tumbled Archaean basalts at Dolphin Island, with a mangrove fringe, typical of the rocky shores of the Dampier Archipelago. *Photo: Barry Wilson.*

quartzitic sandstones, soft mudstones and siltstones, and conglomerate deposits. Accordingly, the rocky shores of those areas are of varied profile. They may be cliffed, e.g., Emeriau Point (Figure 3.7) and Gantheaume Point. Rock platforms are poorly developed on shores of that kind.

FIGURE 3.7   A cliffed headland in Mesozoic sandstone at Emeriau Point on the Dampier
Peninsula. *Photo: Barry Wilson.*

3.1.3 Rocky Shores of Quaternary Limestone

On the north Kimberley coast, there are no Pleistocene rocky shores,
perhaps because of ongoing subsidence.[1] However, throughout the bioregion there are extensive intertidal fringing coral reefs (Chapter 4) with
rock platforms built of Holocene coralgal material over Proterozoic rocks
that provide hard calcareous substrate for burrowing invertebrates.
There are also several examples in the southern part of the bioregion
of massive limestone platforms in the intertidal zone whose age and
origins are undetermined (Chapter 4).

Limestone rocky shores, built of Late Pleistocene limestone of marine or
aeolian origins, are a feature of shores and islands in the Pilbara as far east
as Cape Keraudren (Figure 3.8). These shores may be of low cliffs, commonly with a supralittoral bench 3–6 m above mean sea level and a wide
rock platform (Figure 3.9) or they may form limestone barriers along
sandy shores. The limestone barriers often shelter lagoons and mangals
behind them like that at Cape Keraudren.

Intertidal limestone rock platforms are characteristic of seaward shores
that are exposed to moderate or severe wave action and are discussed in
more detail later (Section 3.1.2). Such limestone rocky shores occur on the
western side of Exmouth Gulf, the inner shelf islands of the West Pilbara,
and around the outer perimeter of the Dampier Archipelago (where they
represent the remnants of the Late Pleistocene shore line). The limestone is
generally massive and hard.

FIGURE 3.8 Cape Keraudren, the most easterly of the Quaternary limestone headlands on the Pilbara coast. It is typical of the limestone coastal barriers of the Pilbara (nearshore) Bioregion with embayments, mangals, and tidal creeks developed behind it. The intertidal rock platform in front has a rich invertebrate fauna including a diversity of corals but there is no reef-building. *Source: Landgate, W.A. Government.*

FIGURE 3.9 A typical West Pilbara Late Pleistocene limestone rock platform with supra-littoral bench on a seaward shore at Kendrew Island, Dampier Archipelago. There is a species-rich fringing reef along the reef-front with a diverse assemblage of reef-building corals (not visible). *Photo: Barry Wilson.*

The limestone shores of western and southern Exmouth Gulf and some of the offshore islands are remnants of Late Pleistocene coral fringing reefs with well-preserved fossil corals. Along the mainland shores of the West Pilbara, some Late Pleistocene limestone shores are of dunal origin with crossbedding and may contain fossil terrestrial snails (camaenids) while others are of lagoonal origin and contain fossil bivalves and gastropods characteristic of shallow inshore marine habitats.

There are also limestone shores, thought to be of Pleistocene age, in the northern part of the Canning Bioregion. They are of varied lithology and have been referred to the Bossut Formation.[1-3] However, a review of the biostratigraphy of the Canning coast[4] has shown that the Bossut Formation may be an artificial mixture of lithologies of different Pleistocene and Holocene ages. Whatever their age, these limestone shores often exhibit well-developed intertidal rock platforms.

On the northwestern coast of the Dampier Peninsula in the southern Kimberley (not to be confused with the Dampier Archipelago in the Pilbara), finely bedded or crossbedded limestone, attributed to the Bossut Formation, outcrops along the coast at several locations.[5] Such limestone shores occur along the seaward side of Packer Island (Figure 3.10) and between Middle Lagoon and North Head (Figure 3.11). This formation is referred to as the North Head Limestone and is described as a "bioclastic and quartzose fine calcarenite."[4] On the shore, it weathers and presents intertidal habitats that are similar to the typical Late

FIGURE 3.10   An algal-dominated rock platform of Pleistocene North Head Limestone at Packer Island (near Lombadina) Dampier Peninsula, southern Kimberley. The upper part of the upper-littoral zone is etched; the lower part with rock-oysters (*Saccostrea cucullata*). The high supralittoral bench suggests an early or middle Pleistocene age. *Photo: Barry Wilson.*

FIGURE 3.11   A rock platform of North Head Limestone on the southern shore of Middle Lagoon on the Dampier Peninsula, southern Kimberley. The upper mid-littoral flat in the foreground is pot-holed with raised algal rims and algal turf. The high rocks in the background are remains of a limestone headland that has been eroded away. *Photo: Barry Wilson.*

Pleistocene limestone rock platforms of the Pilbara and West coasts with high supralittoral benches, wide intertidal rock platforms, and strongly etched surfaces in the supralittoral zone (Figure 3.10). An oyster-zone with its associated invertebrate fauna is moderately well developed in the lower part of the upper-littoral zone. Fronting the cliffs, there is a wide, algal-dominated rock platform with an array of mid-littoral and lower-littoral microhabitats and diverse assemblages of invertebrates including scleractinian corals although the latter are not reef forming.

On the north-facing cliffs at the southwestern entrance of Pender Bay, there are patches of another limestone formation, called the Chimney Rock Oolite, also considered to be the part of the Bossut Formation lying unconformably over Mesozoic sandstone. It is described as a "calcreted, well-cemented oolitic limestone."[4] It does not form intertidal rock platforms but is strongly etched in the high supralittoral zone and there are sections of the shore comprising fallen limestone boulders. Its rocky shore invertebrate fauna is impoverished.

Further south, on the Pindan coast south of Broome, there are cliffs of massive limestone, up to 13 m high (above MSL) along the western shores of Gourdon Bay and southwest of Cape Latouche-Treville as far Cape

FIGURE 3.12   A limestone (calcilutite) cliff, assigned to the Pleistocene Bossut Formation, at Cape Latouche-Treville, Canning Bioregion. This rocky shore has a very restricted invertebrate fauna. *Photo: Barry Wilson.*

Bossut that are also attributed to the Bossut Formation.[5] At Cape Latouche-Treville, the limestone is a fine, horizontally bedded (sometimes slightly folded) calcarenite, and calcilutite with shelly bands (Figure 3.12). There is a high supralittoral bench and the cliff face and edge are smooth (i.e., not etched). There is little or no development of the upper-littoral oyster-zone (*Saccostrea cucculata*) with its characteristic invertebrate associates. A rock platform is poorly developed and the mid-littoral zone below the cliff face comprises a steep sand ramp. There are areas with boulder fields (of limestone and Cainozoic laterite) in the mid and lower-littoral zones that provide habitats for invertebrates but the rocky shore fauna is generally rather sparse.

Holocene limestone known as beachrock commonly occurs as a sloping stony ramp on upper-littoral beaches (Figure 3.13). It is formed by shallow subsurface cementation and consolidation beneath the sand and may be exposed by erosion. Because of its slope and exposure to air, sun, waves, and sand scouring, it is an inhospitable habitat for invertebrates but there may be ledges and shallow pools that provide refuge and some of the upper-littoral rocky shore gastropods may be found there, especially littorinids, nerites, and siphonarians.

FIGURE 3.13   Holocene beachrock slope in the upper intertidal at South Beach, South Maret Island, Bonaparte Archipelago. The rocky shore above the beach is of laterite and there is a wide fringing coral reef of the *Acropora* bank kind in the mid- to lower-littoral zone. In the Maret Islands, beachrock deposits like this are often fossiliferous. *Photo: Barry Wilson.*

3.1.4 Limestone Rocky Shores of Pre-Pleistocene Age

At the western end of the North West Shelf, Miocene limestone underlies Quaternary deposits at shallow depth and outcrops along the shores of Barrow Island. This richly fossiliferous rock, the Trealla Limestone, is crystalline and hard and forms a shore of low cliffs and boulders with deeply undercut notches in the intertidal zone and wide rock platforms. The intertidal habitats thus created are little different to those formed in the Late Pleistocene limestones of that region.

In a very small area at the eastern end of Montgomery Reef in the Kimberley Bioregion, there are "limestone" shores of quite different age and lithology.[6] The High Cliffy complex, with its small islands and reef platform, is an ancient biohermic structure composed of Paleoproterozoic, partially silicified stromatolitic dolomite overlain (high above sea level) by Paleoproterozoic sandstone of the Kimberley Group. The extent of this dolomite formation in the Montgomery Reef structure is not known and it appears to be an unrecognized Paleoproterozoic formation (C. Grey, personal communication, November 2009). This rock is extremely hard and heavy but otherwise forms rocky shore habitats similar to those of the Late Pleistocene limestone shores in the Pilbara, though without an intertidal notch (Figure 3.14).

FIGURE 3.14   A shore of early Proterozoic stromatolitic dolomite at High Cliffy Island on the eastern side of Montgomery Reef, Camden Sound. The dolomite beds are overlain by a quartz sandstone, presently mapped as Pentacost Sandstone, an upper member of the Kimberley Group. *Photo: Barry Wilson.*

3.2 ROCK PLATFORMS

3.2.1 A Definition Issue—Rock Platform v. Fringing Coral Reef

Horizontal or ramped (sloping) rock surfaces in the intertidal zone of rocky shores are known as rock platforms. There is a definition and duplication problem in dealing with the habitat categories "rock platform" (this section) and "fringing reef" (see Section 4.3). There are two types of intertidal platform profiles on the shores of the North West Shelf but they share many geomorphic and biotic features in common

- algal-dominated *rock platform* where there are few corals on the reef-front and little deposition there of biogenic carbonate
- *fringing coral reef* being a rock platform with a coral-dominated reef-front where there is deposition of coralgal limestone and lateral reef growth.

Intertidal algal-dominated rock platforms are a common feature of rocky shores on the Kimberley, Canning, and Pilbara coasts (Figures 3.15 and 3.16). They may occur on rocks of any type but are especially well developed on limestone shores because of their susceptibility to physical and biological erosion processes. Scattered coral colonies are usually common on the mid-littoral reef-flat, especially in shallow pools and moats, and

FIGURE 3.15 The reef-front of the Pleistocene limestone rock platform on the northwestern shore of Barrow Island, dominated by macroalgae; corals sparse and small. This reef is only about 5 km north of Biggada Reef which is a major fringing reef where the reef-front is dominated by corals. There is no immediate explanation why one reef is algal dominated and the other coral dominated. *Photo: Barry Wilson.*

FIGURE 3.16 Reef-front of a limestone rock platform dominated by macroalgae, Packer Island, on the northwestern coast of the Dampier Peninsula, Canning Bioregion. *Photo: Barry Wilson.*

may contribute to the vertical growth and leveling of the platform but the lower-littoral zone is not coral dominated and there is no coral limestone wedge at the reef-front and no lateral reef growth. The platform is usually gently sloping and is often terraced but there is rarely a steep reef-front ramp or a reef crest with a boulder zone.

Kimberley and Pilbara fringing coral reefs share some features of rock platforms. In these bioregions, fringing reefs have developed by vigorous Holocene coral growth in the lower-littoral zone along the reef-front of pre-existing rock platforms and the development there of a biogenic limestone wedge and lateral reef growth. On these shores there is a wide mid-littoral reef-flat with growth of macroalgae dominant over scattered corals. The upper-littoral zone is a noncoral, rocky shore habitat (except where there are impounded perched pools).

On both rock platforms and fringing reefs of the North West Shelf, biotic communities of the mid-littoral reef-flat comprise mixed assemblages of both coral reef and rocky shore species. In the field, distinction between algal-dominated rock platforms and fringing coral reefs is not always clear and it is necessary to describe what one sees rather than attempt rigorous application of a classification. Category overlap and duplication are unavoidable. As a working procedure, the upper-littoral rock-face and mid-littoral rock flat parts of fringing reefs are treated in this section as rock platform habitat. The lower-littoral part, when it is coral dominated, is dealt with in the coral reef section (see Chapter 4) and again here but briefly.

3.2.2 The Formation of Rock Platforms

Rock platforms are sometimes referred to as "wave-cut," the implication being that they are essentially erosional features. However, these geomorphic features may be the result of several different biogeomorphic processes. They may be structural (bedding plane), erosional or constructional features, or combinations of these. Whether the intertidal level of a rock platform is primarily an outcome of wave erosion or biogenic construction is not always easy to determine in the field.

The level of rock platforms in Proterozoic sandstones of the Kimberley Bioregion generally has structural geological origins, determined by the position of bedding planes that have no causal relation to sea level. Levels of contemporary limestone rock platforms of the Canning and Pilbara Bioregions may also be determined by preexisting structures that, in this case, are benches of still-stand sea levels of previous interglacial phases. In both cases, the contemporary level of the platform surface may be modified by Holocene erosional and constructional processes.

The principal erosional processes in the upper-littoral zone of rock platforms are wave and chemical erosion in the upper part of the upper-littoral and, on limestone shores, wave and bioerosion in the lower part. Edwards[7] described rock platforms that he regarded as wave-cut in Proterozoic igneous rocks around the shores of islands in the Buccaneer Archipelago and the mainland of the Yampi Peninsula. He noted that such rock platforms are best developed where the shore rocks are quartz-feldspar porphyry or schists that weather easily, and poorly developed where the rocks are resistant quartzites. Rates of erosion are considerably greater on limestone shores, largely due to their propensity for both chemical and bioerosion, especially by undercut of the upper-littoral shore face and progressive extension of the mid-littoral reef-flat in the shoreward direction. The suite of boring invertebrates that are the principal bioerosion agents on limestone shores of the North West Shelf are discussed in Section 3.3.

The principal constructional processes on intertidal rock platforms are the redistribution and deposition of sediment and the *in situ* deposition of biogenic carbonate derived directly from growth of living organisms with calcareous skeletons, especially corals and calcareous algae. Preexisting intertidal rock platforms occurring at suitable levels and in suitable ecological conditions provide an ideal basis for the growth of these organisms. Carbonate accretion by growth of crustose algae on the platform pavement and deposition in low areas of carbonate sediment derived from breakdown of the corals, algae, and other organisms may raise and level the platform surface.

3.2.3 Geomorphic Features of Rock Platforms

Three primary geomorphic zones may be recognized on rock platforms.

3.2.3.1 The Upper-Littoral Zone of Rocky Shores

The upper-littoral zone at the back of rock platforms is not always rocky shore habitat. It may be a mangal (Chapter 6) or sloping beach (Chapter 5). When the shore is rocky, the profile of the upper-littoral zone varies. It may be a sloping beachrock ramp, an eroded rock ramp, a boulder field, or a cliff.

The upper-littoral profile of limestone rock platforms is typically very different to that of those on igneous and metamorphic rocks because of the vulnerability of limestone to erosion, especially biogenic erosion. The supralittoral fringe (splash zone) and the upper part of the upper-littoral zone of limestone shores, the rock is exposed to aerial physical and chemical erosion as well as wave action, and the rock is characteristically "etched" with sharp-edged facets and turrets (Figure 3.10).

On limestone shores, it is usual for there to be a supralittoral bench, a few or many meters wide. In the Pilbara, like the West Coast, where the limestone is of Late Pleistocene age, the supralittoral bench is usually 3-6 m above MSL and is thought to represent an intertidal rock platform developed at the time of the Last Interglacial high sea level (Figure 3.9). In the Canning bioregion, there are limestone shores with benches much higher than this that may represent earlier phases of Pleistocene high sea level (Figure 3.10).

On most limestone shores exposed to wave action, the shore face in the lower part of the upper-littoral becomes undercut by physical action (wave, sun, and wind erosion) and bioerosion. There is a suite of invertebrates (mainly barnacles and bivalved molluscs) that live in burrows they bore into the rock of this zone, and others (chitons and limpets) that mechanically graze on interstitial algae in the rock surface. Collectively, aided by physical erosion (especially wave force at high tide) these animals may rapidly undercut the rock shore face in the lower part of the upper-littoral. The undercut is known as the *intertidal notch* (Figure 3.17). This feature is commonly conspicuous on limestone shores but poorly developed on shores of igneous or metamorphic rocks.

In the Pilbara, the upper-littoral of limestone rock platforms is typically double-notched with upper and lower notches separated by a band of rock-oysters (*Saccostrea cucullata*). The oyster band is an outward-growing, biogenic constructional zone that may form an overhang (Figure 3.17). Below it, the bioeroded lower notch is typically asymmetrically concave.

FIGURE 3.17   A deep undercut lower notch and oyster overhang on a windward shore, Kendrew Island, Dampier Archipelago. *Photo: Barry Wilson.*

When it is undercut to a critical depth, the overhang of oysters may collapse. Over time, this process leads to progressive shoreward extension of the rock platform. Most spectacular are stacks (remnants) on the rock platform that are undercut all around, creating a mushroom-like structure (Figures 3.18 and 3.19). One such structure was present on the seaward rock platform of Kendrew Island in the Dampier Archipelago, observed by a W.A. Museum survey party in 1971-1972 (Figure 3.18) but had collapsed and vanished in 1973, indicating that this is a powerful and rapid erosion process on Pilbara limestone shores.

3.2.3.2 The Mid-Littoral Reef-Flat

The mid-littoral part of a rock platform (also called the reef-flat) is usually gently sloping or terraced. At its inner margin, the platform meets the upper-littoral rock-face, beach or mangal and the break of slope at that point may be regarded as the boundary between the mid-littoral and upper-littoral zones.

On rock platforms and fringing reefs where there is significant accretion of biogenic limestone on the pavement surface, the mid-littoral flat may be more or less level although algal terraces may be present and there may be shallow pools and a shallow moat at the back where the flat meets the upper-littoral zone. Otherwise, there may be drainage gutters normal to the shore.

FIGURE 3.18   A mushroom-shaped stack on the mid-littoral rock flat at Kendrew Island, Dampier Archipelago, photographed in 1972 but collapsed and vanished in 1973. Note the supralittoral perched pool, impounded by rock-oysters in the top left corner (see also Figures 3.17 and 3.21 of the same structure). *Photo: Barry Wilson.*

FIGURE 3.19   A double-notched stack on a mid-littoral rock flat on the western, seaward side of Barrow Island, with typical development of a broad band of rock-oysters (*Saccostrea cucullata*) at the center of the upper-littoral zone, deeply undercut below and with etched rock surface above. *Photo: Barry Wilson.*

There may be exposures of the underlying rock on the mid-littoral reef-flat, usually as ridges or boulder fields. Such outcrops significantly increase the complexity of the mid-littoral habitat and may act as foci for the deposition of sediment (sand or rubble).

3.2.3.3 *The Lower-Littoral Reef-Front Ramp*

The ramped lower-littoral seaward part of algal-dominated rock platforms generally has an irregular, gently sloping front and no distinct edge (e.g., Figures 3.15 and 3.16). There may be superficial coral growth in that zone but the corals are small and scattered, do not provide a framework for carbonate deposition and are not reef-forming. There is rarely a boulder zone at the top of the seaward ramp on such reefs or any spur and groove drainage system on the fore-reef.

3.2.4 Rock Platform Terraces

A feature of many rock platforms on the shores of the North West Shelf is the presence of terraces that impound water above them at low tide, creating shallow lagoon and pool habitats in the intertidal zone. They may even create pools or lagoons in the upper-littoral zone high above the tidal level normally inhabited by their occupants. Impounded water behind the terraces provides protection for marine plants, corals, other invertebrates, fishes, and turtles against exposure to the air. The terraces are of several major types.

Terraces are sometimes formed by preexisting ridges of hard intrusive igneous rocks or conglomerate deposits, with shallow "perched pools" impounded behind them that eventually become infilled with sediment and biogenic deposits including coral communities. Terraces of this kind are common in the Kimberley Bioregion (Figure 3.20). In the Pilbara, impounded pools high in the upper-littoral zone may also be created by the growth of rock-oysters (Figure 3.21).

On Pilbara limestone platforms, small ringed or lunate pools are very common, especially high on the mid-littoral flat (Figure 3.22). They have surrounding walls of biogenic carbonate a few centimeters high and provide important refuge for marine plants and animals during periods of low tide when the rock platform surface is exposed. The impounding walls are built largely of Holocene carbonate accretions produced by calcareous algae. Similar structures are known from rock platforms throughout the Indo-West Pacific region.

Rock platform terraces with partly biogenic origins are also a conspicuous feature of the Kimberley coast where they are most commonly

FIGURE 3.20 A perched pool high in the upper-littoral zone impounded by the presence of an impervious wall of cemented conglomerate with prolific growth of corals and other invertebrates, North Maret Island, Kimberley Bioregion. *Photo: Barry Wilson.*

FIGURE 3.21   A perched pool high in the upper-littoral zone impounded by an impervious wall of cemented rock-oysters (*Saccostrea cucullata*), Kendrew Island, Dampier Archipelago. *Photo: Barry Wilson.*

FIGURE 3.22   Lunate terraces impounding shallow pools on a mid-littoral rock flat, western shore of Barrow Island, West Pilbara. *Photo: Barry Wilson.*

simple, step-like ledges, relating to bedding plane, and jointing structure of the surface geology but with raised rims built of Holocene bioclastic sediments compounded by algal carbonate secretions. Such terraces are especially common in the Buccaneer Archipelago where they may be up to 1.5 m high and create a "paddy field" effect (Figure 3.23). The carbonate rims develop rapidly in these macrotidal conditions where tidal water flows over an obstruction but the initiation and process of carbonate accretion is not fully understood.[8]

On Montgomery Reef and Talbot Bay on Yampi Peninsula[9] and commonly in the Buccaneer Archipelago terraces formed by banks of living rhodoliths impound shallow lagoons on the intertidal platform on a vast scale (Figures 3.24 and 3.25). The rhodoliths grow on the bed of the shallow lagoons. Near the lagoon perimeters, they are rolled by wind-driven waves and the ebbing tide into coalescing ridges forming crescentic pools (Figure 3.26) and finally up onto the reef-crest where they form the impoundment banks (Figure 3.27A and B). These structures form as a result of the massive tidal flows of this area (range greater than 11 m). They may be a biogeomorphic feature that is unique to the Kimberley Bioregion although rhodolith banks reported from central Indonesia may be of the same kind.[10,11] On the Montgomery Reef platform, rhodoliths appear to be more important carbonate producers and play a more significant role in reef growth than do the corals.

FIGURE 3.23 Terraces constructed by development of carbonate algal rims on the edges of preexisting bedding structures in the mid-littoral zone—Sunday Island, Buccaneer Archipelago, Kimberley Bioregion. The terraces impound pools up to 2 m deep with prolific growth of corals, seagrasses, and associated fauna. *Photo: Barry Wilson.*

FIGURE 3.24   Aerial view of winding rhodolith banks impounding intertidal lagoons on the Montgomery Reef eastern intertidal platform, Kimberley Bioregion. There are two levels, the first on the reef-crest impounding a lower lagoon, and a second one about a meter higher, impounding the vast upper-littoral lagoon of Montgomery Reef. Both terraces may be multiple with coalescing ridges. *Photo: Steve Blake, WAMSI.*

FIGURE 3.25   Northern shore of Molema Island in Talbot Bay, Yampi Peninsula. A multiple bank of rhodoliths along the reef-crest impounds a high, coral-rich lagoon. Note also the mud banks at the mouth of a tidal creek draining a mangal at the base of the sandstone cliffs. *Photo: Steve Blake, WAMSI.*

The ecological outcome of the rhodolith banks on Montgomery Reef is the creation of a vast area (>300 km^2) of shallow, permanently flooded, "benthic primary producer habitat" which may account for the abundance of herbivorous green turtles for which this reef is renowned.

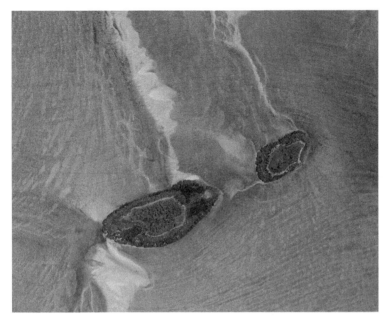

FIGURE 3.26   Egret Islands, Proterozoic structures on the high rock platform at the eastern end of Montgomery Reef. Note the double terraces, upper and lower, formed by banks of loose rhodoliths with sand fans in front and crescentic pools behind. *Image courtesy of Faculty of Science and Engineering , Curtin University, Western Australia.*

FIGURE 3.27   Eastern side of Montgomery Reef: (A) wide impoundment bank of loose rhodoliths on the reef-crest exposed at low tide;

Continued

FIGURE 3.27—cont'd   (B) rhodoliths on a reef-crest bank. The rhodoliths appear to grow in the lagoon impounded above the rhodolith bank and are rolled onto the reef-crest by ebbing tides (>11 m tidal range at spring tide). *Photos: Barry Wilson.*

3.3 THE INTERTIDAL BIOTA OF ROCKY SHORES

3.3.1 Zonation

In the intertidal zone of rocky shores throughout the world, distinct biotic zones (sometimes called "girdles" or "belts") may be recognized, occupied by different invertebrate and algal species arranged in horizontal bands, at successive tidal levels one above another.[12] This vertical zonation of rocky shore invertebrates is particularly well developed in the upper-littoral zone. The width of these species-specific habitat bands varies between regions and locations depending on tidal range and degree of wave exposure and to the varied capacity of the inhabitants to deal with these environmental stresses.

This phenomenon has been known for very many years[13–15] and there has been a great deal written to explain it, to discern common vertical distribution patterns of species, and to devise zonation classification systems that apply world-wide.[12] Generalities can be made about intertidal biotic zonation of rocky shores at generic and family level. Certain invertebrate groups are represented in the upper-littoral zone of rocky shores almost everywhere, their representative species occupying similar biotic zones. However, the specifics of tidal range, wave and wind conditions, and other environmental conditions are so varied, and the species pools available in different biogeographic regions are so different, that no universal biotic zonation scheme for rocky shores has yet been devised.

Intertidal biotic zonation patterns of invertebrates and algae on temperate Australian rocky shores have been described—for the southern coast,[16–24] southern Queensland,[25] and the southwest.[26,27] There have been no studies of this matter on the tropical shores of the North West Shelf. The following notes are based on field observations made by this author on the zonation of invertebrates in the upper-littoral zone on North West Shelf rocky shores but the subject needs detailed research.

3.3.2 Key Inhabitants and Their Zonation in the Upper-Littoral Zone

A suite of crawling, nestling, cemented and boring molluscs, and barnacles inhabit the upper-littoral and supralittoral of rocky shores, most of them occupying restricted levels within those zones. Patterns of their zonation are basically the same throughout the coastal bioregions of the North West Shelf although there are changes in some key species among the bioregions and some variations in ecological distribution on the shore that relate to tidal range and different shore profiles. On the limited rocky shores of coral islands of the Oceanic Shoals Bioregion the upper-littoral community is severely restricted although elements of it are found where there are sloping beachrock deposits in the intertidal zone.

Four gastropod groups, the families Littorinidae, Neritidae, Siphonariidae, and certain species of the superfamily Patelloidea are characteristic of the upper-littoral rocky shores world-wide. Representatives of the bivalve family Mytilidae may also be conspicuous in the lower part of this zone. In the tropical Indo-West Pacific region, one or more species of the rock-oyster genus *Saccostrea* are key inhabitants that adhere to the rock surface and create a microhabitat utilized by other invertebrates.

The taxonomy of the Indo-West Pacific rocky shore Littorinidae has been extensively reviewed.[28–33] On mainland and coastal island rocky shores of the North West Shelf, there are six species of littorinids, two to four species usually present at any one locality. The six species are *Echinolittorina trochoides*, *E. vidua*, and *Littoraria undulata* which are ubiquitous throughout mainland and coastal island shores of the region, *Tectarius rusticus* which is endemic to the Kimberley Bioregion and *Nodilittorina australis* and *N. nodosa* which are endemic to the West Pilbara and the West Coast. Only one species, *L. undulata*, occurs on the shores of the coral platform reefs and shelf-edge atolls of the Oceanic Shoals Bioregion.

On the North West Shelf, there are eight species of *Nerita*, most of them with Indo-West Pacific and northern Australian distributions.[34] The zonation and geographic distribution of the species that occur on the shores of

North West Cape have been described.[35] *Nerita chamaeleon*, *N. reticulata*, *N. squamulata*, and *Nerita undata* are "continental" species of rocky shores. *Nerita albicilla*, *N. plicata*, and *N. polita* are catholic in their habitats, known on the North West Shelf from the Oceanic Shoals Bioregion as well as the coastal islands and the mainland at Ningaloo. One species, *N. grossa*, is an oceanic species known on the North West Shelf only from coral reefs of the Oceanic Shoals Bioregion.

On rocky shores of the North West Shelf, there at least two species of *Siphonaria* but the taxonomy of the family has not been studied in this region and the identity of these species remains ambivalent.

At least four species of true limpets, *Patella flexuosa*, *Cellana radiata*, *Patelloida mimula*, and *Patelloida saccharina*, are conspicuous in the upper-littoral of North West Shelf rocky shores (see Ref. 34 for illustrations). All are common northern Australian rocky shore species, though only *P. flexuosa* is known from the coral islands of the Oceanic Shoals Bioregion.

There are three species of the rock-oyster genus *Saccostrea* on rocky shores of the mainland and coastal islands, *S. cuccullata* (Figure 3.31), *S. echinata*, and *S.* cf. *commercialis* (S. Slack-Smith, personal communication, April 2010). All of these oysters have northern Australian and Indo-West Pacific distributions. The taxonomy and ecology of *S.* cf. *commercialis* remain ambivalent. This genus may be regarded as "continental" and is not represented on the oceanic coral reefs of the outer shelf and margin.

On temperate shores, the family Mytilidae is usually represented by one or more species that form dense, matted colonies attached to the rock pavement by byssal threads. An endemic Western Australian mytilid with this habit, *Brachidontes ustulatus*, ranges from the South Coast to the Kimberley Bioregion. A second, smaller, species of *Brachidontes* is usually present in the upper-littoral zone also but its identity is not yet determined. Another bivalve commonly present is a small species of the family Isognomonidae, *Isognomon nucleus*.

3.3.2.1 Upper-Littoral Zonation in the Pilbara Bioregion

The invertebrate fauna of the upper-littoral zone of Pilbara rocky shores is uniform throughout the bioregion regardless of rock type although the zonation varies according to local ecological circumstances. The zonation patterns are most distinct on Quaternary limestone shores where there is a conspicuous double-notch, summarized diagrammatically in Figure 3.28. Three biotic units within the upper-littoral zone may be recognized. Species of *Saccostrea* are key species that occupy a band at about the center of the upper-littoral. Above and below the oyster-zone, the rock-faces have different profiles and are occupied by different organisms.

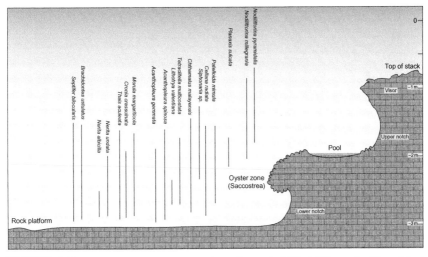

FIGURE 3.28   Diagrammatic representation of a typical West Pilbara double-notched profile on a Quaternary rock platform, showing the vertical distribution zones of characteristic upper-littoral invertebrates.

(a) The high upper-littoral rock-face

Above the oyster-zone, the rock-face of the upper-littoral is often bare, lacking an algal turf, although the red alga *Bostrychia tenella* may be present up to the supralittoral fringe splash zone, sometimes forming a distinct red-orange band at that level. The dominant invertebrates are littorinid gastropods, with up to five species present. In the Pilbara, the highest species is *E. trochoides* (Figure 3.29) that extends from the splash zone of the supralittoral fringe down almost to the level of the oysters. It may be present in vast numbers along the outer edge of the supralittoral bench where there are wave-surge or wave-splash pools. The tiny *E. vidua* is the next one down, overlapping with *E. trochoides* and extending down into the oyster-zone. *N. australis* (Figure 3.30) and *N. nodosa* follow next, both living in the middle part of the upper-littoral among the oysters where they cohabit with *L. undulata* but also extending up onto the bare rock-face above where they cohabit with *E. vidua*.

Of the coastal nerites, *N. undata* is usually highest on the shore, and the only one inhabiting the oyster-zone and rock-face above. However, on rocky shores of the outer islands in the West Pilbara (and the western Ningaloo coast) the oceanic species *N. plicata* may also be common in the oyster-zone and above it. The other species of nerite live in the lower part of the upper-littoral zone.

Another common invertebrate in the upper part of the upper-littoral zone is the small prostrate barnacle *Chthamalus malayensis* that often covers large areas of the rock-face in the lower part of the upper notch of

FIGURE 3.29 The widespread Indo-West Pacific littorinid *Echinolittorina trochoides* occurs on mainland and inner shelf rocky shores where, except in the northern Kimberley, it is the highest marine invertebrate living in the upper part of the upper-littoral and the supralittoral fringe. It clusters in large colonies in splash pools when the tide is low. *Photo: Barry Wilson.*

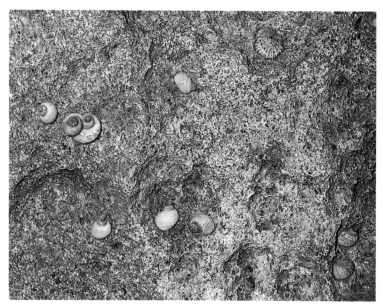

FIGURE 3.30 *Nodilittorina australis* (and one siphonarian limpet) living on bare rock-faces in the upper part of the upper-littoral zone. *N. australis* is an endemic Western Australian species found from the South Coast to the West Pilbara. Two individuals near the left of the picture are copulating, the smaller male in the typical position where it is able to insert its penis into the genital opening within the mantle cavity of the female. After mating, the female stores the fertilized eggs in an egg sac until tide conditions are suitable for spawning. *Photo: Barry Wilson.*

limestone shores. The crab *Leptograpsus* sp. and the long-legged, fast-running isopod *Ligia exotica* live in wet crevices in this zone and below, making forays into the supralittoral.

(b) The rock-oyster band of the central upper-littoral zone

On rocky shores of the mainland and coastal islands, rock-oysters are often abundant, growing on upper-littoral rocks in clusters or spaced apart. *S. cucullata* is the most conspicuous of them on the North West Shelf where it builds thick bands in the middle part of the upper-littoral zone, especially on shores that are exposed to heavy wave action (Figures 3.19 and 3.31). *Saccostrea echinata* is found in more sheltered conditions, generally at a slightly lower level on the shore. On limestone shores of the Pilbara, *S. cucullata* may establish dense, encrusting colonies (Figure 3.17) crowded over each other and forming an outward-growing overhang that is resistant to erosion but may be undermined by bioerosion in the lower notch.

The oyster-zone supports a distinctive microecosystem providing cryptic habitat for a variety of invertebrates, most of them suspension-feeding nestlers. Apart from the oysters, there are three intertidal bivalves common in this zone, *Brachidontes* sp. and *I. nucleus* are nestlers among the oysters while the mytilid *Lithophaga malacana* bores in the rock and oyster

FIGURE 3.31   A dense assemblage of the rock-oyster *Saccoastrea cucullata* forming an overhang at the base of the upper-littoral zone on a limestone shore at Barrow Island (see also Figure 3.19). *Photo: Barry Wilson.*

shells. There are also small amphipods, shrimps and crabs, polychaete and sipunculid worms and sponges, and several kinds of barnacle, including the stalked nestler *Ibla cumingi*. Another barnacle, the large, conical species *Tetraclita squamosa* is common in this zone and may grow attached to the oyster shells (Figure 3.32). In deeper crevices, the alga *B. tenella* may form leafy growths. One of two species of the pulmonate snail genus *Ophicardelus* (*O. ornatus* and *O. sp. indet*) may be present among the *Bostrychia*. The intertidal barnacles in the Pilbara (nearshore) Bioregion are well documented.[36] Bare rock and the mature parts of oyster shells are often covered with the prostrate species *Chthalamus malayensis*.

The gastropods *N. undata*, *Planaxis sulcatus*, *Montfortula variegata*, *P. saccharina*, and *P. mimula* are characteristic of the oyster-zone but may extend slightly above and below it. *Planaxis* often cluster in huge colonies in depressions that retain water at low tide. Two muricid gastropods, *Thais aculeata* and *Morula granulata*, are typical residents, preying on the barnacles, mussels, and oysters. Two large chitons, *Acanthopleura spinosa* (Figure 3.33) and *A. gemmata* (Figure 3.34) are usually conspicuous; the former sometimes extends up into the lower part of the upper notch and the latter downward into the lower notch and onto the inner edges of the reef-flat. These chitons are important bioeroders of limestone shores, grazing on the rock surface to extract interstitial algae.[48]

FIGURE 3.32   The barnacle *Tetraclita squamosa* growing on rock-oysters, *Saccostrea cucullata*, on basalt rocks in the lower part of the upper-littoral zone. Barrow Island. *Photo: Barry Wilson*.

FIGURE 3.33 Two important bioeroders of Quaternary limestone in the intertidal zone. The chiton *Acanthopleura spinosa* is a grazer on surface and interstitial algae. Also visible are plugged burrow entrances of the boring barnacle *Lithotrya valentiana*. Barrow Island. *Photo: Barry Wilson.*

FIGURE 3.34 A very common chiton, *Acanthopleura gemmata*, lives in the lowest part of the upper-littoral zone but extends its range lower down onto bare rock surfaces below the oyster-zone. The gastropod *Thais aculeata* and barnacle *Tetraclita squamosa* are also present in this picture. Cape Latouche-Treville, Canning Bioregion. *Photo: Barry Wilson.*

(c) The lower rock-face of the upper-littoral

The rock-face below the oyster-zone usually bears a cover of algal turf and is habitat for a variety of invertebrates several of which are bioeroders that are largely responsible for the characteristic undercut visor. The invertebrate communities typical of this lowest part of the upper-littoral often occur also, in diminished form, on the mid-littoral rock flat and on boulders on the reef-crest and where high points of the rock surface lie within the upper-littoral level.

Several species of gastropods are characteristic of this part of the upper-littoral zone of rocky shores and are generally confined to it but some may extend above into the oyster-zone or below onto the reef-flat. They are *Turbo cinereus, Monodonta labio, N. albicilla, N. polita, N. squamatula, N. chamaeleon,* one or two species of the pulmonate limpet genus *Siphonaria* and the true limpets *C. radiata, P. flexuosa,* and *Patelloida saccahrina* (Figure 3.35). The lowest limit of most of these gastropods may be taken as an approximate marker of the mid-littoral/upper-littoral boundary. *T. cinereus* is especially useful in this regard as it has a very narrow vertical range and it may be used as a sharp marker of that boundary.

FIGURE 3.35   The true limpet *Patelloida saccharina* and the pulmonate limpet *Siphonaria* sp. on the bare rock-face below the oyster-zone, Barrow Island. True limpets are broadcast spawners with short-lived pelagic larvae. Like other pulmonates, siphonarians are hermaphroditic; their pelagic eggs hatch from gelatinous egg masses that are laid on the rock substrate (visible in the picture) but little is known of larval development of tropical species. *Photo: Barry Wilson.*

Two other large gastropods that commonly feature in the lower upper-littoral zone but which also occur across most of the intertidal rock platform are the trochids *Trochus hanleyanus* and *Tectus pyramis*. Their upper limit is a sharp marker of the mid-littoral upper boundary. The two large chitons *A. spinosa* and *A. gemmata* are usually common in the zone, though extending upward into the oyster-zone. At least one species of the slug genus *Onchidium* is common here. A heavy bivalve, *Chama* sp. may be present, cemented on the rock surface.

At some localities, there is a mat-forming vermitid gastropod, *Dendropoma* sp. These colonial animals construct encrusting masses of tubes that cover large areas of the rock surface. The mytilid bivalve *B. ustulatus* may form sand-embedded mats in the lowest part of the upper-littoral. The biomass of these animals may be substantial and they are the prey of several species of muricid gastropods, especially *Cronia crassulnata* and *Cronia margariticola*.

There are usually two or three species of cemented barnacles in the lower notch. The medium-sized, erect species *Newmanella vitatia* is universally common under ledges and stones. A smaller, prostrate species *Tetraclitella multicostata* is usually very common under stones.

On Pilbara double-notched limestone shores, the depth of the lower notch may be exaggerated in situations where there are burrowing (boring) invertebrates. The most important of these are the barnacle *Lithotrya valentiana* and the mytilid bivalve *Lithophaga nasuta*. Also present and playing an important role may be the boring bivalves *Gastrochaena gigantea*, *Petricola lacipida*, *Lithophaga malaccana*, *L. obesa*, and *L. teres*. The burrow openings of these animals may be a conspicuous feature of the rock surface. The rate of bioerosion produced by these borers, complementing the work of the grazing chitons, has not been estimated on the shores of the North West Shelf but it is certainly significant and must play a key role in cutting back the rock-face and creating the rock platform.

3.3.2.2 *Upper-Littoral Zonation in the Canning Bioregion*

Between the long stretches of sandy shores and mangals that characterize the Canning coast there are rocky headlands of Mesozoic sedimentary rocks and some Quaternary limestone shores. The latter sometimes have a wide mid/lower-littoral rock platform and a cliffed or ramped upper-littoral margin but the conspicuous double-notch seen in the Pilbara is not present.

The upper-littoral zonation patterns of the invertebrate species present on rocky shores of the Canning coast are basically the same as in the Pilbara except that the fauna is diminished. The only nerite is *N. undata* and there are only three littorinids, *E. trochoides*, *E. vidua*, and *L. undulata*. Each of these species occupies the level it does elsewhere on the North West Shelf. The oyster band (*S. cucullata*) with its associated fauna of nestling invertebrates is weakly developed. The absence or shallowness of the lower notch may be related to the absence of the boring barnacle *Lithotrya* and the boring bivalves that are such a feature of this zone in the Pilbara.

3.3.2.3 Upper-Littoral Zonation in the Kimberley Bioregion

On upper-littoral rocky shores of the Kimberley Bioregion there are significant changes in the species present. Notably, there are four species of littorinid. One of these is the endemic littorinid *Tectarius rusticus*, which is the highest invertebrate inhabitant of the zone in the Kimberley. It lives at the very top of the upper-littoral and also in the supralittoral fringe in conditions that appear to be rarely wet by seawater. The biology of this species has not been studied and its position in the shore profile may vary according to tidal and wave-splash conditions. As in the Pilbara, *E. trochoides*, *E. vidua*, and *L. undulata* are all present and arranged in the same order as they are in the Pilbara. However, neither of the two western species of *Nodilittorina* is present.

The assemblage of nerite species in the Kimberley Bioregion also differs from that of the Pilbara. Four species are commonly found in the upper-littoral zone. As usual *N. undata* is common in the oyster band and slightly above. The widespread species *N. polita* and *N. albicilla* are usually abundant, though patchily distributed in the lower part of the upper-littoral. However, neither *N. chamaeleon* nor *N. squamulata* is common and their place in the lower-littoral is usually occupied by the northern Australian endemic species *N. reticulata*. The oceanic species *N. plicata* is not seen on the Kimberley coast (but is common on beachrock of the oceanic reefs offshore).

In the Kimberley Bioregion, the byssate bivalve *I. nucleus* is very conspicuous in rock crevices above the oyster-zone, filling jointing cracks and sometimes giving them the appearance of black lines criss-crossing the upper-littoral rock-face. The oyster-zone, while present, is never as thickly developed as it is in the Pilbara and it generally lacks the associated fauna of nestling invertebrates. As the rock types do not allow boring, there is no suite of boring invertebrates in this zone. (However, barnacle and bivalve borers are usually common boring in coralline limestone boulders and the crustose algal veneer of the reef platform.)

A very conspicuous barnacle, *T. squamosa*, is very common and seems to do well on even the smooth basalt rocks exposed to moderate wave action (Figure 3.36). In the lower part of the upper-littoral zone, there are often matted colonies of the mytilid *B. ustulatus* attached to the rocks among rock-oysters, both probably the prey of the thaid gastropod *T. aculeata* (Figure 3.37).

FIGURE 3.36 A dense colony of the barnacle *Tetraclita squamosa* on a basalt block in the upper part of the mid-littoral zone at North Maret Island. This smooth rock surface seems to suit this species but not others. *Photo: Barry Wilson.*

FIGURE 3.37 The carnivorous muricid gastropod *Thais aculeata* in the lower part of the upper-littoral zone at North Maret Island. The rock-oyster *Saccostrea cucullata*, the barnacle *Tetraclita squamosa*, and the mat-forming mussel *Brachidontes ustulatus* are all present but which of them the thaid is eating was not determined. *Photo: Barry Wilson.*

3.3.3 Key Inhabitants of the Lower- and Mid-Littoral Zones

A distinction was made in Section 3.2 between algal-dominated rock platforms and fringing reefs where corals dominate the lower-littoral reef-front. Although the lower littoral biotic assemblages of these two bio-geomorphic reef platform types are profoundly different, there is usually little difference between their mid-littoral biotic assemblages, except for the lesser prevalence of corals in the latter. In this section, biotic assemblages of algal-dominated rock platforms and the mid-littoral reef-flats of fringing reefs are considered together.

It is not clear why the reef-front of some rock platforms is dominated by algae and others by corals. The presence of typical algal-dominated rock platforms (Figure 3.15) and a coral fringing reef (Biggada Reef) within a few kilometers of each on the western shore of Barrow Island, in seemingly identical sea conditions, is especially interesting. It may be that algae and corals are in a state of competitive interaction and the dominance of one ecosystem type over the other at any one place is a transient thing. But it seems more likely that there are subtle habitat differences, or connectivity factors, that determine which ecosystem is established, at least over relatively short ecological time scales.

On both rock platforms and fringing reef platforms of the North West Shelf coastal zone, the basic biogeomorphic units are

- sloping lower-littoral reef-front
- elevated mid-littoral reef-crest
- near-horizontal mid-littoral reef-flat.

Within these units, five primary microhabitat categories may be distinguished: the rock pavement surface, upstanding boulders, crevices and loose stones that provide refuge for cryptic species, sand sheets and cays, and pools and shallow lagoons. Intertidal biotic assemblages are discussed in these terms. Many of the species in these mid-littoral habitats also live in the subtidal zone but some lower-littoral species that inhabit the reef-front are confined to that zone.

By virtue of its near-horizontal surface, populated by photosynthetic plants and autotrophic corals, the reef platform is usually a zone of high primary productivity. But it is also a zone of high stress, being subjected to wave action and alternate submersion and exposure. There is a stress gradient across the flat in regard to the duration of exposure to air and sun. Even in pools of the middle part of the reef-flat where animals may find submerged refuge at low tide, they must deal with increasing temperature and deoxygenation as the low tide period proceeds. In accord with exposure and immersion gradients and microhabitat variability, there is a high degree of habitat specialization on intertidal reef platforms.

Although there is usually little vertical relief across the flat, just a few centimeters may make a big difference in terms of exposure time and there are changes of biota from the inner margin to the reef-crest as a consequence. Even so, species distribution patterns on the reef-flat tend to be mosaic and not zoned. They relate to the complex interaction of hydrodynamics with distribution of physical microhabitats such as elevation, the extent of guttering, pools and loose stones, and whether the rock pavement is bare, vegetated, or with a gravel, sand, or mud veneer. It is well known that local biodiversity is directly related to microhabitat complexity and some rock platforms are very much more species-rich than others.

3.3.3.1 Intertidal Marine Plant Assemblages

The mainland and coastal flora of the North West Shelf has been estimated to comprise over 350 species, the majority of them found in the intertidal zone and many restricted to it.[37] There are published taxonomic checklists of the flora of the Dampier Archipelago[38,39] and unpublished reports on the marine flora of the Kimberley coast.[40–47] The following generalized summary is gleaned from these reports.

Common mid-littoral plants on rock platforms of shores along the North West Shelf are listed in Table 3.1.

To date, there is no information that compares the floras of mainland and coastal island localities within the coastal bioregions. Reef-front and fore-reef zones of rock platforms of the North West Shelf are dominated by dense beds of thalloid macroalage, mainly species of *Sargassum*. In scour channels, gutters, and shallow, sandy pools, there may be extensive beds of *Halimeda* and *Caulerpa*. On exposed rock surfaces of the reef-front ramp, there are patches of crustose corallines (*Hydrolithon onkodes* and *Lithophyllum* sp.) and turf algae. The algal turfs are generally stunted or grazed-down red algae comprising various combinations of genera such as *Hypnea*, *Laurencia*, *Chondria*, *Ceramium*, *Centrocersa*, *Gelidiella*, *Pterocladia*, and *Gelidiopsis*.

In all the coastal bioregions, there is little difference in the mid-littoral reef-flat floras of rock platforms and fringing coral reefs. On reef-flats of both geomorphic types, macroalgae are dominant and corals, when present, are sparsely distributed and mostly located in shallow pools. Sand-scoured areas on mid-littoral rock pavements of rock platforms may be bare, have a thin veneer of mobile sand or a cover of crustose algae and a sparse algal turf but most often the rock surface has a patchy cover of leafy brown macroalgae that grow attached firmly to the rock base. Macroalgal growth tends to be seasonal with greater biomass in the summer months, especially on the Kimberley coast. On Kimberley rock platforms, mid-littoral *Sargassum* may appear as short fronds centimeters long in winter months but as dense growth with fronds more than a meter long in the summer wet-season.

TABLE 3.1   Common Mid-Littoral Plants on Rocky Shores of the North West Shelf

Sea grasses	*Thalassia hemprichii*
	Halophylla ovalis
	H. decipiens
Green algae	*Entromorpha* spp.
	Cladophora spp.
	Anadyomene brownii
	Caulerpa spp.
	Halimeda cylindracea
	Halimeda cuneata
	Boodlea composita
	Udotea argentea
Red algae	*Galaxaura marginata*
	Galaxaura rugosa
	Jania adhaerens
	Portieria hornemannii
	Eucheuma spinosum
	Asparagoposis taxiformis
	Tricleocarpa cylindrica
Brown algae	*Sargassum polycystum*
	Sargassum olygosystum
	Colpomenia sinuosa
	Cystyoseriea trinodis
	Dictyopteris australis
	Lobophora variegata
	Padina australis
	Padina tenuis
	Spatoglossum macrodontum
	Turbinaria ornata
	Turbinaria gracilis

FIGURE 3.38   The seagrass *Thalssia hemprichti* in a shallow sandy pool on a mid-littoral rock flat at Barrow Island. *Photo: Barry Wilson.*

Several species of seagrass, most commonly *Thalassia hemprichii, Halophylla ovalis*, and *H. decipiens*, may also be common on mid-littoral reef-flats, mainly on sandy patches in shallow pools and gutters (Figure 3.38) but they do not form dense beds as seagrasses do on southern temperate shores. Close to shore, especially where there is freshwater seepage, *Enteromorpha* and *Ulva* may form bands along the inner rock platform. Rock pools on the reef-flat often contain rhodoliths formed by growth of *H. onkodes*. In the Buccaneer Archipelago, most notably on Montgomery Reef and Turtle Reef within Talbot Bay on the Yampi Peninsula, banks of rhodoliths form containment barriers impounding intertidal lagoons and pools and are thereby key species of those rock platforms (Figures 3.24 to 3.27).

3.3.3.2 *Assemblages of Invertebrates on Rock Platforms*

Although there are many species in common, intertidal assemblages of invertebrates of all the coastal bioregions are very different to the assemblages of species found on the coral reefs of the Oceanic Shoals Bioregion. There are also differences between the intertidal rocky shore invertebrate faunas of the Kimberley Bioregion and the Pilbara and Canning Bioregions.

(a) Lower-littoral assemblages

The lower-littoral reef-front invertebrate faunas of algal-dominated rock platforms are not rich in invertebrate species. The large grapsid crabs *Plagusia* cf. *squamosa* and *Eriphia sebana* (Figure 3.39A and B) may be conspicuous foraging among the plant fronds at the reef-edge. The echinoid *Echinometra mathaei* is often present burrowing in the rock surface (Figure 3.40). Where there are overhung ledges in the gutters there may be moderately diverse assemblages of filter-feeding epifauna attached to the walls—mainly sponges, ascidians, alcyonarians, and hydroids. The large barnacle *Megabalanus tintinnabulum* is sometimes present on elevated bare rock surfaces.

FIGURE 3.39   Two grapsid crabs that inhabit the reef-front of algal-dominated rock platforms on the west coast of Barrow Island: (A) *Plagusia squamosa and (B) Eriphia sebana. Photos: Barry Wilson.*

FIGURE 3.40 The echinoid *Echinometra mathaei* lives in the reef-front zone of limestone rock platforms and makes shallow burrows, sometimes meandering channels, in the reef surface. In doing so, it is a significant bioerosion agent on limestone reefs. West Coast Barrow Island. *Photo: Barry Wilson.*

The variety of surface-dwelling molluscs in lower-littoral rock platform habitats is generally low. Sometimes the bivalves *Tridacna maxima* and *Chama* sp. are found on the pavement surface and the mytilid *Septifer bilocularis* may be common nestling in crevices. A few gastropods may be present, including the herbivorous trochid *T. pyramis* and the omnivorous cowries *Cypraea caputserpentis* and *C. moneta*. Carnivorous gastropods, including the muricids *Thais alouina*, *Cronia avellana*, and *Drupina ricinus* and the cones *Conus coronatus*, *C. ceylonicus*, and *C. flavidus* may be common but this functional group is never as species-rich or as abundant as it is in the equivalent habitat zones of the oceanic coral reefs.

One reef-front gastropod predator of special note in this habitat in the western Pilabra is the thaid *Thais orbita* which is a conspicuous member of the reef-front invertebrate communities of the Montebellos, Barrow Island, and Ningaloo Reef. This is a barnacle-eating, temperate, Southern Australian species. Its presence in this marginal tropical area is enigmatic but may be explained if it were found to breed in spring-summer and delivered onto the western North West Shelf by the seasonal Ningaloo Current, or it could be a self-maintaining population relictual since a time of greater connectivity with the West Coast.

In sheltered areas, there is usually no reef-edge and the reef surface simply slopes from the lower-littoral into the sub-littoral zone. The invertebrate fauna of reefs such as this is usually more diverse, along with the more diverse substrate that includes patches of stones, rubble, and sand over the rock pavement. In this situation, the lower-littoral fauna is little different to that of the shallow sublittoral. Mixed communities of algae and sessile invertebrates, especially filter-feeding sponges and scleractinian corals, are present in this zone.

(b) Mid-littoral invertebrate assemblages

The majority of the mid-littoral species on rock platforms and fringing reefs are not coral reef species but are typical of continental shores with or without the presence of corals. Most of the key species of mid-littoral reef-flat invertebrates also occur in the sublittoral zone.

These intertidal invertebrate assemblages are essentially secondary producers. There is a huge variety of feeding strategies and morphologies including suspensory feeders, herbivorous grazers, detrital feeders, predators, and parasites. Microhabitats are diverse and biodiversity is high, extremely so in the case of mid-littoral rock flats in the Pilbara. Many of the species are endemic to the North West Shelf or to Northern Australia more broadly.

Surface dwellers on mid-littoral rock pavement

A large variety of invertebrates live as surface dwellers in these diverse habitats. They include active foragers and grazers that crawl on the surface, burrowers that live in the sediment veneer, borers that burrow in the rock itself (when it is limestone), and fixed, sedentary species that are either cemented to the rock surface or attached by byssal threads. There are also many epiphytic and epizoic species that live among the fronds of the macroalgae or corals.

Attached invertebrates

In bare rock pavement situations, there may be several kinds of suspensory-feeding molluscs that live cemented to the rock base such as the bivalve *Chama* sp. and the irregularly coiled vermitid gastropods *Serpulorbis* sp. and *Dendropoma* sp. On mainland and inner island shores, higher surfaces may be covered by colonies of the byssal-attached mytilid mussel *B. ustulatus*, matted with algae and sand, along with their muricid gastropod predators. *Tridacna maxima* is a conspicuous mid-littoral species throughout the region, living attached to the rock base by a strong byssus, especially on the outer part of the reef-flats. The larger species *T. squamosa* is also common throughout the region but normally in pools and lagoons rather than exposed on the reef-flat pavement. In the Kimberley (and Oceanic Shoals) Bioregions, another tridacnid, *Hippopus hippopus*, is also common in pools but this species does not occur further south in the Canning or Pilbara Bioregions.

Epiphytic invertebrates

Beds of leafy macroalgae on the reef-flat provide protection for small invertebrates and food for some herbivorous species. Many bryozoans, hydroids, isopods, small crabs and gastropods such as *Jujubinus polychromus*, *Phasianella solida*, *Turbo haynesi* and the columbellids *Pyrene bidentata* and *Anachis miser* live as epizooites on the algae of this habitat.

Vagile (crawling) invertebrates

Gastropods are the dominant surface-crawling invertebrates on mid-littoral rock pavements. Herbivores such as *Strombus urceus* (on coastal shores) or *S. variabilis* (on offshore island shores) and the trochids *T. pyramis* and *Angaria delphinus* may be abundant on rock pavements with an algal turf. In the Kimberley Bioregion, the large strombid *Lambis lambis* may be very common but it is not found on coastal rock platforms further south. Also, the omnivorous cowries *Cypraea annulus* and *C. moneta* are sometimes common in this habitat. Carnivorous gastropods (mainly worm eaters) *Mitra scutulata*, *Cronia avellana*, *Conus musicus*, *C. doreensis*, and *C. ebraeus* live half-embedded in the algal-matted sand in the day-time but are active foraging across the rock flat at night. The two giant carnivorous gastropods, *Melo amphora* and *Syrinx aruanus*, also inhabit these intertidal rock platforms, feeding on other molluscs. Both lack a pelagic larval stage and thereby are vulnerable to local extinction by collectors.

Several kinds of detritivorous cerithiid gastropods live in this habitat, *Clypeomorus* among stones and gravel nearshore, *Rhinoclavis bituberculata* and *R. brettinghami* in sandy patches throughout and the large *Pseudovertagus aluco* in slightly muddy pools.

On most rock platforms, echinoderms are conspicuous elements of the surface-dwelling fauna, especially in areas of mixed rocky and sandy habitats in the lower-littoral zone. The detritivore *Holothuria atra* is often both common lying in the open (Figure 3.41). A number of other large holothurians are common on rock flats (Figure 3.42). *H. hilla*, *H. impatiens*, *H. leucospilota*, *H. pardalis*, and *Stichopus horrens* usually shelter under stones while the very long and thin-skinned synaptid *Synaptula macra* crawls on the surface of the reef pavement.

Several species of large asteroids are common in the open among rocks and corals, e.g., *Culcita schimideliana*, *Protoreaster nodulosus*, and *Pseudoreaster obtusangulus* (Figure 3.43). Species of echinoid are also commonly present in rock platform habitats—the venomous *Diadema setosum* shelters under ledges in deep pools while *Tripneustes gratilla* and *Temnopleuris alexandri* conceal themselves among tufts of macroalgae. Many kinds of ophiuroid and several species of comanthid crinoids live under stones (Figure 3.44).

FIGURE 3.41   A species-rich community of filter-feeders on a lower-littoral stony sand flat in Bouguer Passage, Nickol Bay, Pilbara. The very common detrital-feeder *Holothuria atra* also lives in this habitat. *Photo: Barry Wilson.*

Among this multitude of intertidal rock platform invertebrates, there are many examples of close or obligatory predator-prey associations. For example, on mainland and nearshore island rock platforms, mats of the byssal-attached mussel *B. ustulatus* are preyed upon by the muricids *Cronia crassulnata*, *C. margariticola*, and *Hexaplex stainforthi*. Sand-matted colonies of an unidentified ascidian are the prey of the cymatiid gastropod *Cymatium sarcostomum*. However, the complex food chains among these diverse reef-flat invertebrate communities are not yet studied and remain a fruitful field for future research.

The Cryptic Fauna

Probably the most species-rich habitats on mid-littoral reef-flats of the North West Shelf are those that provide hiding places beneath stones and ledges where small "cryptic" fishes and invertebrates find refuge from predation and protection from wave action and exposure. The cryptic invertebrates include sedentary species that live permanently in these

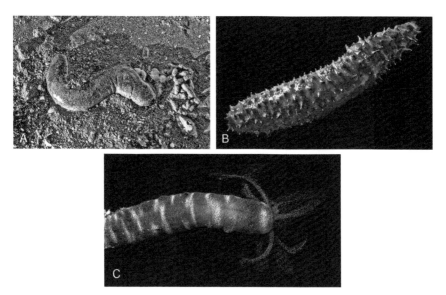

FIGURE 3.42 Some of the holothurians that live on intertidal rocky sand flats in the Pilbara region: (A) *Holothuria edulis*, (B) *Stichopus horrens*, and (C) *Synaptula macra*. Photos: *Barry Wilson*.

FIGURE 3.43 Three large sea stars that are common in the mid- and lower-littoral zone of rock platforms in the Pilbara, usually where there is mixed sand and stony habitat: (A) *Protoreaster nodulosus*, (B) *Culcita schimideliana*, and (C) *Pseudoreaster obtusangulus*. Photos: *Barry Wilson*.

FIGURE 3.44   A common comanthid crinoid that lives semi-attached under stones in the mid- and lower-littoral zone of rock platforms in the Pilbara. *Photo: Barry Wilson.*

refuges and mobile species that use them for daytime refuge and emerge to forage over the reef-flat at night.

The sedentary species that live attached to the rock substrate under stones are mostly suspension-feeders relying on plankton and detrital material delivered to them by water currents. They include tubiculous polychaetes, byssal-attached and cemented bivalves, crinoids, barnacles, sponges, bryozoans, ascidians, corals, anemones, and hydroids. There are also some suspension-feeders that bury themselves in the sand beneath the stones with just their feeding apparatus exposed. This rich assemblage of sedentary invertebrates, all secondary producers, comprise a food resource for a range of predatory and parasitic species.

An array of mobile invertebrates also finds transient refuge under stones and in ledges on the reef-flat during the periods they are not foraging. They include a large number of grazing herbivorous and carnivorous gastropods and predatory crustaceans.

Sand Sheets and Cays

The mid-littoral zone of many intertidal reef-flats contains sand sheets and cays on the rock pavement, providing habitat for infauna. Most of these species are also found on intertidal sand flats (Section 7) but some specialize in habitats with a thin sand veneer over rock pavement.

References

1. Brunnschweiler RO. *The geology of the Dampier Peninsula, Western Australia*. Bureau of Mineral Resources. Report 13; 1957.
2. Linder AW, McWhae JRH. Definitions of new formation names. Appendix C. In: Johnstone MH, editor. *Geological completion report*. Samphire Marsh No. 1 Bore of West Australian Petroleum Pty Ltd. Bureau of Mineral Resources Search Subsidy Acts, Publication 5; 1961.

3. Johnstone MH. *Geological completion report.* Samphire Marsh No. 1 Bore of West Australian Petroleum Pty Ltd. Bureau of Mineral Resources Search Subsidy Acts, Publication 5; 1961.

4. Semeniuk V. Holocene sedimentation, stratigraphy, biostratigraphy, and history of the Canning coast, north-western Australia. *J R Soc West Aust* 2008;**91**(1):53–148.

5. Gibson DL. *Explanatory notes on the Pender 1:250000 geological sheet.* Canberra: Bureau of Mineral Resources, Geology and Geophysics; 1983 p. 1–17, map sheet SE/51-2.

6. Wilson BR, Blake S. Notes on the origins and biogeomorpholgy of Montgomery Reef, Kimberley, Western Australia. *J R Soc West Aust* 2011;**94**:107–19.

7. Edwards AB. Wave-cut platforms at Yampi Sound, in the Buccaneer Archipelago, W.A. *J R Soc West Aust* 1958;**41**:17–21.

8. Brooke BP. Geomorphology, part 4. In: Wells FE, Hanley JR, Walker DI, editors. *Survey of the marine biota of the southern Kimberley Islands.* Perth: Western Australian Museum. Unpublished report no. UR286; 1995. p. 67–80.

9. Wilson BR, Blake S, Ryan D, Hacker J. Reconnaissance of species-rich coral reefs in a muddy, macro-tidal, enclosed embayment—Talbot Bay, Kimberley, Western Australia. *J R Soc West Aust* 2011;**94**:251–65.

10. Weber M. Introduction et description de l'Expedition. *Siboga-Expeditie. Monograph 1*; 1902.

11. Kuenen H. Geology of coral reefs. In: *Geological results of the Snellius expedition in the eastern part of the Netherlands East-Indies 1929–1930.* Utrecht: Kemink en Zoon N.V.; 1933. p. 1–125.

12. Doty MS. Rocky intertidal surfaces. In: Hedgpeth JW, editor. *Treatise on marine ecology and paleoecology: Geological Society of America, Memoir 67.* Washington, DC: Geological Society of America; 1957. p. 535–85 [Chapter 18].

13. D'Orbigny C. Essai sur les plantes marines des côtes du Golfe de Gascogne, et particu-lièrment sur celles du department de la Charente-Inferiéure. *Mus Nat Hist Memoir* 1820;**6**:163–203.

14. Lamouroux JVF. *Corallina: or a classical arrangement of flexible coralline polypidoms.* Bath, UK: Sherwood & Co.; 1824.

15. Gislén T. Epibioses of the Gulolmar Fjord I: Kristinebergs Zoological Station 1877–1927 No. 4: 380. Skr. utg. av. Svenska Vetenskapakad; 1930.

16. Bennett I, Pope EC. Intertidal zonation of the exposed rocky shores of Victoria, together with a rearrangement of the biographical provinces of temperate Australian shores. *Aust J Mar Freshw Res* 1953;**4**:105–59.

17. Bennett I, Pope EC. Intertidal zonation of the exposed rocky shores of Tasmania and its relationship with the rest of Australia. *Aust J Mar Freshw Res* 1960;**11**:182–221.

18. Edmunds SJ. The commoner species of animals and their distribution on an intertidal platform at Pennington Bay, Kangaroo Island, South Australia. *Proc R Soc Vic* 1948;**72**:167.

19. Guiler ER. The intertidal ecology of Tasmania. *Pap Proc R Soc Tasman* 1950;**1949**:135–202.

20. Guiler ER. The nature of intertidal zonation in Tasmania. *Pap Proc R Soc Tasman* 1952;**86**:31–61.

21. Guiler ER. Intertidal classification in Tasmania. *J Ecol* 1953;**41**:381.

22. Guiler ER. Australian belt-forming species in Tasmania. *J Ecol* 1955;**43**:138.

23. Knox GA. The biogeography and intertidal ecology of the Australian coasts. *Oceanogr Mar Biol Annu Rev* 1963;**1**:341–404.

24. Womersley HBS, Edmunds SJ. A general account of the intertidal ecology of South Aus-tralian coasts. *Aust J Mar Freshw Res* 1958;**9**:217–60.

25. Endean R, Stephenson W, Kenny R. The ecology and distribution of intertidal organisms on certain islands off the Queensland coast. *Aust J Mar Freshw Res* 1956;**7**:317–42.

26. Hodgkin EP. Patterns of life on rocky shores. *J R Soc West Aust* 1960;**43**:35–45.

27. Marsh LM, Hodgkin EP. Survey of the fauna and flora of rocky shores on Carnac Island, Western Australia. *West Aust Nat* 1962;**8**:62–72.

28. Rosewater J. The family Littorinidae in the Indo-Pacific. *Indo-Pacific Mollusca* 1970;**2**(11):417–533.

29. Reid DG. *The littorinid molluscs of mangrove forests in the Indo-Pacific region. The genus Littoraria.* London: British Museum (Natural History) Publication No. 978; 1986.

30. Reid DG. The genus *Nodilittorina* Von Martens, 1897 (Gastropoda: Littorinidae) in the Indo-Malayan region. *Phuket Mar Biol Cent Spec Publ* 2001;**25**:433–49.

31. Reid DG. Morphological review and phylogenetic analysis of *Nodilittorina* (Gastropoda: Littorinidae). *J Molluscan Stud* 2002;**68**:259–81.

32. Reid DG. The genus *Echinolittorina* Habe, 1956 (Gastropoda: Littorinidae) in the Indo-West Pacific. *Zootaxa* 2007;**1420**:1–161.

33. Williams ST, Reid DG. Speciation and diversity on tropical rocky shores: a global phylogeny of snails of the genus *Echinolittorina*. *Evolution* 2004;**58**(10):2227–51.

34. Wilson BR. *Australian marine shells.* , vol. 1. Perth: Odyssey Publishing; 1993, 408 pp., 44 col. pls.

35. Wells FE. Ecological segregation among nerites of North West Cape, Western Australia. *J Malacol Soc Aust* 1979;**4**:135–43.

36. Jones DS. Barnacles (Cirripedia: Thoracica) of the Dampier Archipelago, Western Australia. In: Jones DS, editor. *Marine biodiversity of the Dampier Archipelago, Western Australia 1998–2002.* Records of the Western Australian Museum Supplement No. 66; 2004. p. 121–57.

37. Huisman JM. Marine benthic plants of Western Australia's shelf-edge atolls. In: Bryce C, editor. *Marine biodiversity survey of Mermaid (Rowley Shoals), Scott and Seringapatam Reefs.* Records of the Western Australian Museum Supplement No. 77; 2009. p. 50–87.

38. Huisman JM. Marine benthic flora of the Dampier Archipelago, Western Australia. In: Jones DS, editor. *Marine biodiversity of the Dampier Archipelago, Western Australia 1998–2002.* Records of the Western Australian Museum Supplement No. 66; 2004. p. 66–8.

39. Huisman JM, Borowitzka MA. Marine benthic flora of the Dampier Archipelago, Western Australia. In: Wells FE, Walker DI, Jones DS, editors. *The marine flora and fauna of Dampier, Western Australia.* Perth: Western Australian Museum; 2003. p. 291–344.

40. Huisman JM. In: RPS Bowman Bishaw and Gorham 2007. *Inpex environmental impact assessment studies, technical appendix—marine and intertidal.* Unpublished report.

41. Walker DI. Seagrasses, Part IX. In: Morgan GJ, editor. *Survey of the aquatic fauna of the Kimberley Islands and Reefs, Western Australia.* Report of the Western Australian Museum Kimberley Island and Reefs Expedition, August 1991. Perth: Western Australian Museum. Unpublished report UR286; 1992. p. 75.

42. Walker DI. Seagrasses and macroalgae, Part 5. In: Wells FE, Hanley JR, Walker DI, editors. *Survey of the marine biota of the southern Kimberley Islands.* Western Australian Museum. Unpublished report no. UR 286; 1995. p. 67–80.

43. Walker DI. Seagrasses and macroalgae, Part 5. In: Walker DI, Wells FE, Hanley JR, editors. *Marine biological survey of the Eastern Kimberley, Western Australia.* Western Australian Museum. Unpublished report UR353; 1996. p. 36–8.

44. Walker DI, editor. *Marine biological survey of the Central Kimberley Coast, Western Australia, November–December, 1996.* Perth: University of Western Australia. Unpublished report, WA Museum Library No. UR377; 1997. p. 91–5.

45. Walker DI. Seagrasses and macroalgae, Part 5. In: Walker D, editor. *Marine biological survey of the Central Kimberley Coast, Western Australia.* Perth: University of Western Australia. Unpublished report, WA Museum Library UR377; 1997. p. 40–5.

46. Walker DI, Wells FE, Hanley JR, editors. *Marine biological survey of the Eastern Kimberley, Western Australia.* Perth: University of Western Australia. Unpublished report, WA Museum Library UR353; 1996.

47. Walker DI, Prince RIT. Distribution and biogeography of seagrass species on the northwest coast of Australia. *Aquat Biol* 1987;**29**:19–32.

48. Barbosa SS, Byrne M, Kelaher BP. Bioerosion caused by foraging of the tropical chiton *Acanthopleura gemmata* at One Tree Reef, southern Great Barrier Reef. *Coral Reefs* 2008;**27**:635–9.

Coral Reefs of the North West Shelf

The North West Shelf is rich in corals and coral reefs, at least at the northern and southern ends of the shelf where there is extensive reef development, a diversity of reef forms of varied origins, and a high diversity of coral species. In this study, five North West Shelf coral reef bioregions are recognized based on biogeomorphic criteria, each characterized by distinctive reef types discussed in Section 4.3 (Figure 4.1).

Although corals and coral communities are common on suitable subtidal substrata in coastal waters along most of the North West Shelf, coral reefs, in the sense of coral and algal communities undergoing structural reef-building, are developed only in the north (on the Sahul Shelf, the edge of the Rowley Shelf, and the Kimberley coast) and in the south (on the West Pilbara and Ningaloo coasts). Coral reefs are absent on the mainland coast and inner shelf of the central North West Shelf between Cape Leveque and the Dampier Archipelago—a distance of nearly 800 km. The northern part of that reefless coast (Cape Leveque to Cape Keraudren) has a sandy shore of low profile and few rocky headlands that could be regarded as a suitable base for reef development. In the southern part (Cape Keraudren to the Dampier Archipelago), there is a Quaternary sedimentary fringe along the margin of the Pilbara Craton with some limestone headlands and rock platforms with diverse coral communities, but it nonetheless is also reefless.

4.1 HISTORY OF DISCOVERY AND DESCRIPTION

The resources of coastal coral reefs of the Kimberley and Pilbara have been utilized by indigenous people for millennia. Similarly, the large oceanic reefs along the margins of the North West Shelf have been long known to Asian seafarers, especially fishermen from the islands of eastern

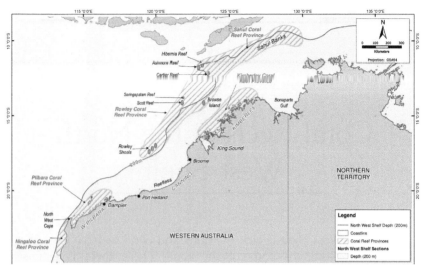

FIGURE 4.1   Coral reef bioregions of the North West Shelf. *Drawn by URS (Australia).*

Indonesia who have fished there for many centuries, taking invertebrates such as holothurians, clams and other molluscs, and fin fish.[1] Dutch navigators were familiar with Ningaloo Reef as a navigation hazard in the seventeenth century, and the shelf margin reefs further north were seen and their positions noted by other European navigators during the seventeenth, eighteenth, and nineteenth centuries.[2–5]

Scientific interest in the coral reefs of the North West Shelf began with a series of hydrographic surveys of the Kimberley coast in the nineteenth century carried out by ships of the British Admiralty (*Mermaid* and *Bathurst* 1818-1822; *Beagle* 1837-1838; *Penguin* 1890-1891). References to corals and coral reefs were published by officers of each of these three surveys.[6–8]

On his voyage north along the coast of Western Australia in the *Mermaid* (March 1818), Phillip Parker King charted and named the three Rowley Shoals (Imperieuse, Clerke, and Mermaid Reefs) referring to them as "of a coral formation."[6] There are also many references to fringing reefs in the Kimberley in his narrative. Referring to reefs around what are now known as the Heyward and Champagny Islands in the Kimberley, he specifically referred to coral reefs: "To the westward of Augustus Island is a range of islands extending for five leagues; on their north side, they are fronted by considerable coral reefs, which at low water are dry." And again "Off the north end of Byam Martin's Island are several smaller islets and coral reefs; the latter extend from it for more than six miles." These appear to be the first published references to the Kimberley coral fringing reefs (King, Appendix of vol. 2, Sailing Directions, p. 346-7).[6]

King also collected coral specimens at the entrance to Munster Water (vol. 2, p. 55-6 and Appendix B).[6] He wrote, "... we landed on the reef off the east end of the Midway Islands, which we found to be more extensive than had been suspected ... being low water when we landed, the reef was dry. Upon it we found several species of coral, particularly *Expanaria mesenterina*, Lam.; *Caryophyllia fastigiata*, Lam.; and *Porites subdigitata*, Lam." These specimens must have been among the first corals collected in Australian waters for scientific purposes. Many of the natural history specimens collected by the King expedition were later deposited in the British Museum of Natural History, but these corals have not been located and their identities are not verified.

The *Beagle* survey of northern Australia in 1837-1838 was initially commanded by Captain Wickham, but when he retired because of ill health, John Lort Stokes took over and it was he who wrote the subsequent narrative account of the expedition.[7] This voyage of the *Beagle* was the third scientific expedition undertaken by that historic vessel, the second being the famous one in which Charles Darwin had participated, leading to his monumental works on coral reefs[9] and the origin of species.[10] Just prior to the departure of the *Beagle* from Plymouth in July 1837, Darwin had delivered his controversial lecture to the Geological Society of London, proposing the significance of subsidence in the formation of coral atolls and barrier reefs. This was the genesis of the "coral reef problem" that prevailed for many years. The matter was of great moment to English natural scientists of the time and, in our present context, it is interesting to note reference to it in the sailing instructions given to Captain Wickham by the Lords of the Admiralty that included the following passage written by the Hydrographer F. Beaufort:

> It has been suggested by some geologists, that the coral insect, instead of raising its superstructure directly from the bottom of the sea, works only on the summits of submarine mountains, which have been projected upward by volcanic action. They account, therefore, for the basin-like form so generally observed in coral islands, by supposing that they insist on the circular lip of extinct volcanic craters; and as much of your work will lie among islands and cays of coral formation, you should collect every fact which can throw any light on the subject.

Stokes had been a junior officer during the second *Beagle* expedition and thereby a shipmate of Darwin. He was well informed on coral reefs, and in his narrative (Stokes, vol. 1, p. 331),[7] we find the following: "Previous to my departure from England, I had the pleasure of hearing a valuable paper by my friend Mr. Darwin, on the formation of coral islands, read at the Geological Society; my attention being thus awakened to the subject ..."

Earlier in the narrative (p. 171-172, entry for March 25, 1938), Stokes gave an account of an effort to find a passage for the *Beagle* from King

Sound through the islands of the Buccaneer Archipelago into Collier Bay. He described the view from a cliff top on the eastern side of Irvine Island (referring to it as Bathurst Island) saying "... we proceeded to examine that channel more minutely, and were sorry to find the extensive coral reefs which fronted the islands, left a space of only half a mile between..."

A little further on (p. 185, entry for April 5), Stokes described the discovery of Beagle Reef (see Figures 4.4 and 4.32), one of a cluster of major, open sea platform reefs in the Adele complex, as "... well marked by a bank of very white sand and dead coral from which the reef extends two miles and a half, in a N.N.W. and one mile in a S.S.E. direction; and which rising some 15 feet above the mean level of the blue surrounding water, became a conspicuous object from our deck, even at a distance of six miles. We gave our discovery the name of Beagle Bank, as another memorial of the useful services in which our little vessel had so frequently been engaged ..."

During the third Admiralty survey on the *Penguin*, Bassett-Smith made collections of corals at Holothuria Bank and on the fringing reefs of Troughton Island and other islands of the Bonaparte Archipelago on the northern Kimberley coast. Several years later,[8] he published a brief note on his observations on the formation of the fringing reef of Troughton Island, commenting, "Another peculiarity of this region was the great turbidity of the water near the coast and the large amount of slimy mud deposited on the flats, which as fringing reefs were everywhere present ..." The corals collected by this expedition were also lodged by the Admiralty in the British Museum of Natural History in London where they became an important resource for coral taxonomists.

Geologists were the first to describe the structure and geomorphology of coral reefs of the North West Shelf. Based on the interpretation of military aerial photographs, Curt Teichert and Rhodes Fairbridge discussed the geomorphology and origins of the Sahul and Rowley Shelf reefs and noted the presence of large platform reefs and fringing reefs on the Kimberley coast.[2,11,12] Jones[13,14] presented the results of a geological reconnaissance of the northwest Australian continental shelf in 1967 and 1968 in which he discussed the structure and origins of the Rowley shelf-edge atolls.

Biological research on the coral reefs of the North West Shelf began in the 1970s when the Western Australian Museum initiated a study of the crown-of-thorns starfish (*Acanthaster planci*) in the Dampier Archipelago[15–18] and environmental survey work began that was associated with the development of Dampier Port.[19–25]

In 1987, a summary of available information at that time on Western Australian coral reefs was produced for the IUCN Directory of Coral Reefs.[26] Since then, a great deal of new information has become available from taxonomic and ecological surveys by the Western Australian

Museum and the Australian Institute of Marine Science and many local environmental surveys conducted by and for the North West Shelf petroleum and gas industry. As a result of these studies, the coral fauna of the region is now fairly well known taxonomically.[27] There are lists of corals of the Oceanic Shoals Bioregion,[28,29] the Kimberley Bioregion,[30–32] and the Pilbara Bioregions.[33–37] However, to date there is little information on functional assemblages of coral species in the reef ecosystems of the North West Shelf. Ongoing research will address this knowledge gap.

4.2 CONDITIONS FOR CORAL REEF GROWTH

The environmental conditions that foster the establishment and growth of coral reefs may not be the same as those that foster settlement and growth of corals. Benthic communities that include corals are commonplace in intertidal and subtidal habitats along the whole coastline of the North West Shelf but the deposition of framework and the construction of biogenic reefs do not occur unless a set of suitable conditions prevail. The controlling factors of the reef growth are reviewed by Montaggioni.[38] They include changing sea level and tectonism, antecedent topography, substrate, wave energy, sea surface temperature and salinity, nutrient levels, light levels and turbidity, aragonite saturation, atmospheric CO_2, and sea surface salinity. To this list, we may add connectivity with an available source of larvae. These factors are universal within the low-latitude coral provinces of the world. In this section, antecedent surfaces for reef development and light and turbidity are discussed in the North West Shelf context.

Antecedent Topography. Modern reefs are Holocene structures developed on the sea bed or rocky shores following the Post Last Glacial rise of sea level. There is a common view that coral reefs grow best on preexisting firm substrata but this is not always the case. While many reefs have grown on rock surfaces, coral communities commonly establish on sediment banks and mounds and may consolidate, eventually forming a reefal structure with or without a solid base.

The North West Shelf has been a depocenter of carbonate sediments throughout the Tertiary and biogenic limestone structures developed on topographic high points along its subsiding margin at least as early as the Miocene (Figure 2.1 of Chapter 2). Some of these are ancient coral reefs with layers of limestone built during successive phases of reef growth, forming atolls and platform reefs on deep terraces on the continental slope and along the outer shelf. Late Pleistocene reefs on the tops of these structures were left exposed by lowered sea level during the Last Glacial period but provided ideal surfaces for renewed coral growth when inundated by the Holocene sea level rise. The original topographic high points on which coral growth began may have been geological features resulting from tectonic activity. On the Rowley Shelf, the series of slope atolls appear to be arranged along antecedent anticlinal structures. On the Sahul Shelf, there is evidence of carbonate sediment mounds built by chemolithotrophic organisms around hydrocarbon leaks on the seabed and these also may have been starting points for reef growth in some cases. In either case, coral reefs of the outer continental slope are bioherms with modern reefs growing on older biogenic reef structures.

There are no mid-shelf platform reefs on the North West Shelf like those of the Great Barrier Reef in Queensland. However, on the inner shelf of the Kimberley, there are several platform reefs including a cluster of elongate platform reefs in the vicinity of Adele Island, around 90 km from the coast (Figure 4.4). These may be regarded as bioherms insofar as their tops are caps of Quaternary reefal limestone but they are built on inundated, rocky hills of the dissected Kimberley Basin margin over a Proterozoic basement. The elongate form of these modern reefs, aligned with geological trends of the region, suggests geological control of their initial establishment and subsequent growth.

On the Pilbara and Kimberley coasts, the inner shelf is characterized by Holocene fringing reefs of several kinds. Some are built on antecedent rock platforms—Quaternary limestone platforms in the Pilbara and mainly Proterozoic sandstones and igneous rocks in the Kimberley. Fringing reefs built on rock platforms on shores of rocky islands generally occur on the sides facing the prevailing swell. Fringing reefs are also commonplace on sandy leeward shores of these coastal bioregions, where they appear to be built on unconsolidated sediment banks. In both bioregions, there is extensive development of patch reef and coral shoals on preexisting topographic high points of the seabed.

Levels of Light and Turbidity. Since the publication of Charles Darwin's epic study of coral reefs,[9] the view that coral reefs flourish best in clear oceanic waters has been held by many and still prevails but it needs to be qualified. It is true that the world's largest barrier reefs and atolls occur in those conditions but species-rich coral reefs with well-developed coral framework also occur in turbid coastal waters and there appears to be no correlation between the diversity of coral species and turbidity.

There is extensive development of biodiverse fringing and platform reefs in turbid, coastal waters of the Coral Triangle[39–41] and on the inner continental shelf of northern Australia. Species-rich reefs are commonplace in the turbid coastal waters of central Queensland coast.[42–46] Bassett-Smith[8] and Davis[47] both commented on what they considered to be the unusual presence of coral reefs in the turbid conditions of the Kimberley coast of Western Australia. However, fringing, patch, and platform reefs with high species richness are not unusual but prolific on the

Kimberley and Pilbara coasts and the idea that coral reef communities are stressed by these turbid conditions may be misplaced.

Turbidity is a measure, principally, of the amount of particulate material in suspension. The material may be organic (living microplankton or detrital) or inorganic sediment. Particle suspension is highest in nutrient-rich coastal water where planktonic primary production is high, runoff from rivers delivers terrestrial sediment, and the substrate is muddy. It may have two quite different impacts on benthic organisms. High turbidity is associated with attenuated light penetration—light penetrates deeper in clear oceanic water than in turbid coastal water. Light attenuation has detrimental impacts on benthic invertebrates that depend on phototrophic nutrition and may affect their depth range. Also, settlement of the suspended particles (sedimentation) may have detrimental physical impacts. Yet suspended organic particles may have high nutritional value for heterotrophic suspensory-feeding invertebrates. Nutrient-rich habitats generally support benthic communities characterized by high diversity and abundance of heterotrophic organisms, but these communities must achieve a balance between the advantages of abundant food supply and the problem of sedimentation.

Most scleractinian corals have dual modes of nutrition, phototrophic through photosynthetic activity of commensal zoothanthellae[48] and heterotrophic by means of filtration of suspended organic particles.[49–51] Spatial distribution and reef-building capacity of corals that are dependent on phototrophic nutrition may be limited by light attenuation resulting from increased turbidity but this effect will be less significant if it is compensated for by heterotrophic nutrition. Some phototrophic coral species that live in oligotrophic oceanic conditions of offshore reefs adapt to inshore turbid conditions by greater reliance on heterotrophic nutrition, taking advantage of the abundant suspended food particles.[51] Enhanced heterotrophic capacity inshore may provide a partial physiological explanation for the success of many reef corals in high-turbidity nearshore habitats. In fact, a case can be made that heterotrophic life in the turbid water of inner shelf habitats is the natural circumstance for many corals, rather than an adaptation to abnormal stressful conditions. Perry and Larcombe[52] put it this way, ". . . it is more appropriate that these [reefs in "marginal settings"] be considered not as restricted or disturbed communities, but as alternative states of coral community development." In fact, an argument could be made that, on evolutionary principles, phototrophy based on derived commensalism is likely to be an acquired character and that the ancestral condition of corals is heterotrophic filter-feeding in nutrient-rich habitats.

It follows that the concept of corals being adapted to clear water and stressed by increased turbidity by virtue of light attenuation may be misplaced if there are species naturally adapted to nutrient-enriched coastal conditions that advantage heterotrophy. Perhaps the argument might even be put the other way—heterotrophic corals that thrive in turbid coastal water may be stressed in clear, nutrient-poor oceanic conditions and compensate by a capacity for phototrophic nutrition. However, the damaging effects of sedimentation are another matter.

Settlement of suspended particles, either organic or nonorganic, poses physical problems for benthic organisms, especially suspension feeders living in muddy nearshore conditions like those that are common along the Kimberley coast. Sessile benthic invertebrates have morphological and behavioral adaptations for dealing with sediment that clogs up their respiratory and filter-feeding systems, but this activity requires significant expenditure of energy to deal with it. Some benthic species deal with sedimentation very effectively, while others do not manage this so well and failure to do so may be fatal.

Dependence of corals on phototrophic and heterotrophic nutrition and their ability to deal with sedimentation is likely to vary greatly among the coral taxa, and we should expect that both light attenuation and sedimentation are major factors determining coral spatial distributions. These environmental variables are primary controllers of coral community species composition.

On the North West Shelf, nearshore fringing and platform reefs have higher coral species diversity than the oceanic shelf-edge reefs. Over 60 kinds of corals, including many reef-building species, are found in the turbid waters of the Kimberley and Pilbara but not on the outer oceanic reefs. These corals comprise a suite of "continental" reef-building species that appear to be well adapted to nutrient-enriched, turbid coastal waters and such conditions may be regarded as their natural environment. The existence of "continental" and "oceanic" invertebrates in the Indo-West Pacific Realm has been recognized for many years as a biogeographic feature of the region.[53–58]

The contrasts between coral and other invertebrate assemblages of inner shelf reefs of the Kimberley and Pilbara and those of the oceanic reefs of the shelf margin will be discussed in a later section. The matter needs further investigation as it has important relevance to environmental management issues. The conditions in coastal waters enhancing (or inhibiting) heterotrophy, and perhaps even calcification, reef-building, and resilience to stress, are unlikely to be the same as those that operate on the open ocean reefs along the shelf margin.

4.3 CATEGORIES OF CORAL REEF ON THE NORTH WEST SHELF

Coral reefs of the North West Shelf include some of the types that occur in Queensland (described by Hopley[42,59]) and also continental slope atolls of the kind that occur in eastern Indonesia and in the South China Sea. Fringing reefs are numerous and diverse, but there are few platform reefs on the North West Shelf and they do not represent the range of this category seen in Queensland. There is no barrier reef at the present time although Teichert and Fairbridge[2] noted that the series of submerged banks along the margin of the Sahul Shelf may have comprised such a feature earlier in the Quaternary. That notion will be discussed later in this section.

4.3.1 Slope Atolls

Atolls are more or less circular reefs enclosing a central lagoon. They are relatively old structures that originated as fringing reefs growing around an island or patch reefs on a topographic high area of the sea bed and owe their present annular form to ongoing vertical and lateral growth as the central foundation has subsided. The "typical" atolls whose origins and development have been the subject of so much debate and controversy since the epic work of Charles Darwin[9] developed on subsiding volcanic peaks in mid-ocean.[60] Such structures are prolific in the Western Pacific and Indian Oceans. However, reefs of the same form are also common on subsiding margins of continental shelves of South East Asia and are present also on the continental margin of north-western Australia.

There are many examples of large annular reefs with central lagoons on the Sunda Shelf of Indonesia to which the term "shelf atoll" has been applied.[40,41,61–63] Wang et al.[64] described atolls in the South China Sea, drawing distinctions between three categories they called oceanic, continental slope, and continental shelf atolls. Tomascik et al.[40,41] noted that while shelf atolls are common in Indonesia, continental slope atolls "... are quite rare and described thus far only from the South China Sea." However, Karang Muaras (Muraras Atoll) located 90 km off the coast of East Kalimantan is an Indonesian example of a slope atoll. This large atoll arises from a plateau on the margin of the continental slope at a depth of 300 m.[40,41,63]

There are five coral reefs on the continental slope of the Rowley Shelf that are similar to the continental slope atolls of the South China Sea and Muraras Atoll of Kalimantan. The most studied of them is Scott Reef (Figure 4.2). They originated, probably in the Middle or Late Miocene, on anticlines on subsiding terraces of the continental slope.[65,66] There have been varied opinions among reef scientists whether these reefs should be considered to be atolls in the strict sense. Davis[47] described them as "normal atolls." Teichert and Fairbridge[2] commented that "these reefs all conform to the general annular pattern of the oceanic atoll," and noting their position along the margin of a continental shelf, they suggested that they might be thought of as "bank-atolls" except for the fact that they rise from depth on the continental slope and not on the shelf. Jones[14] called them "true atolls." Fairbridge[11] and Berry and Marsh[3] referred to them as "shelf atolls." Veron[28] called them platform reefs.

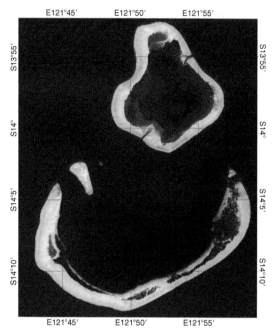

FIGURE 4.2   Scott Reef. North and South Scott Reef are separated by a deep channel. North Scott is annular with two narrow entrances; South Scott is a lunate reef, open to the north. *Landsat Image, courtesy Department of Land Information, Western Australia.*

The Rowley Shelf margin reefs owe their geomorphic form to ongoing subsidence matched by rapid coral growth. The process has been the same as the Darwinian process that produces "typical" atolls on subsiding volcanic pinnacles and there have been similar geomorphic and ecological outcomes. In this author's opinion, subsidence of the base and the consequent development of the atoll form, with peripheral reef platforms, deep central lagoons, and diverse atoll habitats, are sufficient to determine these reefs as atolls rather than shelf platform reefs.

With this background, Seringapatam Reef, North and South Scott Reef, and the three Rowley Shoal reefs, all arising from terraces on the continental slope, are referred to here as continental slope atolls or slope atolls for short.

4.3.2 Platform Reefs

There are several platform reefs on the North West Shelf, four at the shelf margin in the eastern part of the Oceanic Shoals Bioregion that are true bioherms (i.e., built of accumulated carbonate skeletal remains of marine organisms) and seven of quite different character on the inner shelf of the Kimberley.

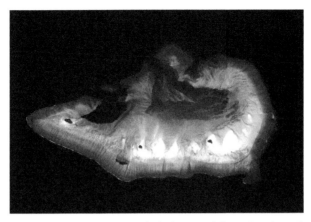

FIGURE 4.3   Ashmore Reef. A lagoonal platform reef that arises from the continental slope. *Source: Google Earth.*

Ashmore Reef (Figure 4.3) is the largest platform reef of the bioregion. It rises from the continental slope close to the continental shelf margin. It has two lagoons open on the leeward side and resembles the lagoonal platform reefs of Queensland. Hibernia Reef is a much smaller, coral-rich intertidal platform with a shallow lagoon atop a pinnacle rising from the continental slope. Cartier Reef and Browse Reef are small planar platform reefs located adjacent to the shelf margin. They are oval, lack lagoons and have small sand islands. These platform reefs of the outer shelf and slope are thought to be entirely of biohermic origins.

The Kimberley platform reefs of the inner shelf, in open sea conditions, are built on antecedent Proterozoic features that are remnants of topographic high points of the submerged Kimberley Basin margin. Their form is aligned to the underlying geological structure although contemporary reef growth on their tops is controlled by wind, swell, and tidal conditions. In the far north Kimberley, East Holothuria Reef and Long Reef are elongate platform reefs of this kind with a hard seaward edge and a sloping leeward side without a distinct edge. In the outer Buccaneer Archipelago of the southern Kimberley, there is a cluster of ovate platform reefs separated by deep channels presumed to be ancient river canyons (Figure 4.4). They are Adele, Churchill, Albert, Beagle, Mavis, and Brue Reefs. The very large Montgomery Reef in Collier Bay (Figure 4.5) is also an inundated terrestrial topographic feature with a surface veneer of Holocene coral and calcareous algae, and although there is vigorous coral growth in its high lagoon, it is not a typical coral platform reef (see below).

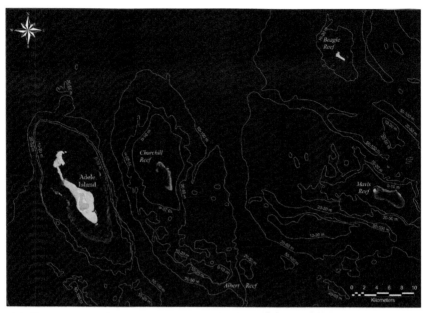

FIGURE 4.4   The cluster of platform reefs of the Adele complex, southern Kimberley, showing the bathymetry of the now inundated land surface. Brue Reef is out of the picture to the south. *Source: Landsat Image, courtesy Department of Land Information, Western Australia.*

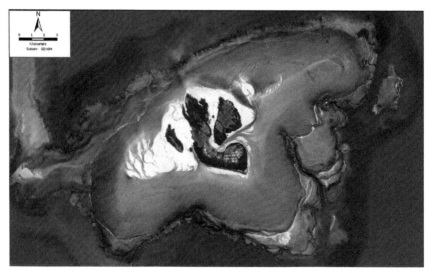

FIGURE 4.5   Montgomery Reef. A vast rock platform (area ca. 400 km²) of Proterozoic rocks with a Holocene coral and algal veneer. *Source: Google Earth.*

4.3.3 Fringing Reefs

As their name implies, fringing reefs grow as a fringe around the shores of islands and mainland shores. They are highly variable in geomorphology, determined by underlying geological structure and degrees of exposure to wave action, tidal currents, salinity, and turbidity.

Hopley *et al.*[67] classified Queensland fringing reefs into four main categories:

(1) headland-attached fringing reefs that develop on rocky headlands;
(2) bayhead fringing reefs that develop in embayments and prograde out from the head of the bays;
(3) narrow beach-based fringing reefs that develop along stretches of sandy coast;
(4) nearshore shoals that are not directly attached to the shoreline.

Versions of all four of these fringing reef categories occur in the Kimberley and Pilbara Bioregions, but judging from field observations and photo interpretation, there appear to be many variations in geomorphology that are different to the typical fringing reefs of Queensland. Of the four categories, "bayhead fringing reefs" are the least distinguishable on the shores of the North West Shelf because they tend to grade into fringing reefs of the other categories and will not be considered further in this account. The "nearshore shoal" category is treated here as a variation of patch reef. They occur in large numbers and great geomorphic diversity on the inner shelves of the Kimberley and Pilbara Bioregions. Reefs of the "headland-attached" category are referred to here as fringing reefs on rock platforms. The rock platforms may be of preexisting rocks with a coralgal veneer or biogenic reef limestone built on preexisting foundations, or a combination of the two. Reefs of the "narrow beach-based fringing reefs" category are referred to as fringing reefs without rock platforms and they may have rocky shore or sedimentary foundations.

In both the Kimberley and the Pilbara, rock platforms built of biogenic Holocene limestone develop on west and north-facing shores that are exposed to the prevailing swell and they are characterized by reef-front assemblages of domal faviid and other massive and robust corals. The biogeomorphic processes that lead to the development of rock platforms in these circumstances need investigation.

In positions on leeward shores and in bays that are protected from the swell, extensive rock platforms do not develop and fringing reefs are generally dominated by *Acropora* assemblages, sometimes with clusters of massive corals and fungiids. This basic division of fringing reef types into seaward reefs with rock platforms and leeward reefs without is illustrated by a diagram based on field observations in the Maret Islands (Figure 4.6). In the Pilbara, similar situations prevail (Figure 4.33) although in that

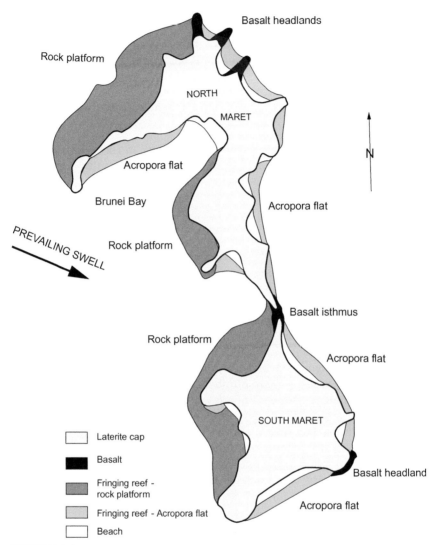

FIGURE 4.6 Diagram illustrating the distribution of the two main types of fringing reefs at the Maret Islands, Bonaparte Archipelago, Kimberley. Wide rock platforms with fringing reefs are developed on the western and northern shores that are exposed to the prevailing swell, with the reef-fronts dominated by domal faviids. On leeward shores and shores protected from the swell, rock platforms are poorly developed and the fringing reefs are primarily banks of *Acropora* growing on banks of *Acropora* rubble.

bioregion the circumstances are complicated by the fact that the anteced-ent rocks may be limestone prone to the development of rock platforms through erosional processes, for example, on the eastern shore of Barrow Island. However, there is no lateral development along the reef-fronts of such rock platforms by means of coral growth and reef-building.

Significant exceptions to this generality occur in the southern Kimberley where there are several examples of high limestone platforms with rich coral communities in high, impounded lagoons and on peripheral lower-littoral ramps and flats. The origin of these reefs is unknown and they may be older Quaternary structures.[68]

4.3.4 Patch Reefs

These are simply "unattached" reefs of varied size and form, established on antecedent topographic high points of varied nature on the sea floor. Some are submerged and may be regarded as coral shoals rather than coral reefs. They are abundant in the coastal waters of both the Pilbara and the Kimberley Bioregions. The foundations may be hard rock surfaces or sediment mounds and the extent of coral framework building is very variable.

4.4 CORAL REEF BIOREGIONS ON THE NORTH WEST SHELF

The five North West Shelf coral reef bioregions recognized in this study (Figure 4.1) equate, in part, to the IMCRA mesoscale bioregions. From north to south, they are

1. Submerged banks and platform reefs of the Sahul Shelf margin (eastern Oceanic Shoals Bioregion).
2. Atolls and platform reefs along the Rowley Shelf margin (western Oceanic Shoals Bioregion).
3. Fringing reefs and inner shelf platform and patch reefs of the Kimberley coast (Kimberley Bioregion).
4. Fringing and patch reefs of the West Pilbara coast (western Pilbara (offshore and nearshore) Bioregions).
5. Ningaloo Reef Tract (Ningaloo Bioregion) that separates the North West Shelf and the Dirk Hartog Shelf.

The distinctive characteristics of these five coral reef bioregions are founded in differences in tectonic history, ocean currents and connectivity, tidal regimes, and oceanographic (metocean) conditions (Table 4.1). The geomorphology and invertebrate reef assemblages of these five coral reef bioregions are discussed in the following sections.

A helpful recent development has been the posting online of a spatial database of coral species by J.E.N. Veron called *Coral Geographic* (http://www.coralreefresearch.org/html/crr_cg.htm). Coral species of the world are grouped by Veron into "Coral Ecoregions," based on their geographic distributions. These ecoregions equate only approximately with the coral reef bioregions proposed in this account.

TABLE 4.1   Characteristics of the Five Biogeomorphic Coral Reef Provinces of the North West Shelf

Feature	Sahul Shelf	Rowley Shelf	Kimberley Coast	West Pilbara	Ningaloo
Bioregion	Oceanic Shoals (East)	Oceanic Shoals (West)	Kimberley	Pilbara (offshore and nearshore)	Ningaloo
Reef category	Platform reefs and submerged banks (drowned reefs)	Shelf-slope atolls and platform reefs	Fringing and platform reefs	Fringing and patch reefs	Fringing/ barrier
Age of origin	Pliocene	Miocene	Holocene	Pleistocene/ Holocene	Pleistocene/ Holocene
Tectonic activity	Subsidence	Subsidence	Subsidence	No subsidence	No subsidence
Built on	Biogenic limestone	Biogenic limestone	Proterozoic igneous and metasediments	Quaternary[a] coastal limestone	Quaternary reef limestone
Wave regime	High energy	High energy	Low energy	Low energy	High energy
Water conditions	Clear oceanic	Clear oceanic	Turbid coastal	Moderately turbid coastal	Clear oceanic
Currents	ITF/ Holloway	Holloway	Local tidal	Local tidal mixing with Holloway along the shelf margin	

[a] *Biggada Reef on Barrow Island may be an exception—it may be on Miocene limestone.*

4.4.1 Quaternary History of the Coral Bioregions

The three northern coral reef bioregions occur in a region of ongoing continental shelf subsidence associated with downwarping along the margin of the Timor Trough (Section 2.1.5 of Chapter 2). Tilted regional subsidence is believed to have been a fundamental factor in the development of the reefs of the Sahul Shelf and northern Rowley Shelf.[2,14,69] Sandiford[70] described a pronounced tilt of the Australian continent, along a NW/SE axis, with ongoing subsidence of the northern continental margin. He estimated vertical motion to be 250-300 m downward since the mid-Miocene, significantly greater than the rate of Quaternary eustatic sea level change.

In contrast, there is no evidence of Quaternary tectonic subsidence in the Pilbara. There, and on the Ningaloo Reef tract, coral reef development appears to be associated with interglacial climatic periods and eustatic transgressions, the modern reefs having developed during successive high sea level stands on stable preexisting reef structures.[71,72]

The IMCRA Oceanic Shoals Bioregion occupies the outer shelf and continental slope in the northern and central part of the North West Shelf and lies more than 300 km from the Australian coastline, far from terrestrial influences. It has been a coral reef bioregion for a very long time. Northward drift of the continent into tropical latitudes in the Miocene was associated with the development of platform reef complexes and atolls as a scatter of shallow marine environments on the outer shelf (Figure 2.1 of Chapter 2) and modern reefs and submerged banks of the bioregion are their biogeomorphic descendants. Coral reef growth has been ongoing in this bioregion since the late Tertiary, although distribution of the reefs, the geomorphic types present, and their species composition may have varied in response to the dramatic tectonic, climatic, and sea level change events that have affected the region throughout this period. The interaction of shelf subsidence and eustatic sea level change is clearly of the utmost relevance to the development of coral reefs in the Oceanic Shoals Bioregion.

The history of coral reef development in the three coastal bioregions, Kimberley, Pilbara, and Ningaloo, has been very different. Coral reef development in all the three has been profoundly affected by the eustatic sea level change, but subsidence appears to have been a factor in the Kimberley and not in the other two bioregions. It is possible that, for different reasons, coral reef growth may have ceased or been severely restricted in these bioregions during the Last Glacial Maximum requiring the regional reconstruction of coral reef ecosystems when sea level rose again and the along-shelf current system was restored in the Holocene.

Ningaloo Reef lies near the margin of the optimal temperature zone for coral reef development. It remains an open question whether there were coral reefs on the lower limestone ridges on the seabed off the Cape Range Peninsula during the LGM period (Figure 4.39). Not only was sea level lower but also the ocean was colder and there was no warm Leeuwin Current, providing connectivity with reefs further north (if there were any). In the Kimberley and Pilbara Bioregions, especially the former, when sea level was below the present −50 m mark, there would have been limited rocky substrate suitable for coral reef development. In neither of these bioregions could there have been a profusion of coral reefs like those of the present day. Obviously, coral reef ecosystems had to be entirely reconstructed on the coastal surfaces that had been exposed during the LGM. The question is where did the recruits come from? Were there enough, and diverse enough, reef communities in rocky refuges along the shelf margin? Or were the Kimberley and Pilbara coastal coral reefs rebuilt with recruits from a source further north—the Sahul Shelf perhaps. The current interest in coral genetics on the North West Shelf will need to take this open question into account.

The Kimberley Bioregion is a coral reef region in a macrotidal, turbid, coastal water environment. Fringing, platform, and patch reefs are developed as Holocene veneers on Proterozoic basement rocks and on Holocene sea floor sediments. The Pilbara Bioregions (offshore and near-shore) are also essentially coastal with moderately to severely turbid con-ditions. Pilbara coral reefs are Holocene structures built on Late Pleistocene limestone platforms, bare pavement of the sea floor, or Holo-cene sediment banks. The fringing and patch reefs of the outer islands (Pilbara (offshore) Bioregion) are close to the shelf margin and are influ-enced by the Holloway Current. Their biota includes oceanic elements derived from mixing of the oceanic and coastal water. Fringing and patch reefs are also prolific on the inner shelf of the Pilbara (nearshore) Bioregion where the coastal water is turbid and presumably less influenced by mixing with the oligotrophic water of the Holloway Current.

4.4.2 Oceanic Shoals Bioregion

In this account, the Oceanic Shoals Bioregion is divided into an eastern Sahul Shelf section with platform reefs and submerged banks thought to be of Pliocene origins and a western Rowley Shelf section with shelf slope atolls and platform coral reefs of Miocene origins. The boundary between the two is taken to be the southwestern face of the Londonderry Rise which also marks the boundary between the Sahul and Rowley Shelf divisions of the North West Shelf (Figure 1.1 of Chapter 1).

4.4.2.1 Sahul Shelf Coral Reef Bioregion

The eastern section of the Oceanic Shoals Bioregion, between longi-tudes 123°20'E and 130°46'E, is characterized by three platform reefs and a series of submerged carbonate banks and shoals lying along the margins of the Sahul Shelf and directly in the path of eastern elements of the Indonesian Through Flow (Figure 2.14 of Chapter 2). These struc-tures are bioherms, built by the accumulation of marine biogenic carbon-ate during periods of submergence. Teichert and Fairbridge[2] suggested that earlier in the Quaternary the Sahul Banks and reefs may have been a major barrier reef complex.

The Sahul Shelf terminates at the Ashmore Reef in a prong projecting west into the Timor Sea (Figure 4.7). Its northern margin borders the Timor Trough that narrowly separates the Australian continent from Indonesia. At Ashmore Reef, the continental margin recurves on the southern side of the prong along the edge of the Londonderry Rise that represents the location of a deep-seated basement fault or accommodation zone.[73]

FIGURE 4.7   Reefs and banks of the Sahul Shelf. *Image courtesy Geoscience Australia.*

The paleogeographic history of the Sahul reefs and banks is dominated by their position in the subsidence zone associated with subduction of this part of the Australian continental margin beneath the Eurasian plate and the formation of the Timor Trough. They are structures of high relief, arising steeply from the sea bed along the shelf-edge and continental slope. They may be considered in three categories.

- The Sahul Banks, along the north-facing margin of the continental shelf bordering the Timor Trough, comprising three main clusters rising from the continental slope between the 200- and 300-m isobaths, and a chain of smaller shoals along the 200-m isobath.
- Johnson and Woodbine Banks and a series of smaller submerged banks along the 200-m isobath on the southern side of the Londonderry Rise.
- Three platform reefs at the western end of the shelf, Ashmore (Figure 4.3), Hibernia, and Cartier, all rising from the continental slope between the 200- and 300-m isobaths.

The three platform reefs are emergent at low tide and corals are the main reef builders, each supporting a diverse and vigorous reef fauna. Ashmore is the largest of them and is located on an isolated platform separated from the continental shelf by a narrow trough. Glenn[74] described a model for the Holocene development of the modern Ashmore Reef showing the resumption of coral growth over an antecedent Pleistocene structure, subsequently keeping up with late Holocene sea level rise. Hibernia and Cartier are small platform reefs, the latter quite like the planar platform reefs of the Queensland continental shelf with a central sand cay.

The submerged Sahul banks appear to be largely accumulations of skeletal remains of the calcareous alga *Halimeda* although there is coral growth on many of them. Their structure and origins have been described.[75] The tops of the banks have irregular topography, often more or less level at a depth of 20-25 m but with high points as shallow as 7 m. Through much of the period that followed the Late Pleistocene high sea level and during the lowest sea level period of the Last Glacial (Figure 2.13 of Chapter 2), they would have been a chain of islands close to the shore of the mainland but they were inundated by the post-LGM transgression. The rate of inundation would have been exaggerated by ongoing subsidence of the shelf margin. (Estimates of subsidence rates along the margin of the Sahul Shelf are discussed in Section 2.1.5 of Chapter 2 and the rate of the transgression is discussed in Section 2.1.6 of Chapter 2).

It has been proposed that these conspicuous topographic features are drowned coral reefs that have been unable to "keep up" with the rapid rise of relative sea level brought about by the joint effects of subsidence and the post-LGM transgression.[76] However, their history may be more complicated than this and involve contemporary ecology as well the response of the reef builders, or lack of it, to sea level change.

The reefs and banks of the Sahul Shelf are probably younger than the Rowley Shelf slope atolls. There is seismic evidence that there were no reefs on the Sahul Shelf in the Miocene, reef development beginning there in the Pliocene.[77] Origins on seabed topographic highs are likely. Extensive oil seepages in the Sahul and Karmt Shoals have been associated with the development of carbonate mounds on the seabed built by chemolithotrophic organisms.[77,78] It was suggested that the reefs and banks on the Sahul Shelf may have been initiated by seafloor hydrocarbon seepage stimulating the deposition of authigenic carbonates. Such topographic high points are thought to provide ideal bases for reef development.

If carbonate deposition leading to the development of the Sahul Banks began in the Pliocene, we could assume that the growth of the bioherms was intermittent through the Quaternary, stalling during periods of low sea level like that of the Last Glacial Maximum when the banks were islands above the sea level. Cores to a depth of 55 m at Big Bank revealed sediment derived from disarticulated *Halimeda* skeletal material,[75] but it is not known whether this material continues all the way to the basement and whether there is a lower Pleistocene sequence. If the entire sequence is Holocene, it would indicate an extraordinary rate of deposition. It remains to be determined whether *Halimeda* has been the main reef builder throughout the history of these bioherms or whether there have been earlier phases of coral framework building. What can be said is that *Halimeda* growth and reef-building was initiated or resumed during the mid-Holocene, supplemented by moderate coral growth.

The idea that the Sahul Banks are "drowned" coral reefs needs to be reconsidered. It presumes that the banks were once coral reefs that failed to "keep up" with sea level rise, as Ashmore, Hibernia, and Cartier Reefs did, and that their reef-building role was taken over by *Halimeda*. However, tops of the banks are mostly well within the optimal depth for coral growth. Healthy coral growth is present on many of the banks (A. Heyward, personal communication) and their failure to grow up to the present sea level is not satisfactorily explained by rapid sea level rise alone. It is more likely that the corals have failed because they cannot compete with vigorous growth of *Halimeda* in the circumstances that prevail at most of these banks.

Bioherms built primarily by *Halimeda* have been reported from the Great Barrier Reef,[79–85] Indonesia,[86–88] and the Caribbean.[89–91] They are associated with localized areas of upwelling and nutrient enrichment.

The position of the extensive *Halimeda* bioherms on Kalukalukuang Bank in the Java Sea relates to places along the margins of the bank where there are episodic upwellings of cold Pacific water.[87,88] It has been suggested that upwelling and nutrient overloading may advantage these algae at the expense of corals. Littler *et al.*[92] have shown that *Halimeda* species are tolerant to low light levels and are able to take advantage of nutrient pulses. Both these facilities would provide these algae with a competitive advantage over corals in deep water. At South Scott Reef (see Section 4.4.4), there is evidence of episodic upwelling over the edge of the channel into the deep lagoon with *Halimeda* forming prolific beds along that edge and overgrowing coral assemblages there. Episodic upwelling up the very steep continental slopes of the Sahul Shelf is possible and variations in its frequency and intensity due to local topography and current flows could well explain the variability in the relative success of corals and *Halimeda* among the Sahul Banks and the success of coral reef growth at Ashmore, Hibernia, and Cartier Reefs and its failure (to reach the surface) elsewhere.

If this explanation has merit, it would follow that the present depth of the Sahul Banks may be a natural circumstance and they have not been "drowned" at all. It would also follow that continuing growth of the *Halimeda* depends on upwelling and that, in turn, may depend on the strength of the ITF and continuity of the monsoonal wind system. Cores on the banks may show whether the present circumstances that favor *Halimeda* have prevailed through the Quaternary or whether they turn on and off with phases of climate change. More information is needed on the living and past biota of these bioherms before this very interesting matter is resolved.

Ashmore Reef is a large ovoid platform reef, approximately 26 km long by 14 km wide (area approximately 239 km^2) with three low vegetated sand islands (Figure 4.3). Its long axis is oriented east-west. Its southwestern, windward side rises abruptly from a depth of more than 400 m, along the escarpment of the Londonderry Rise. On the northeastern side, there is a wide platform beyond the reef with a depth less than 100 m. In this regard, Ashmore Reef is very different to the Rowley slope atolls that arise abruptly from greater depths on the continental slope and have vertical perimeter walls all around.

Ashmore Reef does not have an annular outline and the windward (southwestern) side has a much wider intertidal reef flat than the leeward (northeastern) side. The windward reef flat is 2-6 km wide with a moderately developed reef-crest boulder zone and a wide back-reef that merges with wide tidal sand flats. There is a sloping (ca. 5°) fore-reef about 150 m wide with a strong spur and groove system and a cover of mostly soft corals. The reef flat on the northeastern side has a steep fore-reef, dropping off from 3 to 20 m before a 35° slope, and supports prolific scleractinian coral growth. Unlike the windward side, there is little sand flat development in the leeward back-reef.[4] There are two lagoons, both broadly open to the sea, one with a maximum depth of 46 m and the other, while much larger in area, has a maximum depth of 15 m.

Cartier Reef is a small, elongate-ovate, planar platform reef, about 4.5 km long with an area of 10.85 km². It is situated on the continental slope, beyond the 200-m isobath. It has a small, central sand island surrounded by a wide reef platform with a well-developed boulder zone on the weather (western) side and a wide fore-reef slope with a spur and groove system. There are extensive shallow pools on the reef platform but no real lagoon. Like Ashmore, the southwestern windward side has poor coral growth on the fore-reef slope while the northeastern side has prolific growth of scleractinian corals.

The invertebrate fauna of these three reefs is characteristic of oceanic coral reefs throughout the Indo-West Pacific. For example, there is an assemblage of herbivorous and predatory gastropod molluscs that are characteristic of the reef-front zone of oceanic coral reefs but are not found on the reef-fronts of fringing reefs of the Kimberley coast (Section 4.5.3). There are many sand cays on the reef platform of Ashmore and wide areas of shallow sandy back-reef pools. These sandy habitats support species-rich infaunal invertebrate assemblages of burrowing bivalves, gastropods, and echinoderms that, like the reef-front species, are widespread on oceanic coral reefs throughout the Indo-West Pacific but few of them occur on the Kimberley coast.

Several molluscan species are endemic to Ashmore Reef with affinities to endemic North West Shelf groups.[93] The presence of these molluscs may be explained by the proximity of the reef platform to the continental shelf. During the lowest sea level period of the Last Glacial Maximum, the Ashmore platform would have been connected to the main continental shelf by benthic shelf habitat less than 100 m deep. Post-LGM sea level rise would have isolated the benthic populations of Ashmore and allowed several thousand years of genetic divergence. Some of the Sahul Banks have a similar history and study of their benthic fauna may discover new populations, or sister species, of the Ashmore endemics.

Habitats of Ashmore Reef and Cartier and Hibernia Reefs have been mapped and the major taxa of the fauna at the two reefs have been documented.[4,5,94,95] The geology and geomorphology of Ashmore Reef have been described.[78,96–99]

4.4.2.2 Rowley Shelf Coral Reef Bioregion

This is the western part of the Oceanic Shoals Bioregion. It has emergent coral reefs of two quite different kinds, slope atolls and shelf platform reefs, and some submerged banks.

Slope Atolls. The five slope atolls of the Rowley Shelf lie on subsiding terraces of the continental slope up to 100 km beyond the shelf-edge[14] with depths of 230-500 m on the upslope side and as much as 800 m on the downslope side. North and South Scott Reef (Figure 4.2) and their neighbor Seringapatam Reef (Figure 4.8) arise from the Ashmore Terrace. The three reefs comprising the Rowley Shoals, Mermaid, Clerke, and Imperieuse Reefs rise from the Rowley Terrace (Figure 4.9).

A detailed account of these oceanic reefs may be found in Collins.[14] They are large carbonate structures with more-or-less annular form and deep central lagoons. Although they have typical atoll geomorphology and habitats, they have come about by biogenic growth on anticlinal

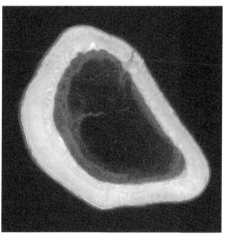

FIGURE 4.8   Seringapatam Reef—a slope atoll of the Rowley Shelf. *Landsat Image, courtesy Department of Land Information, Western Australia.*

FIGURE 4.9   Mermaid Reef, Rowley Shoals. *Landsat Image, courtesy Department of Land Information, Western Australia.*

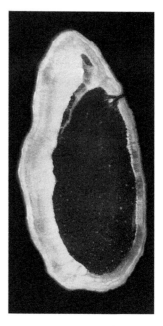

ridges of a subsiding continental shelf margin in a subduction zone, not on subsiding volcanic oceanic ridges and seamounts like the classical atolls of the Pacific and Indian Oceans.

Geological Structure and Reef Growth. The subsurface structure of the Scott Reef is known. Exploratory hydrocarbon bores in the lagoon (BOC Scott Reef Nos. 1 and 2) drilled to depths of nearly 5000 m into late Triassic and early Jurassic shelf and deltaic sediments.[65] The bores found a potentially commercial gas resource (the Torosa field) in the anticline on which Scott Reef is located and basalt intrusions within deltaic Jurassic sediments at a depth of 4279 m. The sequence of Paleocene-Quaternary carbonate sediments below the Scott platform and details of its Holocene reef growth history have been described from study of shallow cores.[100] Contemporary reef growth was initiated at 10.6 ka BP and was followed by up to 40 m of vertical growth, with accretion rates varying (between cores) from 2.65 to 3.57 m/ka.

Geomorphology. All five atolls of the Rowley Shelf slope have more-or-less annular outlines with an intertidal reef platform around the perimeter enclosing the central lagoon. South Scott, Clerke, and Imperieuse Reefs have sand islands. The perimeter reefs have an outer reef-front ramp, a prominent boulder zone on the reef crest, and a wide mid-littoral reef flat with some bare pavement and sand cay development. The inner edge of the platform is ragged with prolific coral growth and shallow pools and a steep back-reef sloping into the lagoon. There is a narrow fore-reef slope with a well-developed subtidal spur and groove system. The upper part of the fore-reef slope, at 20-30 m depth, has prolific growth of scleractinian corals. Below the fore-reef slope, the outer walls of these atolls are steep to vertical down to the top of talus slopes that descend on to the deep sea floor.[14,94,101]

Teichert and Fairbridge,[2] Wilson,[101] and Berry and Marsh[3] provided notes on the geomorphology of Seringapatam atoll, the most northerly of the group, located about 100 km from the shelf margin. This reef is trapezoidal in outline with dimensions of about 8.5-5.5 km (Figure 4.8). It has a moderately deep (to 30 m) silty lagoon. The wide reef flat has a distinct boulder zone on the reef rest (Figure 4.10). There is prolific coral growth on the fore-reef slope which has a slope of about 15° and ends abruptly at a depth of 40 m at the edge of a vertical wall. During a survey of reef invertebrates, a party from the Russian research vessel R.V. *Bogorov* found that the outer wall descends to a depth of around 200 m, below which there is a steep rubble slope to the sea floor which is 750 m deep at the northwestern side of the atoll and 500 m on the southeastern side.[101] The outer wall, to depths accessible to divers, bears a sparse cover of alcyonarians and sponges (Figure 4.11).

The dual Scott Reef is the largest of the Rowley shelf-edge atolls lying about 65 km beyond the shelf margin. Its geomorphology has been described in detail.[72] It has an area of more than 250 km^2 in total with the two parts separated by a channel more than 200 m deep (Figure 4.2). South Scott is the larger part and is crescent shaped, open to the north with

FIGURE 4.10   The boulder zone on the reef crest of Seringapatam Reef with a remnant of a large boulder that has been bioeroded by the barnacle *Lithotrya valentiana*. *Photo: Barry Wilson.*

FIGURE 4.11   An alcyonarian colony and other soft corals and sponges on the vertical outer wall at 55 m; Seringapatam Reef. There are few scleractinian corals at this depth but a prolific growth of other attached filter feeders. *Photo: Barry Wilson.*

a deep lagoon (to 60 m). North Scott is annular with an enclosed a shallower lagoon (to 20 m) and two narrow entrances. The fore-reef slope has rich scleractinian coral assemblages to depths of around 20 m, but the lower parts of the slope, to 60 m, are populated mainly by sponges and large nepthiids and other soft corals. Studies of Scott Reef (unpublished URS report to Woodside) have revealed a similar profile to that of Seringapatam. The perimeter wall has vertical (sometimes undercut) buttresses and steep, unstable talus slopes to the sea floor at around 800 m depth on the western side. ROV surveys show that the wall fauna is very sparse, comprising mainly alcyonarians, sponges, and crinoids.

The general morphology of the three Rowley Shoals reefs, Mermaid, Clerke, and Imperieuse Reefs, has been described.[3,100] They are elongate-oval (slightly pear-shaped) with deep lagoons and narrow entrance channels on their northeastern sides. The three of them are aligned with their long axes parallel, roughly in a north-south direction. There is decreasing lagoon infill northward in the series of the Rowley Shoals reefs indicating differential rates of Holocene subsidence (18 m for the southern Imperieuse and 37 m for the northern Mermaid).[100]

Fifty kilometers further south of Imperieuse Reef there is a fourth structure of a similar kind but submerged, rising abruptly 60-100 m from the seafloor to within 287 m of the surface. This appears to have been a shelf-edge atoll that was drowned by subsidence/sea level rise.[14]

The lagoons of all five atolls have a floor of silty sand and coral rubble and extensive patch reefs. The deep South Scott lagoon is exceptional. Its deepest part has a limestone pavement floor that is bare in patches, but elsewhere it has a dense community of branching and foliose corals that does not seem to be replicated elsewhere on the North West Shelf

FIGURE 4.12   The plate coral *Pachyseris speciosa*, a dominant species in the assemblage of foliaceous and delicate branching corals that inhabit the deep (30-50 m) lagoon of South Scott Reef. *Photo courtesy: AIMS.*

(Figure 4.12). On the open northern side, there is a steeply sloping sandy ramp forming the southern side of the channel separating the two reefs. On the South Scott lagoon floor bordering the ramp, there are large areas of dense *Halimeda* (unpublished URS report to Woodside). These are likely to grow so vigorously because of episodic upwelling bringing nutrient-rich bottom water from the deep sea into the northern part of the South Scott deep lagoon. ROV surveys revealed that the *Halimeda* overgrows the foliaceous corals in places suggesting that these two carbonate producers may be competing for space, with the alga most successful when and where the lagoon bottom water is enriched with deep sea nutrients.

Behind the back-reef on the western and southern sides of South Scott, there is a band of deep sand/mud habitat with infaunal communities comprising dense populations of spatangoid echinoids, benthic foraminifera, and other invertebrates. This community type also does not seem to be replicated elsewhere on the North West Shelf.

The intertidal and shallow subtidal reef communities of the shelf-edge atolls have been sampled by Western Australian Museum and AIMS surveys. Algae, coral, sponge, mollusc, echinoderm, and fish taxa present in these oceanic reef systems are now moderately well documented,[102] but the smaller taxa remain undescribed. Coral and fish communities at the Scott Reef have been mapped in detail and are regularly monitored by AIMS biologists.

The invertebrate fauna of the Rowley shelf-edge atolls, like that of Ashmore, is essentially oceanic in character. Scott Reef, in particular, has a diverse fauna in keeping with its diverse reef and sandy habitats. Significantly, almost all the invertebrates are widespread Indo-West

Pacific species and (with the exception of some sponges, J. Fromont, personal communication) none are endemic to the region.

Shelf Platform Reefs and Banks. Some 150-200 km northwest of the Kimberley coast, there are three carbonate bioherms arising steeply from the shelf margin. Two of these, Heywood Shoals and Echuca Shoals, are submerged with depths at their summits around 13 m. The third and most southerly one, Browse Reef, is a typical planar platform coral reef.

Results of a survey of Browse Reef by RPS Australia in 2007 are available in an unpublished report and are being prepared for publication.[33] The following notes are based on that survey. Browse Reef is almost circular and arises steeply from the outer shelf at a depth of around 150 m. Teichert and Fairbridge[2] described its geomorphology based on aerial photo interpretation (Figure 4.13) showing it to be similar to Cartier Reef on the margin of the Londonderry Rise a little further north. It has a central sand island surrounded by a wide intertidal reef platform (Figures 4.13 and 4.14) with no lagoon, although there are extensive shallow pools on the reef platform and a sandy moat around the island's shore. There is a well-developed boulder zone on the weather (western) side and a wide fore-reef slope with well-developed spur and groove systems.

The fore-reef and reef platform have poor to moderately rich coral growth. Coral assemblages appear typical of oceanic reefs exposed to heavy weather conditions. The reef-front ramp and mid-littoral reef flat on the western, windward side is high and relatively barren (Figure 4.16). It has a veneer of crustose algae over the pavement and very little live coral growth except close to the reef-edge where the coral life forms are robust and prostrate. On the leeward eastern side, the reef-platform is lower and there is prolific coral growth of diverse species on the lower-littoral reef-front. The mid-littoral reef flat is characterized by prolific growth but little diversity of corals. In the outer zone of the eastern reef flat, adjacent to the reef crest, *Isopora palifera* and *Porites lobata* are the dominant species, both forming large, coalescing colonies with mainly dead centers and a living fringe. Close to the shore, there are extensive areas of *Porites lutea* micro-atolls, also coalescing. The flat tops of these mid-littoral corals are exposed at low tide and it is clear that coral growth almost has reached its limit and that further sedimentation will soon complete the process of consolidating a new rock platform. Taking account of this and the high barren reef platform on the western side, Browse Reef may be regarded as senile, or almost so. This would be consistent with the interpretation of Hopley[42] that planar reefs represent a final stage of platform reef development.

While detailed faunal inventories are needed, provisional indications are that the reef fauna of Browse is like that of the shelf-edge atolls but perhaps less species rich, in keeping with the more restricted habitats and the late stage of reef development. There is very little intertidal sand

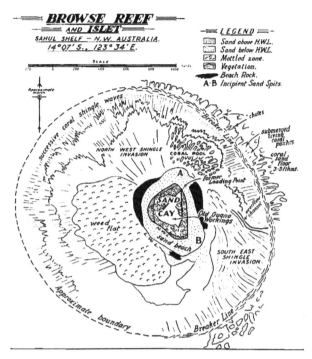

FIGURE 4.13   Diagram of Browse Island and reef—a planar platform reef near the shelf margin with a central sandy island and no lagoon, based on interpretation of a wartime aerial photograph. *Reproduced from Ref. 2.*

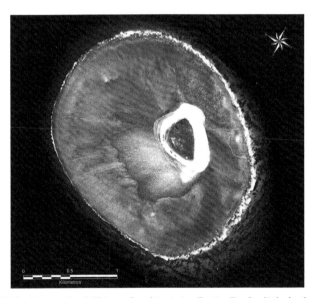

FIGURE 4.14   Browse Reef. This reef and its twin, Cartier Reef, a little further north, are the only planar platform reefs on the North West Shelf. Taken in July with the prevailing SE trade wind blowing so that the heaviest wave action is on the low eastern side of the reef, whereas the western side is relatively calm. Compare with Figure 4.13. This new satellite image shows how accurately Teichert and Fairbridge interpreted the geomorphology of the reef and how little has changed in the 70 years since their military photograph was taken. *Source: Digital Globe Quickbird2 image, July 18, 2006; supplied by Landgate, Western Australian Government.*

FIGURE 4.15   NE side of Browse Reef. A shallow sandy moat surrounds the central island and the inner reef flat is dominated by *Porites* and *Isopora palifera*. *Photo: Barry Wilson.*

FIGURE 4.16   The RPS survey team at work on the ramped reef-front on the western side of Browse Reef. Note the Indonesian praus at anchor alongside the survey team's charter vessel. The Indonesian crew came ashore at low tide to fossick for invertebrates and collect turtle eggs. *Photo: Barry Wilson.*

habitat on the reef platform of Browse and no back-reef or lagoon habitats. Its intertidal reef assemblages are quite different to those of the fringing and platform reefs of the Kimberley coast 150 km to the southeast.

There is little information on the submerged Heywood and Echuca Shoals. Video tows by consulting firm RPS over parts of Echuca Shoal at depths of 13 m and more revealed abundant coral debris and some large but dead coral bombies, suggesting that this shoal may be a failed (drowned) planar platform reef.

4.4.3 Kimberley Coral Reef Province—Kimberley Bioregion

The coral reefs of the Kimberley Bioregion remain the least known in Australia. They include several inner shelf, open sea platform reefs, and a large number of fringing and patch reefs and subtidal banks around mainland shores and coastal islands. They occur in a macrotidal regime with a tidal range as much as 11 m and high but variable turbidity (TSS measures ranging from 0.8 to 13 mg/L—Section 2.2.1 of Chapter 2). Newly emerging evidence indicates that Kimberley reefs are far richer in coral species than was once supposed.

Bassett-Smith[8] and Davis[47] both commented on the "unusual" presence of coral reefs in these turbid conditions. Davis dealt with the Kimberley in a chapter on "reefless coasts," mistakenly referring to the large nearshore platform reefs (Long Reef, Holothuria Bank, Adele Reef, Lacapede Reefs) as "bank atolls" and to the coast and islands as "generally without fringing reefs." This misconception was corrected by Teichert and Fairbridge[2] who observed that there are "ordinary fringing reefs … in great profusion around most of the offshore islands and on many parts of the mainland coast."

The location, extent, and form of coral reefs on the Kimberley coast appear to be controlled by aspect in relation to swell and wind conditions and the preexisting geology. Although wind direction switches from westerly (summer) to easterly (winter), the main impact on the development of fringing reefs in the region is by the persistent prevailing westerly swell.

The preexisting foundations upon which they are built are mainly proterozoic metasedimentary or igneous rocks or Holocene sediment banks but there is little information on their structure. There are some large limestone platforms in the southern part of the Kimberley Bioregion that may have a Pleistocene base.

The northwestern escarpment of the Kimberley Plateau, along which the modern and Quaternary coastline lies, has been strongly dissected, mainly by subaerial erosion during its long terrestrial history and its outer parts now lie submerged beneath the sea. The post-LGM rise in sea level and ongoing subsidence have resulted in its present ria character (Section 2.1.5 of Chapter 2).

Throughout much of the Quaternary, the coastal geomorphology would have been much the same as it is now. In interglacial periods of maximum sea level like the present and during interstadial periods of intermediate sea level that averaged around −50 m, the coastline lay over the western margin of the Proterozoic Kimberley Basin with its complex, dissected topography and large-scale relief including high ridges and plateaux and deep paleo river channels. During those times, high points and rocky substrata would have been widespread along the inner shelf with a range of geomorphic situations suitable for the development of coral reefs. However, during Quaternary glacial periods of lowest sea level, the coastline would have been very different. For example, during the Last Glacial Maxima with sea level at −125 m, the coastline would have lain far beyond the margin of the Proterozoic Kimberley Basin with its complex sea floor topography, over the relatively featureless sedimentary sea floor plains of the Browse Basin.

The modern reefs of the Kimberley coast are Holocene, with growth initiated following inundation by the Holocene transgression on preexisting terrestrial land surfaces but whether they developed directly on Proterozoic rocks remains to be determined. There are no reports of supralittoral exposures of Late Pleistocene eolianite or reefal limestones on the Kimberley coast north of Cape Leveque, like those of the West and Pilbara coasts of Western Australia. That such deposits occurred along Late Pleistocene shores of the Kimberley may be assumed and the absence of them on modern shores is evidence of ongoing subsidence. There are, however, several large intertidal reefs in the southern part of the Kimberley Bioregion, that is, in Collier Bay and the Buccaneer Archipelago, that comprise massive intertidal platforms of hard consolidated limestone, e.g., Montgomery Reef,[68] Turtle Reef in Talbot Bay,[103] Bathurst-Irvine Islands Reef, and reefs of the Sunday Island complex. These reefs may be of entirely Holocene construction on antecedent Proterozoic geology but drilling or seismic study is needed to test the possibility of them being Pleistocene structures with a Holocene veneer and fringe.

4.4.3.1 Fringing Reefs

Fringing reefs are well developed on the shores of almost all the islands of the Kimberley Bioregion from the Bonaparte Archipelago (e.g., West Montalivet Island, Figure 4.17) to the Buccaneer Archipelago. They also occur on mainland shores, for example, on the shores of Cape Londonderry (Figure 4.18) and Cape Voltaire in the far north and at One Arm Point and the Yampi Peninsula in the south. This stretch of coast is more than 600 km long.

Montaggioni[38] has described the varied processes involved in the growth of fringing reefs. Much work is needed to effectively account for the range of geomorphic form of Kimberley fringing reefs. The

FIGURE 4.17   West Montalivet Island, Bonaparte Archipelago, Kimberley Bioregion, looking NW. This is a sandstone island with a laterite cap. The seaward fringing reef on the windward northern and western shores is developed on a rock platform of flat-bedded King Leopold Sandstone. It is subject to persistent swell while the leeward side (foreground) lacks a rock platform. *Photo: Barry Wilson.*

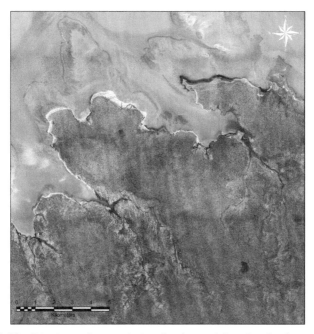

FIGURE 4.18   Fringing reefs around a headland east of Cape Londonderry in the North Kimberley. These are the most northerly of the fringing reefs in the Kimberley Bioregion. *Photo courtesy: Department of Land Information, Western Australia.*

following notes are based on field surveys at the Maret Islands and nearby islands of the Bonaparte Archipelago done by consulting firm RPS Australia[33] and this author's field observations in the southern Kimberley.

(a) *Fringing Reefs on Rock Platforms.* Hopley *et al.*[67] found reefs with intertidal rock platforms are best developed in Queensland on headlands exposed to moderate wave action. In the Kimberley, fringing reefs are also developed in this situation but commonly occur along stretches of rocky shore that are exposed to the prevailing westerly ocean swell. Figure 4.6 is an illustration derived from field observations showing wide limestone fringing reefs developed along northwest- and west-facing rocky shores of the Maret Islands. Such intertidal rock platforms on Kimberley shores appear to be prograding Holocene structures with vigorous lateral growth originating from antecedent coral and algal communities established along rocky shores. The western rock platforms of the Maret Islands are up to 400 m wide but there is no information on their internal structure or the extent to which their width is due to lateral biogenic growth. Brooke[104] inferred the thickness of some Kimberley reefs by measurement of the height of the reef-edge above the sea floor, but no reefs have been drilled and there is no information on the thickness of the Holocene limestone of the reef flats or the nature of the foundations.

Many Kimberley fringing reefs are Proterozoic rock platforms with Holocene biogenic surface veneers and limestone wedges built by coral and crustose algal communities along the reef-fronts. The antecedent rock platforms are often near-horizontal, flat-bedded, metamorphosed sandstones of the Kimberley Group lying within the intertidal zone. The seaward fringing reef of West Montalivet Island is of this kind where emergent sandstone rocks near the reef-front suggest that the wide intertidal platform has been created by infill of biogenic growth and sediment deposition (Figure 4.17). Biogenic reef structures around the nearby East Montalivet Islands are built on a basement of Proterozoic basalt (Figures 4.19 and 4.20). There is no evidence at this time that the kind of basement rock affects the kind of coral assemblages that grow on them or the spatial distribution of community types or species. However, fringing reefs around the shores of these islands with different geologies exhibit similar geomorphology, apparently because the same Holocene reef-building processes are superimposed over the antecedent rock surface whatever its nature.

In the southern part of the Kimberley Bioregion, around the islands and bays of the Buccaneer Archipelago and Yampi Peninsula, the basements are of complex igneous and metamorphic rocks and the shores and reefs are extremely variable in form. There are many reef flats on shores facing the open sea where the fringing reefs are narrow, e.g., the fringing reef at the northern end of Bathurst Island (Figure 4.21). This may be because flat-bedded surfaces are not common on these igneous or intensely folded rocky shores.

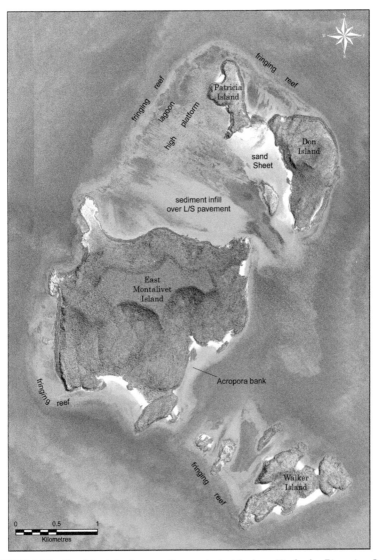

FIGURE 4.19 East Montalivet group, Bonaparte Archipelago, Kimberley Bioregion. Here the base rock is basalt but there is a laterite cap. Again there is a wide rock platform with fringing coral on the windward northern and western shore. The complex of fringing reefs around Walker Island in the south was found to be extremely rich in coral species by a RPS survey team. *Photo courtesy: Department of Land Information, Western Australia.*

On the reef-front, coral species diversity and live coral cover may be very high, e.g., on the seaward reefs of the Maret Islands where domal faviids dominate (Figures 4.22 and 4.23). In such cases, the reefs appear to be vigorously prograding, although there is significant variation in species composition of the coral communities. On seaward shores in the

FIGURE 4.20   Northwestern shore of Patricia Island, East Montalivet group, Bonaparte Archipelago. Along the western shore of Patricia Island, there is a mid-littoral terrace where the force of waves is reestablished during flooding tide and a band of domal faviids occurs along its margin. The shore here comprises a columnar basalt cliff. *Photo: Natalie Rosser.*

FIGURE 4.21   Fringing reef on a narrow rock platform exposed to prevailing swell. Northern peninsula of Bathurst Island, Buccaneer Archipelago. Note the fragmented reef-edge suggesting rapid lateral growth. *Image by Whelans, Mount Hawthorn, Western Australia; courtesy Pluton Resources Ltd, Perth.*

FIGURE 4.22   Species-rich reef-front coral assemblage on a seaward fringing reef exposed to prevailing swells, NW side of North Maret Island, Bonaparte Archipelago. Dense cover of live coral at low spring tide—domal faviids dominant on the reef-front and foliaceous corals on the vertical fore-reef of a wide rock platform. *Photo: Barry Wilson.*

FIGURE 4.23   Seaward fringing reef on a wide rock platform exposed to prevailing swells. West side of South Maret Island, Bonaparte Archipelago. Field of domal faviids and diverse other corals on the reef-front—a dawn photograph at low spring tide. *Photo: Barry Wilson.*

Bonaparte Archipelago, the reef-front is dominated by a diversity of domal colonies, mainly faviids, while acroporids are mainly small, robust, more-or-less prostrate species. In the Buccaneer Archipelago, the narrow fringing reef on the northern headland of Bathurst Island, exposed to moderate ocean swell (Figure 4.21), has similar diverse coral assemblages with domal faviids dominant on the reef-front (D. Blakeway, personal communication).

At the East Montalivet Island group, the seaward rock platform on the western side is wide but with a distinct terrace on the mid-littoral reef flat where there is a second band of domal faviids, presumably there because the terrace is also affected by heavy swell (Figure 4.20).

Mid-littoral reef flats of seaward reefs that do not face the direct action of heavy swells but are nevertheless benefited by vigorous water circulation, may support highly diverse coral assemblages dominated by small, robust, and tabular *Acropora* colonies (Figure 4.24). In contrast, rock platforms on sheltered shores with little wave action, like those of the mainland reef at One Arm Point, support reef-front assemblages that are dominated by ramose and plate *Acropora* and large colonies of the mussid *Lobophyllia hemprichtii* (Figure 4.25). There is not yet enough information on Kimberley reefs to interpret this variability in reef-front assemblages.

Typically, Kimberley fringing reefs of the rock platform category have a steep fore-reef lacking a spur and groove drainage system, a distinct reef-edge, and a gently or steeply sloping lower-littoral reef-front ramp with or without a boulder zone along the reef crest. There is usually a wide, sometimes terraced mid-littoral reef flat and a shallow moat near the shore. While corals may be diverse and reef growth vigorous along the fore-reef and reef-front zones, the mid-littoral zone of the reef flats is generally dominated by macroalgae, mainly *Sargassum*, and corals are sparse except around the fringes of pools, moats, or lagoons impounded by terraces.

FIGURE 4.24   North side of Patricia Island, East Montalivet Group, Bonaparte Archipelago. Diverse *Acropora* assemblage on the low reef-front, protected from the direct force of the prevailing swell. *Photo: Natalie Rosser.*

FIGURE 4.25   Fringing reef on a sheltered shore—*Acropora* bank at the edge of a rock platform at low spring tide. Scattered domal faviids on the rock platform behind the reef-front. One Arm Point, Buccaneer Archipelago. *Photo: Barry Wilson.*

The fore-reef slope of rock platform fringing reefs is usually steep, sometimes vertical in the upper few meters, often with prolific coral growth of massive and foliose species, to around 30 m where it gives way to mud or silty sand. Below a depth of around 20-30 m, scleractinian corals become scarce and, on exposures of hard substrata, are replaced by communities of other filter-feeding epifaunal invertebrates, especially sponges and alcyonarians.

A striking feature of some reefs in the Bonaparte Archipelago (e.g., Albert Reef, East Montalivet reef) is the porous nature of the poorly consolidated material forming the matrix between the domal corals of the reef-front and reef-edge and the many cavities and tunnels therein (personal observations). The material appears to be largely of algal origin. It is friable and easily collapses underfoot. It creates extensive and unusual cryptic habitat for small invertebrates. The tunnels provide the drainage normally achieved by a spur and groove system. The impression given is of rapid carbonate deposition and outward reef growth, coupled with weak or slow consolidation and without a strong coral framework. Domal coral assemblages are usually considered to result in slow reef growth,[38] but the circumstances would be different if there were vigorous growth of the calcareous algal matrix.

There are rock platforms in the Buccaneer Archipelago that do not fit the usual form of fringing reefs. These are intertidal rock platforms that connect high islands, presumably through coalescence of fringing reefs, e.g., the reef between Bathurst and Irvine Islands (Figure 4.26) and twin Woninjaba Islands (Figure 4.27). They comprise hard, consolidated limestone with a high, terraced reef flat impounding pools and lagoons, a steep reef-front ramp, and a species-rich coral fringe along the reef-front. It is common for there to be a terrace, a meter or more high, with a sand fan

FIGURE 4.26   High-rock platform between Bathurst and Irvine Islands, Buccaneer Archipelago. Note the remnant pools captured by reef growth and the sand fan below a terrace on the western (left) side. *Image by Whelans, Mount Hawthorn, Western Australia; courtesy Pluton Resources Ltd, Perth.*

FIGURE 4.27   Woninjaba Islands, northern side of the Yampi Peninsula, Collier Bay, southern Kimberley. This high rock platform connects the two islands and is fringed with vigorous coral growth around the lower-littoral margins. *Photo: Tim Willing, Pearl Sea Cruises.*

on the downside and impounding a high shallow lagoon on the upside (Figure 3.26 of Chapter 3 and Figures 4.26 and 4.27), perhaps indicating a late Holocene sea level change event. It is difficult to categorize these reefs and further study is needed to reveal their origins.

On limestone platforms in the Buccaneer Archipelago, e.g., Montgomery Reef, Turtle Reef, and Bathurst Reef, there are deep holes that are spaces left by coalescing reef growth and do not appear to be karst features (Figure 4.26). They have vertical or undercut walls with profuse growth of foliose corals even though the reef-flat surface around them may be barren. (A friend of the author's tells a story about jumping into one of these deep holes at low tide to see what was there—and found that it was "full of reef sharks"!)

There are similar high-reef platforms between islands within the almost land-locked Talbot Bay on the north side of the Yampi Peninsula where moderately species-rich coral reef growth occurs in spite of the extreme muddy conditions.[103] Turtle Reef at the center of the bay (Figure 4.28) is a vast structure with a high rock platform and a shallow lagoon high in the intertidal zone, contained by banks of rhodoliths on the reef crest (Figures 3.26 and 3.27 of Chapter 3) and rich coral growth on narrow lower littoral flats around its margins (Figure 4.29). The Turtle Bay reefs are in close proximity to (sometimes overlapping with) extensive mud banks and the corals are frequently covered by slimy mud. There is virtually no wave action in this bay and it appears that the corals are able to deal with the extreme level of intermittent sedimentation because of the cleansing effects of the very strong tidal currents.

Some rock platforms in the Kimberley Bioregion lack significant modern coral growth, even on the reef-front. These reefs may be, and may always have been, algal-dominated rock platforms, but it is possible that they are coral fringing reefs whose growth has stalled, macroalgal growth replacing prior coral communities.

(b) *Fringing Reefs Without Rock Platforms.* On Kimberley shores that are not exposed to the prevailing westerly swell, that is, on the eastern and southern sides of islands, behind protective headlands and in sheltered bays, there are often dense intertidal coral banks composed primarily of *Acropora* with occasional clusters of massive faviids. They may occur in front of either rocky shores or beaches. These reefs appear to be roughly equivalent to the category of Queensland fringing reef as "narrow beach-base fringing reefs that develop along stretches of sandy coast."[67] On these fringing reefs, there is no rock platform and it is presumed that the reefs are built on unconsolidated sediments.

There is no defined reef-edge on these fringing reefs; the fore-reef simply slopes into the subtidal zone where there is usually extensive growth of *Porites* bommies and other massive and foliose corals. The coral banks may be separated from the beaches or boulder shores by shallow sandy moats that are usually just a few meters wide. These reefs often have high live coral cover but relatively low species diversity and the associated biota of other invertebrates and algae is very restricted. Fringing reefs of this kind may be seen in the Maret Islands, along the leeward eastern and southern shores and on the sheltered sides of western peninsulas (Figures 4.6, 4.30, and 4.31).

FIGURE 4.28   South Turtle Reef in Talbot Bay, Yampi Peninsula; mud and coral banks coexist in an enclosed, macrotidal embayment. This high rock platform has a steep reef-front with banks of rhodoliths along the reef crest that impound a wide, shallow, coral-rich lagoon on the top of the platform. The mud bank in the foreground is soft and dangerous to walk on. *Photo: Steve Blake, WAMSI.*

FIGURE 4.29   Fringing reef in the mud-dominated, almost enclosed gulf of Talbot Bay on the Yampi Peninsula, exposed at extreme low spring tide. Communities of this kind grow on narrow lower littoral reef-front ledges at the base of the steep reef-front slope. *Photo: Steve Blake, WAMSI.*

4.4.3.2 Platform Reefs and Banks in Open Sea

On the inner shelf off the Kimberley coast, there are many large "unattached" reefs and banks in the open sea with flat-topped platforms that are emergent at low tide. Most of the nearshore examples appear to be Proterozoic rock platforms veneered to varied extent with Holocene biogenic limestone, sometimes with emergent outcrops of the basement rocks. There is also a cluster of offshore reefs in the vicinity of Adele Island near the outer margin of the Kimberley Bioregion that are thought to be constructed on submerged Late Pleistocene coral limestone platforms, with no surface evidence of the Proterozoic basement (Figure 4.4).

Very little is known of the biota of the inshore platform reefs. They are usually dominated by brown algae (primarily *Sargassum*) and the extent of coral growth is variable. Coral growth is usually sparse but there may be more or less vigorous growth along the reef-front margins building limestone wedges at the periphery. Examples in the central Kimberley Bioregion are Colbert, Rob Roy, and Wildcat Reefs. Wildcat Reef has built around a sandstone platform core that outcrops intertidally up to 1.5 m above the mean sea level.[104] It has a steep coralline algal reef-front ramp (with few corals) and a subtidal fore-reef "cliff" to −18 m on the west side, while on the eastern side, the reef-flat slopes gently into the subtidal zone. Rob Roy Reef has a similar structure, but the steep side faces east and the west side slopes into the intertidal zone. These are not coral platform reefs in the classical sense and may be best described as rock platforms with varying degrees of Holocene biogenic veneer and fringing coral reef

FIGURE 4.30   Fringing reef adjacent to a sandy beach on a sheltered shore without a rock platform—*Acropora* bank with patches of massive faviids. Southern shore of South Maret Island, Bonaparte Archipelago. *Photo: Barry Wilson.*

FIGURE 4.31   Fringing *Acropora* bank in front of a beach in a sheltered situation behind a peninsula within Brunei Bay, North Maret Island, Bonaparte Archipelago. *Photo: Barry Wilson.*

development. Their geomorphic form, for example, the steep cliff-like fronts, is the most likely construct of their preinundation terrestrial erosional history.

The reefs of the Adele complex are more akin to true coral platform reefs. (Figure 4.4) They comprise Adele Reef and its neighbors Churchill, Mavis, Albert, Beagle, and Brue Reefs, all arising from depths of less than 100 m on elongate, flat-topped banks. Teichert and Fairbridge[2] provided photo interpretation of the geomorphology of Adele Reef and suggested that it and the other reefs of the group are Pleistocene limestone structures built on a Proterozoic base, with rejuvenated Holocene reef growth following from regional subsidence. Jones[14] figured the complex of platform reefs associated with Adele Island as limestone bioherms. During the LGM period, these reefs would have been high, flat-topped hills close to the coast, separated from each other by deep river canyons. The depth and age of biogenic limestone on their tops and the extent of coral reef development need to be tested by drilling or seismic study.

The long axes of these reefs are aligned with prevailing geological trends of the region, suggesting that the Proterozoic basement may be at shallow depth. Otherwise, they have geomorphic surface characteristics of shelf platform reefs with wide limestone reef flats. The platforms have ramped reef-fronts, coral boulder zones at the reef crests. Adele has a central vegetated sand island and there are emergent sand cays on Mavis and Beagle Reefs. The sand cay on the Beagle Reef (Figure 4.32) is the one noted by Stokes.[7] Adele has a shallow lagoon with a wide entrance on the northeastern side and its platform is separated from the close by bank on which Churchill and Albert Reefs are situated by a channel up to 100 m deep.

A survey of the biota of Adele Reef by the Western Australian Museum found a zone of live domal coral along the reef crest on the western side (C. Bryce, personal communication), suggesting the presence of coral communities similar to those of seaward fringing reefs exposed to the prevailing swell elsewhere in the Kimberley Bioregion.

FIGURE 4.32   The sand cay on Beagle Reef, described by John Stokes in 1846.

Montgomery Reef is a unique case.[68] It is a very large, nearshore platform reef (area ca. 400 km^2) in Collier Bay (Figure 4.5). It is not a coral platform reef in the strict sense. Rather, it is a flat-topped, inundated terrestrial topographic feature (a mesa), and its major geomorphic characteristics are inherited from its ancient terrestrial history. Conversion of the terrestrial Montgomery mesa to a marine platform would have occurred in the Holocene during the last phases of the post-LGM transgression. Geomorphic features of the reef surface today, including the central mud islands, sand sheets, lagoon sediments, and marine coral and algal communities, are products of contemporary marine sedimentary and reef-building processes superimposed over preexisting terrestrial geomorphology. This is not a high energy shore and the prevailing extreme tidal range, not wave action, appears to be the dominating force dominating the sediment production and dispersal processes.

The reef platform of Montgomery is very high. At low spring tide, the reef-edge stands 4 or 5 m above the water level and water cascades over it in dramatic waterfalls. There is little coral growth along the steeply ramped, lower-littoral reef-front, the rough reef surface being mainly a coralline algal crust with small algal terraces; nor is there much coral growth in the fore-reef zone of the outer perimeter or, apparently, any significant lateral reef growth.

At the western end of Montgomery Reef, there are series of small lower-littoral terraces of fixed algal construction on the reef-front ramp but elsewhere major terraces comprise wide, mobile banks of loose, golf-ball-sized rhodoliths (Figures 3.26 and 3.27 of Chapter 3). These rhodolith banks impound shallow mid- and upper-littoral pools and lagoons that occupy most of the surface area of the reef. Live corals are common in the lagoons along with algal rhodoliths, carbonate sand, and coral rubble. The extensive development of rhodolith impoundment banks and the obvious key role they play in reef growth are a feature of this reef and may be a characteristic of reefs in the Buccaneer Archipelago with its extreme macrotidal conditions.

The vast, shallow lagoons on the platform of Montgomery Reef never dry and primary production on the reef top is likely to be very high. This may explain the enormous numbers of herbivorous green turtles and dugong that may be observed feeding there. Perhaps, the Montgomery platform could be thought of as a "marine Serengeti"—a huge shallow expanse of marine herbivore habitat.

This remarkable structure has exceptionally high importance for its cultural heritage and geoheritage as well as its ecological values. Excavated material from sites on the small High Cliffy Islands at the eastern end of Montgomery Reef, dated at 6700 years B.P., shows that these islands were inhabited then by people with a specialized maritime economy, relying principally on resources of the reef.[105] During the Last Glacial, the

Montgomery mesa and its surrounding plains and river valleys would have been many kilometers from the coast and the switch from a terrestrial economy to a marine one must have been a dramatic period of cultural change for the inhabitants of the region, in addition to the loss of territory.

On its eastern side, the rock platform of Montgomery Reef comprises a bed of stromatolitic dolomite (apparently an unrecognized formation of the early Proterozoic Kimberley Group—K. Grey, personal communication, 2009). If this is found to be the case for the whole reef platform, Montgomery Reef may be thought of as a bioherm, but one that is 1.8 billion years old with a thin Holocene veneer of modern reef builders. It is perhaps a unique situation where, on the shore of the High Cliffy Islands, it is possible to stand on a supralittoral bench built of well-preserved stromatolites representing the reef builders of the time when life on Earth began, and look down on a modern coral reef a few meters below, built by an entirely different life form (scleractinian corals and crustose algae). Between the periods when these two reef builders did their work, other kinds have come and gone.

4.4.4 Fringing and Patch Reefs of the Pilbara Bioregions

There are extensive species-rich coral reefs of various kinds in both the offshore and nearshore Pilbara Bioregions. To date, the most studied coral reefs of the Pilbara have been those of the Dampier Archipelago and the Montebello-Lowendal-Barrow Island complex. Most of them have been associated with environmental studies for the petroleum-gas industries that operate in that area.[106–108] Accounts of the coral reefs and fauna of the Montebello Islands were produced as background for the Montebello-Barrow marine reserve management plan by the Department of Conservation and Land Management.[17109110] A map of the complex marine habitats prepared by the Department as a management tool is reproduced here (Figure 4.33).

The marine fauna and aspects of the ecology of reefs in the Dampier Archipelago have been extensively studied and there is now a moderately comprehensive list of corals from the bioregion (see Section 4.1). The prevalence of the coral-eating asteroid *Acanthaster* in the archipelago has been examined.[16,18,25,111,112] A comprehensive account of the marine fauna and flora of the Dampier Archipelago, including its corals, has been published in 2004 by the W.A. Museum.[113]

4.4.4.1 Fringing Reefs

Fringing reefs occur along the seaward shores of offshore islands throughout the Pilbara (offshore) Bioregion. There is no evidence of subsidence in this region. Most contemporary biogenic reef growth takes place on preexisting Late Pleistocene limestone rock platforms that are remnants of coastal limestone barriers or coral reefs. There is no

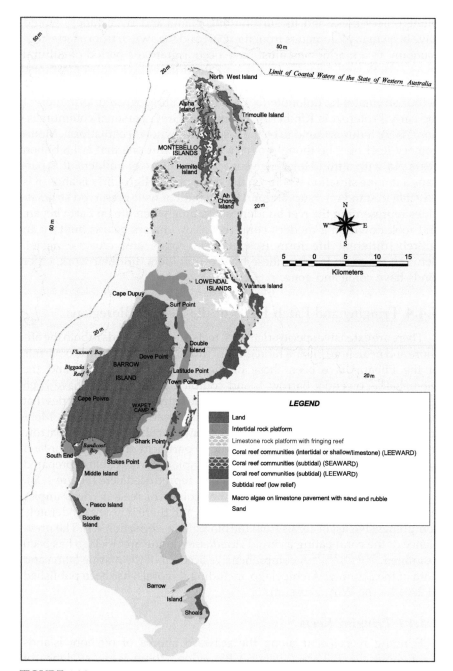

FIGURE 4.33   Major marine habitats of the Montebello/Barrow Islands Marine Conservation Reserves. *Courtesy Western Australian Department of Environment and Conservation.*

information on the depth of Holocene limestone deposits on these fringing reefs but they are probably veneers over the rock platforms and thickest and outward growing along the reef-edges where most of the modern reef-building takes place.

The best developed fringing reefs of the region lie off the western side of the Montebello-Lowendal-Barrow complex of islands that are built on a long shallow bank over an anticline of the Miocene Trealla Limestone. These are high-energy reefs facing the open sea with heavy surf breaking on the reef-front and they probably lie within the mixing zone of the Holloway Current and coastal waters.

Barrow is a large island comprising Miocene rocks (Trealla Limestone) with some Quaternary coastal sediments and limestone rock platforms in the intertidal zone. There is a small but species-rich fringing reef, known as Biggada Reef, on its central west coast. The nearby Montebello and Lowendal groups include a large number of low islands of Quaternary limestone with many sheltered bays, lagoons, and channels. The prominent fringing reef off the western shore of the Montebello group has a total length of about 18 km, including an isolated section at the southern end (Figure 4.34).

The Montebello fringing reef consists of a limestone rock platform separated from the western chain of islands by a shallow lagoon that is narrow at the northern end and wide and open to the sea at the southern end.[110] The fore-reef is not steep and lacks a spur and groove system. The gently sloping sea floor in front of the reef is rocky with only patches of coral. There is said to be sparse coral growth along the reef-front although, for accessibility reasons, this habitat remains poorly studied. Much of the wide reef flat is pavement limestone with an algal turf and few corals. There are many coral slabs and, in places, a moderately developed boulder zone along the crest. The wide back-reef has rich and diverse coral assemblages. Breaks in the reef, forming channels into the lagoon, are also fringed with dense living coral—tabular *Acropora* usually dominant. There are patch reefs in the lagoon, including colonies of the coral *Porites cylindrica*, but for the most part, its bed is sandy and dominated by algae. In deeper parts, these are mainly *Sargassum* but species of the brown alga *Turbinaria* dominate in the shallower northern part of the lagoon.

Many of the limestone islands of the West Pilbara also have fringing reefs on their seaward sides. Kendrew Island is a small outer limestone island in the Dampier Archipelago that is a remnant of the Late Pleistocene shoreline (Figure 4.35). The reef is developed on the northern side of the island along the margin of a 200-m wide limestone rock platform. There is no lagoon or moat and only limited coral growth in shallow pools on the mid-littoral reef platform but on the reef-front there is extensive growth with more than 50% live coral cover in places. At the western end of the reef, tabular *Acropora* cover a broad area. There is a prominent

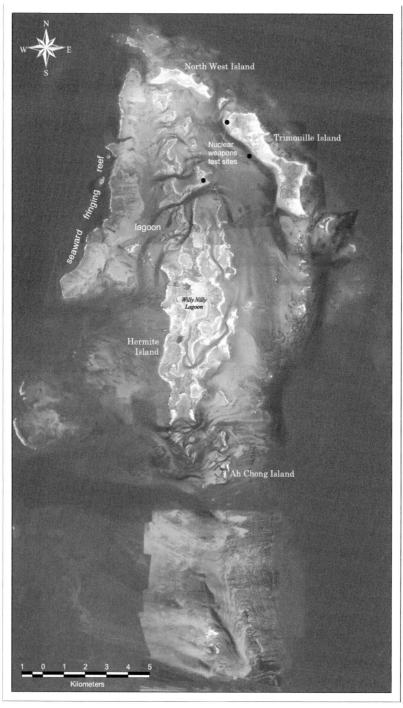

FIGURE 4.34 The Montebello Islands—a cluster of low Quaternary Islands with a complex of lagoons and channels and a major fringing reef on the western side. *Photo courtesy: Apache Energy Ltd, Perth.*

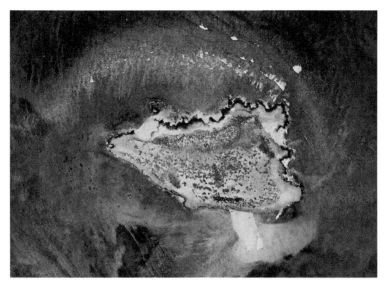

FIGURE 4.35   Fringing reef on the seaward shore of Kendrew Island, outer Dampier Archipelago, Pilbara. This island is a limestone remnant of the Late Pleistocene shore line. Coral growth is active along the reef-front zone of a wide limestone rock platform, with a strong spur and groove system in the fore-reef. *Photo courtesy: Department of Land Information, Western Australia.*

reef crest with a pronounced boulder zone and strong development of a spur and groove system along the fore-reef.

There are similar fringing reefs on rock platforms along the seaward sides of other outer islands in the Dampier Archipelago, also growing on limestone remnants of the Late Pleistocene shoreline. In the case of Rosemary Island, there is a shallow sandy lagoon rather than a rock platform between the reef and the shore and the island itself is built of Proterozoic basalt.

The easternmost coral fringing reef in the Dampier Archipelago surrounds Delambre Island (Figure 4.36). Like Kendrew, the core of this island is an east-west remnant of the Late Pleistocene limestone shore but it has a long, north-south, vegetated sand cay behind it so that the island has a T-shape. There is a narrow rock platform with a fringing reef along its northern, seaward shore, and wide tidal flats on both sides fringed by subtidal coral communities. The fauna of this reef has not been surveyed but personal observations indicate that it has a rich and diverse coral and associated fauna.

Fringing intertidal flats with extensive coral communities also occur around many of the small nearshore limestone islands of the Pilbara (nearshore) Bioregion, for example, Weld Island off Onslow. The water conditions in this nearshore zone are generally turbid and the sea floor sediment is subject to resuspension by wave action, especially during cyclonic storms. Whether the corals are reef-building to the extent that

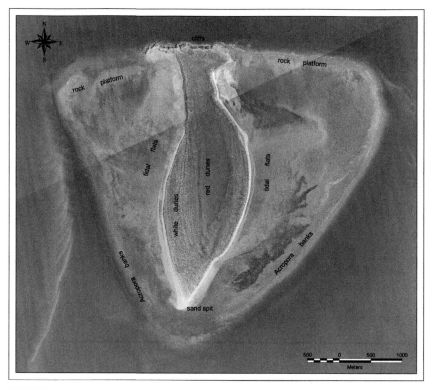

FIGURE 4.36 Delambre Island, outer Dampier Archipelago, Pilbara. There is a rocky seaward shore, remnant of the Late Pleistocene shore-line, and a large, vegetated sand spit on the leeward side, surrounded by a wide tidal flat with rich coral growth around its margin. *Courtesy Department of Land Information, Western Australia.*

the flats might be called coral reefs is a moot point. None of them have been studied in any detail.

4.4.4.2 Patch Reefs

Development of subtidal patch reef occurs along the leeward sides of many islands on the inner shelf of the West Pilbara (between the Dampier Archipelago and the Exmouth Gulf) and on topographic high areas of the rock pavement of the sea floor. Regionally, they are poorly studied to date but available information indicates that they are species rich and very variable in their coral assemblages. They occur in water conditions ranging from moderately clear (eastern Montebello reefs) to turbid (nearshore Onslow reefs).

Patch reefs are extensive in moderately clear water conditions at the Hamersley Shoals in the Dampier Archipelago. The W.A. Museum 1998 survey recorded 50 species at sites in this area with *Acropora* (11 species) and *Montipora* (7 species) the dominant genera.[35]

There is a chain of subtidal and intertidal patch reefs along the eastern margin of the Montebello-Lowendal-Barrow bank, from the Montebellos to the Barrow Island Shoals (Figure 4.37) and large areas of bommies scattered across the shallow subtidal platform between the Lowendal Islands and the Montebellos. The Gorgon coral monitoring study has shown that

FIGURE 4.37   Coral assemblages on a subtidal patch reef off the east coast of Barrow Island, West Pilbara: (A) *Porities* bombie with associated corals; (B) a branching *Acropora* assemblage;

Continued

FIGURE 4.37—cont'd   (C) *Porities* bombie with associated corals; and (D) a plate *Acropora* assemblage. *Photos: Natalie Rosser.*

these patch reefs are species-diverse (229 species recorded) with communities dominated by *Acropora* or *Porites* assemblages.[36,37]

Throughout the Pilbara (nearshore) Bioregion, there are extensive areas of subtidal patch reefs in turbid, shallow, subtidal habitats. They are well developed on rocky bottom in the inner bays of the Dampier Archipelago[24,34] and even on sandy and gravely seabed along the mainland coast west of Cape Preston and in Exmouth Gulf. Along the 10-m

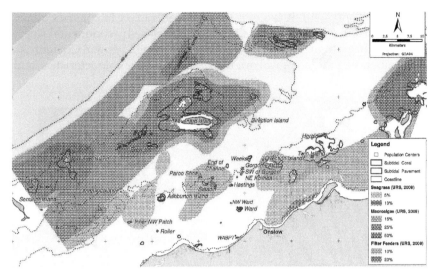

FIGURE 4.38   Nearshore patch reefs on the inner shelf off Onslow on the West Pilbara coast. *Drawn by URS (Australia).*

contour of the inner shelf off Onslow, many such subtidal patch reefs occur on high parts of the limestone pavement on the seabed (Figure 4.38). These reefs have not been studied in detail but a survey report[106] provides an account of the coral assemblages. There is no information to date on their species composition except that those closest to the coast are mostly dominated by *Montipora*. Patch reefs further offshore tend to be dominated by *Acropora* and *Porites* assemblages.

A feature of coral reef invertebrate faunas in the West Pilbara is that the intertidal communities usually comprise a mixture of both coral reef and rocky shore species. The latter group includes a high proportion of species that are not restricted to coral reef habitats but are widely distributed on rocky shores, with or without corals present. In this regard, the Pilbara coral reef faunas are like the fringing and platform reefs of the Kimberley.

4.4.5 Ningaloo Bioregion

Strictly speaking, Ningaloo Reef is not part of the North West Shelf. It marks the western boundary of that shelf, separating it from the Dirk Hartog Shelf to the south. However, by its nature and its position, it plays a crucial role in defining the historical biogeography of the North West Shelf and it is included here for that reason.

Ningaloo Reef is an extensive (ca. 260 km long) oceanic barrier/fringing reef on the mainland coast facing a very narrow continental shelf

(Figures 4.39 and 4.40). It is a high-energy reef, subject to constant heavy wave action and occasional upwelling from the close-by continental slope, and lies in the path of the warm, southward-flowing Leeuwin Current. As a consequence of its geological origins, biogeomorphic characteristics, and

FIGURE 4.39   The Cape Range Peninsula; Ningaloo Reef on the western (open ocean) side, Exmouth Gulf on the eastern side. This peninsula forms a partial biogeographical separating the North West Shelf from the Dirk Hartog Shelf. *Landsat Image, courtesy Department of Land Information, Western Australia.*

FIGURE 4.40   Turquoise Bay; at this locality, Ningaloo Reef is a fringing reef with a shallow lagoon separating it from the arid shore. The reef is close to the shelf-edge and exposed to constant heavy wave action. Note the Late Pleistocene rocky shore—this is a Late Pleistocene fossil reef, antecedent to the modern reef. *Photo: Barry Wilson.*

exposure to oceanic swell and currents, Ningaloo possesses a variety of habitats that are more typical of an oceanic reef than a coastal fringing reef and this is reflected in its biota.

The reef has a Holocene coralgal framework built on a preexisting Late Pleistocene reef along the western side of the anticline that forms the Cape Range Peninsula. It consists of a series of lineate reef platforms, with lengths varying from tens to thousands of meters, lying more or less parallel to the shore, from which it is separated by a shallow lagoon that varies in width from a few meters to several kilometers. The lineate reefs are separated by channels, or passes, that drain the lagoon (Figure 4.40). There is a well-developed spur and groove system along the fore-reef slope. The reef crest is low and rarely has a boulder zone but there is often a band of living domal faviids along the lower-littoral reef-front. Typically, the rear of the reef platform is demarcated by a distinct, undercut edge and a complex back-reef mosaic of sandy pools, large *Porites* head corals and patch reefs with a variety of coral species.

The shallow lagoon, usually merely a few meters deep, has sand and rubble substrata. Extensive banks of ramose *Acropora* are common in deeper parts of the lagoon. In several locations, there are wide subtidal sand sills behind the reef platform. In the outer parts of the lagoon that are most affected by warm water passing over the reef platform, coral colonies dominate the lagoon fauna, but in nearshore parts, the abundance and variety of corals is much less, the area characterized instead by rock pavement with turfs of leafy algae and an invertebrate fauna that includes many temperate elements of the West Coast.

In one significant respect, Ningaloo Reef differs from many other oceanic Indo-Pacific reefs that are similarly exposed to constant heavy surf. The highly speciose genus *Acropora* is less well represented than is usual.[114] *Acropora* species are here confined largely to back-reef, channel and lagoon habitats and small prostrate colonies on the reef-front. Ningaloo Reef's reef-front is dominated by robust corals with massive and encrusting growth forms, mainly massive, rounded colonies of *Platygyra sinensis*, and prostrate *Acopora* that can survive the force of waves.

Ningaloo Reef lies at the boundary between the Indo-West Pacific and the West Coast biogeographic regions. At the generic level, its coral fauna is like that of most reefs in the Indo-Pacific.[27,114]

4.5 INTERTIDAL ASSEMBLAGES CORAL REEF PLANTS AND INVERTEBRATES

This section discusses intertidal distribution patterns of plants, corals, and selected other invertebrates that are associated with coral reefs of the North West Shelf.

Assemblages of species that characterize geomorphic zones of coral reefs are described, and differences observed between the reef communities of the Oceanic Shoals Bioregion and the coastal bioregions are noted.

Like rocky shores, coral reefs provide habitat for a vast array of invertebrates. Many of these are common to both habitats but there are very many species that live associated specifically with one or the other. In the case of fringing reefs of the coastal Kimberley and Pilbara Bioregions, emphasis here is on plant and invertebrate assemblages of reef-front and reef-crest habitats. The plants and invertebrates that live in reef-flat habitats of the mid-littoral zone in these bioregions are mainly rocky shore species and are discussed in Section 3.1.4 of Chapter 3, although corals, when they occur, are also discussed here.

Coral reef invertebrates form functional assemblages adapted to the microhabitats that typify coral reefs, relating to the peculiarities of corals themselves and the habitat structures that they build. In some cases, the determining factors are physical conditions that both the corals and their coinhabitants require.

Protection and restoration of functional groups is the basis for building resilience in coral reefs in the face of change.[115–117] An element essential to this aim is documentation, not only of faunistic lists but assemblages of species in functional groups. Historical biogeography seeks to explain the distribution of flora and fauna in terms of both ecological and historical factors and there is much to be done before North West Shelf species distribution patterns can be understood in this wider context. Developing this phase of coral reef biogeography will be core business in the future.

Regional differences in biogeomorphic, oceanographic, and climatic settings in which coral reefs develop may be expected to result in different coral species assemblages and, thereby, different assemblages of associated fish and invertebrate fauna as well. Such ecological factors are fundamental when we compare, for example, the coral reef faunas of the shelf-edge atolls of the Oceanic Shoals Bioregion with those of the fringing reefs of the Kimberley and Pilbara Bioregions.

Geomorphic structure of coral reef habitats, itself determined by interactions of geological history, coastal and oceanic physical processes, and biological activity, is a key factor in determining spatial arrangement of reef plants and animals. Species richness on coral reefs is strongly correlated with reef topography due to the three-dimensional complexity and diversity of microhabitats it provides.[118] Corals themselves create standing structures that comprise much of reef habitat complexity and this is a major determinant of the faunistic composition of associated fish and invertebrate assemblages.[119–121] Within reef systems, clearly discernable geomorphic zones are distinguishable[67] each with characteristic coral, sessile benthos, and fish communities.[122–125] Yet, observed patterns of ecological distribution are not always easily interpreted. There are often mosaic patterns of plant and coral colonies of varied ages, sizes, and species composition, reflecting the responses of communities to earlier disturbance.[126–128]

Another major determinant of the structure and faunistic composition of coral and associated fish and invertebrate assemblages is the source of available recruits, their abundance, and the timing of their arrival.[129] Since coral reef species are predominantly planktotrophic, this aspect resolves largely into questions of currents, connectivity, and the seasonality of spawning (Section 9.3 of Chapter 9).

In Western Australia, the study of coral reefs and their biota is so new that even a basic inventory of the species present is incomplete and there is very little information about assemblages of species into functional groups. The key species that provide dominant trophic and reef-building roles are not generally known, or at least, are not yet documented. Once the faunistic inventory is more or less settled, the next phase of investigation will be to distinguish assemblages of coral and other reef species and determine to what extent these vary with reef type, tidal and oceanographic conditions, and geographic location.

The Western Australian Museum and the Museums and Art Galleries of the Northern Territory are currently engaged in a program aimed at documenting the biota of coral reefs on the North West Shelf.[102,113,130] The Australian Institute of Marine Science has begun documentation of coral and fish assemblages on coral reefs of the region. From these ongoing studies, it is already evident that the shelf-edge and outer shelf reefs, and Ningaloo Reef, support typical oceanic reef communities that are significantly different to the inner shelf and coastal reefs in both their species composition and the structure of their biotic assemblages.

4.5.1 Coral Assemblages

In this section, coral assemblages of three community types are considered:

(i) non-reef-building benthic communities of the shelf and rocky shores.
(ii) atoll and platform reef communities of the shelf-edge and outer shelf that are essentially oceanic in their ecological characteristics.
(iii) fringing and patch reefs of the mainland and coastal islands that support biotic assemblages that are distinctly continental but with or without oceanic elements.

4.5.1.1 Non-Reef-Building Coral Assemblages

There are many azooxanthellate scleractinian corals that live in soft substrate habitats of the continental shelf and continental slope. Cairns[131] published an account of these Western Australian corals, listing 105 species of which 87 were recorded from the North West Shelf. The majority of them live on the slope where they constitute a significant element in the deep sea suspension-feeding communities. These corals tend to be widespread in the Indo-West Pacific realm although Cairns lists nine species that are endemic to the region. Benthic azooxanthellate corals are not usually included in regional coral species diversity counts (see Section 9.2.2 of Chapter 9).

Some zoothanthellate corals, like solitary species of the fungiid genera *Cycloseris*, *Diaseris*, and *Heliofungia* and the dendrophylliid *Heteropsammia*, also live in sandy benthic habitats including coral reef lagoons. These corals probably do not contribute directly to reef framework construction although they may be locally abundant and sediment derived from their skeletons may contribute to reef-building processes.

Many corals that play important roles in reef-building communities are also common in nonreef benthic habitats as scattered colonies on rocky shores and on rocky pavement in subtidal benthic filter-feeding communities of the inner shelf. Colonial species like the caryophylliid *Euphyllia glabrescens* and the dendrophylliids *Turbinaria frondens* and *Turbinaria bifrons* (Figure 4.41) may be found living attached to shells or stones in benthic soft substrate habitats as well as in reef communities. *T. frondens* is widespread in shallow sandy pavement habitats of the inner shelf of the West Pilbara and is sometimes cast ashore in large numbers during cyclonic storms.

FIGURE 4.41   Coral colonies of *Turbinaria frondens* and *Turbinaria bifrons* and several faviids in the lower-littoral zone of a sandy rock platform in Bouguer Passage, Nickol Bay (Pilbara (nearshore) Bioregion). *Photo: Barry Wilson.*

To date, there have been no studies of the species composition of non-reef coral assemblages of rocky shore and benthic shelf habitats of the North West Shelf. There is no evidence of any species that live solely in these communities but not also in reef communities. Rather, it seems that some reef species have a greater capacity than others to establish themselves and grow in any place with suitable hard surfaces, even on shells or stones on soft substrata, and appear more often in nonreef rocky habitat communities.

Stony corals are a minor feature of the biota of algal-dominated rock platforms. For example, only small and scattered colonies of the scleractinians *Pocillopora damicornis*, *Goniastrea* sp., and encrusting *Montipora* and *Porites* are common among the fronds of the plants on seaward rock platforms of Barrow Island's west coast.[132] On the muddier, leeward, rock platforms on the eastern shores of that island, there are no fringing coral reefs but moderately species-rich coral assemblages occur in the lower-littoral and sublittoral zones of rock platforms with 74 scleractinian species (28 genera) recorded there.[133]

The nearshore, coastal waters of the North West Shelf are generally turbid and subject to episodic heavy sedimentation events and the nonreef coral communities comprise species that are necessarily adapted to these conditions. In muddy conditions of the coastal bays in the Kimberley Bioregion, a variety of coral species may be exposed on rocky shores at extreme low tide, growing directly on Proterozoic basement rocks. For example, 41 species of 24 genera are known from the muddy, rocky shores of Admiralty Gulf.[27] Marsh[30] noted, "A suite of species

FIGURE 4.42   A multispecies assemblage of *Acropora*, mainly *Acropora aspera*, on a muddy rock flat near Anketell Point, a mainland site in the Pilbara.

FIGURE 4.43   In the turbid waters of mainland and island shores in the Dampier Archipelago, Pilbara, there are sometimes extensive colonies of the mussid *Lobophyllia hemprichii* growing on basalt boulders in the intertidal zone.

(*Catalaphyllia jardinei*, *Euphyllia glabrescens*, *Duncanopsammia axifuga*, *Oulastrea crispata*, *Moseleya latistellata* and *Trachyphyllia geffroyi*) known only from turbid lagoons or inshore waters is well represented on Kimberley reefs." These corals are also found in turbid waters of the Pilbara[133] but are not found on oceanic reefs of the shelf margin and may be regarded as a continental element in the North West Shelf coral fauna.

Corals are also common on shores of basalt boulders in the Dampier Archipelago. For example, at Wilcox Island in the Flying Foam Passage[134] and Anketell Point and Poverty Point west of Cape Lambert[135] diverse living corals cover much of the surface of boulders in the lower littoral zone including areas dominated by thickets of *Acropora aspera* (Figure 4.42) or large colonies of the mussid *Lobophyllia hemprichii* (Figure 4.43). These coral communities are moderately species rich, usually with around 40-50 species but it is doubtful whether they qualify as coral reefs.

Griffith[35] listed 112 scleractinians at intertidal stations in the Dampier Archipelago, of which 30 were found only intertidally, many of them in rocky shore communities rather than on coral reefs in the strict sense.

4.5.1.2 Coral Assemblages in the Oceanic Shoals Bioregion

There is now a good taxonomic account of the major plant and animal groups of coral reefs of the Oceanic Shoals Bioregion but limited information on functional assemblages, although significant progress has been made. Atolls and platform reefs in the bioregion share similar fore-reef, reef-front, reef-crest, and reef-flat habitats although there are some local differences among them. However, the small platform reefs (Hibernia, Cartier, Browse) lack large lagoons and back-reefs and thereby lack assemblages characteristic of those habitats.

Coral and fish assemblages of the Rowley Shelf atolls have been described[136] and life-form criteria have been used to characterize the biota of several of the reefs of the bioregion, focusing on the distribution and abundance of key species of particular management concern.[94,95] A brief survey of Browse Reef by consulting firm RPS Australia produced information on the biota of that oceanic reef.[32] There is an ongoing monitoring program on the oceanic reef communities by the Australian Institute of Marine Science.

A quantitative assessment of coral abundance and biodiversity at the Rowley Shelf atolls examined the taxonomic and biogeographical relationships of the complex mosaic of coral communities that exist in the Oceanic Shoals Bioregion.[29] This study found distinct coral communities associated with reef-front, lagoon, and reef-flat habitats and identified the major species that discriminate between these habitat types. Species assemblages of reef-front and lagoon communities showed strong similarities between reefs although there was weaker correlation between their intertidal reef-flat communities.

Reef-Front Coral Communities. On windward shores of the shelf-edge atolls and outer shelf platform reefs, the corals of the exposed reef-front are generally small, robust, or prostrate colonies capable of living in this turbulent, medium to high energy environment, and diversity is generally low. McKinney[29] wrote of reef-front stations at the Rowley shelf-edge atolls "... key groups of taxa were the non-branching *Acroporas* and encrusting or massive non-*Acroporas.*" At Browse, an outer shelf platform reef, the reef-front coral assemblages may be described in similar terms.[32] There, a survey team found the most common reef-front corals were *Pocillopora verrucosa*, a small bushy *Acropora*, encrusting forms of *Porites* and *Montipora*, and small colonies of *Favites*, *Goniastrea*, *Tubipora*, *Heliopora*, and *Millepora*.

Reef-Crest Coral Communities. A reef-crest boulder zone is prominent around the periphery of the atolls and platform reefs, most conspicuous on the windward side. This band, around 20-50 m wide, generally has sandy gutters and pools between its dead coral boulders and slabs and often supports various coral assemblages, again mostly small colonies. On Browse Reef, there is 5-15% live coral cover in this habitat with *Isopora palifera*, *Seriatopora hystrix*, *Goniastrea retiformis*, *G. flava*, *P. lutea*, *P. cylindrica*, *P. damicornis*, *P. verrucosa*, *Goniopora edwardsi*, and species of *Cyphastrea*, *Leptastrea*, *Favites*, *Tubipora*, *Helipora*, and *Millepora* being most common.[32] Also common are soft corals of the genera *Sinularia*, *Sarcophyton*, and *Lobophyton*. On the windward southwestern side of Browse Reef, massive-living *Goniastrea* colonies may be up to 1.5 m high, often with live polyps mainly facing the incoming swells and the back side of the colonies dead.

Reef-Flat Coral Communities. The pavement of the mid-littoral flat on oceanic reefs is often virtually devoid of live corals but there may be small colonies present in shallow pools and gutters. The key species of the reef-flat habitat on the shelf-edge atolls are described as "massive non-*Acroporas*, particularly *Porites* species."[29] The atolls of the shelf-edge frequently have areas of large pools toward the back of their wide, rather barren reef flats. These have extensive and diverse coral communities and there are species-rich coral assemblages along the irregular back-reef bordering the inner margin of the reef platform and the deep lagoon. On the Browse Reef platform, there is no true lagoon although there is a mid-littoral moat around the central island, knee-deep at low tide, almost filled with flat-topped *Porites* micro-atolls and mostly dead colonies of *I. palifera* (Figure 4.15).

Lagoon Coral Communities. The extent of lagoon development varies greatly among the coral reefs of this bioregion. The deep lagoon of South Scott Reef has an area of around 300 km^2 and depths to 60 m. Ashmore Reef and the other shelf-edge atolls of the Rowley Shelf have smaller and shallower lagoons. The Hibernia, Cartier, and Browse platform reefs do not have lagoons but do have areas of shallow sandy pools and moats on their reef platforms.

Shallow lagoon habitats (depths to 12 m) along transects at Mermaid, South and North Scott, and Seringapatam Reefs were surveyed during the Western Australian Museum survey of 2006.[137] The dominant lagoon substrate at all these reefs is sand, rubble, and rock with very low live coral cover but there are scattered coral bommies with abundant reef epifauna.

Coral species richness associated with bommies at lagoon stations of the same survey were moderately high with 72 species at Mermaid (station 12), 71 species at South Scott (station 23), 68 species at North Scott (station 39), and 54 species at Seringapatam (station 43), and "key groups of taxa were the branching *Acroporas*, and massive non-*Acroporas*."[29]

On the floor over most of the deep South Scott lagoon (depths of 40–60 m), there is a patchy coral community, established over coralline limestone pavement and rubble. In the southern part, furthest away from the deep channel, the coral community is categorized as *Deep-Water Foliaceous Coral*.[138] Forty-two species of scleractinian corals are recorded so far from this habitat. The community is dominated by the foliose corals *Montipora aequituberculata*, *Echinopora lamellosa*, and *Pachyseris speciosa* (Figure 4.12), large colonies of *Pavona cactus*, and delicate branching and tabular acroporids including six species that are not also found in the shallows of Scott Reef. No other coral assemblage like this is yet known from the Oceanic Shoals Bioregion.

The northern side of the South Scott lagoon slopes very steeply into the 250-m-deep channel that separates the north and south reefs. The sandy substrate of the slope shoulder (depth 50-100 m) is dominated by a filter-feeding community of alcyonarians (mainly large neptheids) and sponges.

4.5.1.3 *Coral Assemblages of the Kimberley and Pilbara Bioregions*

The Kimberley and Pilbara Bioregions are separated by more than 1000 km of reefless coastline and are characterized by very different geological and metocean conditions. The geomorphology of their fringing reefs and that of their patch reefs vary in detail but there are many similarities between these two bioregions. Most importantly, in both regions fringing reefs built on antecedent rock platforms are best developed on west- and north-facing shores that are exposed to moderate ocean swells and their diverse reef-front coral assemblages are dominated by a band of domal faviids. Coral assemblages of fringing reefs of leeward shores in both bioregions are rarely built on rock platforms and are less diverse and often dominated by *Acropora* species. In these respects, the coral assemblages of reefs of the coastal bioregions are very different to those of the oceanic atolls and platform reefs.

Although the taxonomy of corals in the Pilbara has progressed well, only localized information is available on coral assemblages in the bioregion.[34–37,139–141] The taxonomic account of corals of the Dampier Archipelago[35] includes station data from which information on species assemblages may be gleaned. There is less information on coral assemblages on fringing and platform reefs of the Kimberley. The results of a study of intertidal habitats in the Bonaparte Archipelago include coral communities presently being prepared for publication[32] and ongoing surveys by the WA Museum and AIMS will address this knowledge gap in the Kimberley.

(a) *Fringing and Subtidal Patch Reefs of Leeward Shores.* In both the Kimberley and Pilbara Bioregions, leeward and bay fringing reefs generally slope into the sublittoral zone and there is no rock platform and no distinct reef-edge. Intertidally, such reefs are usually dominated by branching and tabular acroporids, often growing on dense banks of coral rubble. There is usually a variety of massive, foliose, and branching corals and *Fungia* in sandy pools and gutters. Subtidally, the fore-reef slope is moderate or steep with species-rich and abundant coral communities dominated by massive *Porites* and faviids and foliose coral colonies. It seems likely that reefs of this kind, though protected from prevailing swell, may be vulnerable to occasional severe disturbance by cyclonic storms and their community structure may be unstable.

Examples of fringing reefs of this kind have been described at the Maret Islands in the northern Kimberley Bioregion (Figures 4.22, 4.23, and 4.26) where live coral cover exceeds 60%.[32] The RPS survey of the fringing reef on the southern side of the peninsula of North Maret Island (Figure 4.31) found that coral assemblages in the shallow lagoons were diverse, the most common species being *Acropora pulchra*, *A. aspera*, *A. intermedia*, *A. muricata*, *A. yongei*, and *A. brueggemanni*, with patches where some of these species formed virtually monospecific stands. Similar reefs are prolific in sheltered bays in the southern Kimberley Bioregion but they have not yet been studied in detail.

The Gorgon coral monitoring study has shown that patch reefs along the eastern side of the Montebello-Lowendal-Barrow Island chain in the Pilbara (offshore) Bioregion (Figure 4.33A–D) are species rich (around 230 species recorded) with communities dominated by *Acropora* or *Porites* assemblages.[36,37] In the nearby Pilbara (nearshore) Bioregion, there are extensive areas of subtidal patch reefs in shallow, turbid, subtidal habitats. They are well developed on rocky pavement in the inner bays of the Dampier Archipelago.[24,34] There is a chain of patch reefs along the 10-m contour of the inner shelf off Onslow (Figure 4.38) those closest to the coast mostly dominated by *Montipora* while those further offshore tend to be dominated by *Acropora* and *Porites* assemblages.[106]

(b) *Fringing Reefs on Windward Rock Platforms.* On fringing reefs of the rock platform type in the Kimberley and Pilbara, exposed to the prevailing swell, the lower-littoral reef-front ramp is commonly dominated by massive corals, mainly domal faviids, with a variety of small corals, soft corals, and turf algae between them. A reef-front band of large domal faviids is often a spectacular feature of these high-energy reefs, especially in the Kimberley where there is high diversity of scleractinian corals and live coral cover along the reef-front is up to 90% (Figures 4.22 and 4.23). The faviids occupy a band of 10-50 m wide shoreward from the reef-edge with individual colonies up to 40 cm in diameter.[32] The common species are *Goniastrea favulus, Goniastrea aspera, G. retiformis, Favia pallida, Favites halicora, Leptastrea purpurea, Platygyra ryukyuensi, Platygyra pinni,* and *P. sinensis,* and species of *Coeloseris.* Large colonies of the mussids *Symphyllia radians* and *L. hemprichii,* the poritids *Goniopora pendulans* and *G. lobata,* and *Heliopora coerulea* and *Millepora* sp. are also common. Alcyonarians of the genera *Tubipora, Xenia, Nepthea, Sarcophyton, Lobophyton,* and *Sinularia* are all represented.

Domal faviids are also dominant along the reef-front of Ningaloo Reef that is similarly exposed to heavy swell. *Acropora* species are not prominent in this reef-front habitat and those present are usually small robust and encrusting colonies but they may be prolific and diverse where the reef flat is low and semiprotected from the main force of the swell (Figure 4.24).

Variations of this lower-littoral reef-front coral assemblage are common on fringing reefs of high energy shores throughout the Kimberley. Less species-rich and vigorous variations occur in positions that are less exposed to swell. Restricted versions of it often occur along the margins of terraces in the mid-littoral zone, apparently in situations where there is renewed wave energy. There is an example on the western side of Patricia Island in the East Montalivet group where a band of domal *Goniastrea* spp. lies along the edge of a terrace that borders a mid-littoral lagoon, a hundred meters or more from the reef-edge (Figure 4.21).

An exception to the faviid-dominated windward, fringing reefs occurs on the northeastern side of Patricia Island. There the reef-front is covered by profuse growth of small tabulate, corymbose, and digitate *Acropora* colonies (Figure 4.20). This community extends for a distance of about 400 m along the reef-front in a position that is protected from the direct impact of the prevailing northwesterly swell. More than 20 species of *Acropora* were recorded at this site.[32] Common species were *Acropora hyacinthus, Acropora clathrata, Acropora anthrocercis, Acropora spicifera, Acropora cerealis, Acropora millepora, Acropora nasuta, A. aspera,* and *A. muricata.* Small faviids, *Montipora* and *Merulina,* were also common among the acroporids.

Another example of an *Acropora*-dominated reef-front, coral community in semiprotected conditions occurs on the mid-littoral and lower-littoral rock platform reef on the eastern side of Suffron Island in the Albert Island group. In this case, the dominant corals on the reef flat are *A. pulchra* and *A. nasuta* with dense thickets of *A. pulchra* and *A. aspera* in shallow pools.

An entirely different coral reef-front assemblage occurs on a limestone rock platform at One Arm Point near Cape Leveque at the southern end of the Kimberley Bioregion (Figure 4.25). Here, there is a band of branching and tabular *Acropora* along the reef-front, around 10 m wide, with a zone of small domal faviids behind it. The species involved in this assemblage have not been determined. This reef is located in a semiprotected bay and there is very little wave action (except during storms) but there are extremely vigorous tidal currents.

From this very small sample, it seems that on fringing reefs on rock platforms in the Kimberley, reef-front coral communities are very variable with their species composition determined by degrees of wave action.

Species-rich, faviid-dominated reef-fronts like those of Kimberley windward reefs have no counterpart in the Pilbara or Ningaloo Bioregions. The reef-front coral assemblages of the high- and medium-energy, seaward fringing reefs of those bioregions may be dominated by domal faviids but they have relatively low species diversity.

On a windward fringing reef at the Kendrew Island in the Dampier Archipelago, the dominant corals at the reef-edge are massive, domed, colonies of the faviid *P. sinensis*, the agariciid *Pavona minuta*, small robust colonies of digitate *Acropora digitifera*, *Pocillopora eydouxi*, and *P. damicornis*, and encrusting *Montipora* spp.[27,139] At the southwestern end of the Kendrew fringing reef, the rock platform gives way to shoal water with dense cover of *A. hyacinthus* in the shallow sublittoral zone. In the Dampier Archipelago, *P. minuta* and *P. eydouxi* occur only in reef-edge and shallow fore-reef habitats of seaward fringing reefs.[139]

At the edge of the reef-front of a narrow, seaward fringing reef at nearby Legendre Island, just seven coral species were recorded: *A. hyacinthus*, *Coeloseris mayeri*, *G. aspera*, *Leptastrea transversa*, *P. sinensis*, *P. verweyi*, and *P. lutea*.[35] The station notes for this location (DA1/98/05) refer to "coral bommies abundant on the reef edge," but there is no indication which of the massive species was being referred to. *P. minuta* was not recorded there.

P. sinensis is a dominant species on the reef-front at Biggada Reef on the windward side of Barrow Island.[132] *P. minuta* was recorded on the mid-littoral flat at this locality but not at the reef-edge. On the reef-front of the oceanic barrier-fringing reef at Ningaloo (at least in its northern part), the dominant corals are *P. sinensis*, robust prostrate colonies of *Acropora*, and encrusting *Montipora* (Marsh, personal communication, May 2010).

(c) *Coral Assemblages of Mid-Littoral Reef-Flat Habitats.* On the fringing reefs of the Kimberley and Pilbara coasts, mid-littoral flats are most often dominated by prolific growth of macroalgae, especially species of *Sargassum*. Corals tend to be sparsely distributed on rock pavement areas and more common and diverse in pools and gutters. There is little information available on the species composition and structure of these intertidal coral communities. Reporting on coral reefs in the northern Kimberley Bioregion Blakeway[31] noted, ". . . the low diversity of most reef-flat coral communities, and the small size of most colonies. Exceptions occur in the many reef-flat pools, which provide refuge for a variety of corals, most commonly *Montipora*, *Porites*, and *Acropora* species. Large colonies of *Goniastrea* are common on outer reef-flats, but even these tolerant corals seem close to their desiccation limit, as sides facing away from the reef-front are often dead."

An uncommon example where the mid-littoral reef flat is dominated by corals and not macroalgae occurs on the northeastern side of Patricia Island in the East Montalivet Group, north Kimberley. The *Acropora*-dominated reef-front zone of this site was noted earlier. The higher mid-littoral flat between that zone and the shore (beach and columnar basalt) is dominated by the faviids *G. aspera, G. favulus*, and *Favites micropentagona*, and *P. lutea* micro-atolls (Figure 4.24).

A reef-flat survey in the Bonaparte Archipelago[32] reported diverse mid-littoral coral communities on seaward rock platforms with complex mosaics of coral-rich areas and algal-dominated and muddy pavement areas with few corals. Like the reef-front habitats in that region, the dominant coral taxa observed were massive faviids (*Favites, Favia, Goniastrea, Platygyra*) with lesser numbers of acroporids. Amalgamating, flat-topped micro-atolls of *P. lutea* were common on the inner flats. Shallow pools and lagoons encompassed by the mid-littoral flats were often dominated by branching or tabular *Acropora*, sometimes forming dense thickets. The species comprising these reef-flat pool assemblages seem to be highly variable. For example, a shallow pool close to shore on the northern side of Patricia Island in the East Montalivet group has a diverse coral assemblage dominated by monospecific thickets of *A. muricata*.

At Biggada Reef, a small fringing reef on the western side of Barrow Island in the Pilbara, 65 scleractinian coral species and 11 alcyonaceans were listed from the mid-littoral reef flat.[132]

A feature of the Kimberley Bioregion is the occurrence of impounded tide pools high in the upper mid-littoral and upper-littoral zones of rocky shores (Section 3.1.2.4). These frequently contain diverse coral communities but there is no published information on the species involved.

4.5.2 Marine Plant Assemblages

Macroalgae and other plants of mid-littoral rock flats and fringing reefs on shores of the North West Shelf are discussed in Section 3.1.3 of Chapter 3. Those of the atolls and platform reefs of the shelf margin are the subject of this section. There are taxonomic lists and community data for the common marine plants on reef flats of Ashmore, Hibernia, Cartier, Browse, Seringapatam, and Scott Reefs.[94,141]

There are minor differences between the floras of these reefs that may be accounted for by seasonal and local variation. An exception is the mat-forming green alga *Cladophora herspectica* that is common at the Rowley Shoals but apparently absent at Scott and Seringapatam Reefs. Most of the intertidal reef surface bears a crust of coralline algae (primarily *Hydrolithon onkodes*). The most common leafy brown algae on the mid-littoral reef flats of these oceanic reefs are *Turbinaria ornata* and *Lobophora variegata*. In places and at times, the former covers large areas of the reef flats.

Halimeda sp., the green turf *Boodlea vanbosseae* and the red turf *Coelothrix irregularis*, also are usually present. In shallow sandy pools on reef flats, the macroalgae *Halimeda cylindracea*, *Halimeda macroloba*, *Udotea glaucescens*, *Caulerpa cupressoides*, and *C. serrulata* and the seagrasses *Thalassia hemprichi* and *Halophila ovalis* are common. These macroalgal reef-flat assemblages are significantly different to those of the mainland and coastal island fringing reefs. For example, the genera *Sargassum* and *Cystoseira*, which are strongly represented on the coastal fringing reefs and rock platforms, are conspicuously absent on the oceanic reefs.

Extensive beds of the green alga *Halimeda* spp. are reported from the deep banks of the Sahul Shelf where they are clearly key species that provide structural habitat and are major producers of biogenic carbonate.[75] Beds of *Halimeda* also cover large areas of the South Scott lagoon floor along the side of the deep channel above about −50 m where it intermingles with the deep coral community. The presence of *Halimeda* apparently relates to local upwelling of nutrient-rich water and it may be in a competitive state with the corals.

4.5.3 Reef Invertebrates Other Than Corals

The difference between invertebrate assemblages (other than corals) of oceanic reef-front habitats and those of coastal reefs is striking. In particular, many of the invertebrates that live in lower-littoral reef-front and reef-crest habitats of the oceanic reefs do not occur, or are rare, in equivalent habitats of the coastal bioregions where invertebrate communities, other than corals, are severely restricted.

Conversely, invertebrate assemblages of coastal mid-littoral reef flats are species rich. They include species with broad habitat requirements that are common to both coral reefs and rocky shores but also many that are rocky shore species that do not occur in equivalent zones of the off-shore oceanic coral reefs. The latter group includes a significant proportion of regionally endemic species, that is, confined to rocky shores of the North West Shelf and adjacent coastal bioregions, with or without the presence of corals.

Certain functional assemblages of reef molluscs are chosen here to illustrate this situation but the principles seem to apply across the invertebrate taxa.

4.5.3.1 *Reef-Front Assemblages*

Most of the invertebrates of this lower-littoral zone are suspension feeders utilizing the supply of suspended food particles delivered by constant wave action and the trophic system is based mainly on secondary production. While corals, numerically, are usually the principal suspension feeders (although many of them are also autotrophs), other anthozoans, hydrozoans, sponges, ascidians, polychaetes, bryozoans, echinoderms, and bivalve and gastropod molluscs all play important roles. However, the algal turf of the pavement supports a variety of herbivores and there are may be significant numbers of predators that feed on corals and other invertebrates. Molluscs characteristic of these habitats include suspension feeders, grazing herbivores, and carnivores.

Conspicuous suspension-feeding molluscs of this wave-swept zone are sessile, fixed to the substratum in some way. *Tridacna maxima* is very common in this habitat (as well as on mid-littoral reef flats) fixing to the substrate by means of strong byssal threads. Like corals, tridacnid clams are suspension feeders that supplement their diet by means of commensal cyanobacteria in the surface tissues of its mantle. Six species of this family are recorded from the oceanic reefs of the North West Shelf. Of these, *T. maxima* and *T. squamosa* and *Hippopus hippopus* also occur on Kimberley fringing reefs, but only the first two occur in the Pilbara and Ningaloo Bioregions (Figure 4.44).

Another very common suspension-feeding bivalve is *Chama* sp. that cements one valve to the limestone pavement. Several species of the bivalve family Mytilidae also live abundantly in this zone, either byssal-attached like *Tridacna* or living in burrows bored into the pavement. Suspension-feeding, tubiculous gastropods of the family Vermitidae may be common also, as mat-forming colonies cemented to the pavement or boring in corals and coral boulders. Most of these suspension-feeding molluscs are also common in lower-littoral habitats of coastal reefs.

Suites of herbivorous and carnivorous gastropods that characterize the reef-fronts of coral reefs in the region show distribution patterns that are different to those of the reef-front suspension feeders. These molluscan assemblages are species rich on oceanic reefs of the shelf margin but severely restricted on fringing and platform reefs of the coastal bioregions. Many of the key oceanic reef-front species are missing in the Kimberley and Pilbara Bioregions and there is no equivalent functional assemblage of "continental" species to replace them.

(a) *Herbivorous Gastropods of Reef-Front Habitats.* On oceanic reefs of the North West Shelf margin, there is a suite of herbivorous gastropods, including seven large species that are conspicuous in reef-front and reef-crest habitats (Table 4.2; Figure 4.45). They comprise a significant functional assemblage in reef-front habitats on the Oceanic Shoals coral reefs that is reduced at Ningaloo Reef (three species) and further restricted on Kimberley and Pilbara fringing reefs (both with two species).

Five of these gastropods are widely distributed throughout the Indo-West Pacific realm and may be regarded as diagnostic of reef-front habitats of oceanic coral reefs. In Western Australia, *Turbo chrysostomus* and *Cypraea depressa* are known only from reef-fronts in the Oceanic Shoals Bioregion. *Lambis chiragra*, *Trochus maculatus*, and *Turbo argyostomus* are common in those bioregions but are also on Ningaloo Reef.

FIGURE 4.44 Four of the six species of giant clam (Tridacnidae) that are found on the coral reefs of the North West Shelf: (A) *Tridacna maxima*—the most common species that is found on intertidal reef flats as far south as the Abrolhos Islands on the West Coast; (B) *Tridacna squamosa*—common in pools and lagoons of reef flats and in the shallow subtidal zone as far south as Ningaloo Reef; (C) *Hippopus hippopus*—a reef-flat species common on reefs of the shelf margin and the Kimberley coast; (D) *Tridacna gigas*—the largest of the giant clams; on the North West Shelf this species is found only on the oceanic reefs of the shelf margin where its populations have been severely depleted by subsistence fishing. *Photos: (A)–(C) Barry Wilson; (D) Clay Bryce.*

TABLE 4.2 Nine Large Herbivorous Gastropods That Play Key Roles in Reef-Front Ecosystems of Coral Reefs in the Bioregions of the North West Shelf

Species	Ashmore-Cartier	Scott-Sering	Browse	Rowley Shoals	Kimberley	Pilbara	Ningaloo
Trochus niloticus	1	1	1	1	x	x	x
Trochus maculatus	1	1	1	1	x	x	1
Tectus pyramis	1	1	1	1	1	x	1
Angaria delphinus	1	1	1	1	1	1	1
Turbo chysostomus	1	1	1	1	x	x	x
Turbo argyrostomus	1	1	1	1	x	1	1
Cypraea depressa	1	1	1	1	x	x	x
Lambis chiragra	1	1	1	1	x	x	1
Lambis lambis	1	1	1	1	1	x	x
Total	9	9	9	9	3	2	5

The species highlighted in bold are regarded as diagnostic of oceanic reef-front habitats. The stromb *Lambis lambis* is recorded in the Pilbara from a single specimen and is not a significant herbivore on Pilbara coral reefs.

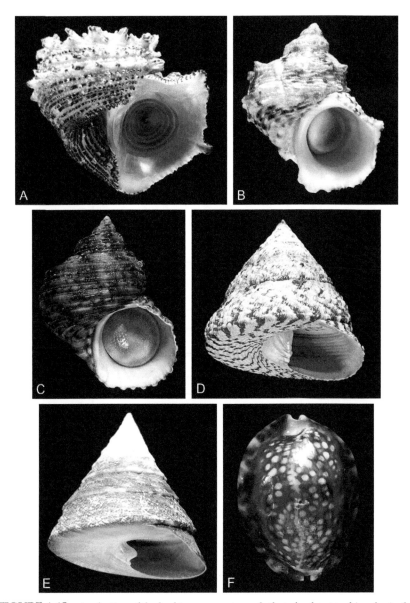

FIGURE 4.45   A selection of the herbivorous gastropods that play key trophic roles in the reef-front zone of coral reefs of the North West Shelf. Of this group, *Angaria delphinus* and *Tectus pyramis* are found on reefs throughout the region but the others are strictly oceanic species of the shelf margin reefs (see also Table 4.2). (A) *Angaria delphinus*; (B) *Turbo chrysostomus*; (C) *Turbo argyrostomus*; (D) *Trochus maculatus*; (E) *T. pyramis*; (F) *Cypraea depressa*.

The two remaining species of the group, *Tectus pyramis* and *Trochus niloticus*, are also common in reef-front (and mid-littoral reef-flat) habitats of the Kimberley Bioregion on rocky shores and fringing coral reefs, with or without the presence of corals, but the latter does not extend further south. *T. pyramis* is geographically and ecologically ubiquitous and is very common on intertidal coral reefs and rocky shores throughout the coastal bioregions (and in the temperate waters further south).

The oceanic reef species *T. argyostomus* is common on the reef-fronts of outer islands of the Dampier Archipelago, the Montebellos, and Barrow Island, and its presence there is anomalous. It may be a result of connectivity with Ningaloo Reef and perhaps with the Rowley Shoals. Its absence in the Kimberley may be because of unsuitable habitat or lack of connectivity with reefs of the Oceanic Shoals Bioregion.

(b) *Carnivorous Gastropods of Reef-Front Habitats*. There are also assemblages of carnivorous gastropods that are characteristic of reef-front habitats on oceanic coral reefs of the Indo-West Pacific realm.[101,145–147] They are mainly species of the families Conidae, Muricidae, and Buccinidae that feed on polychaetes, sipunculids, or tubiculous vermitid gastropods. Cone and buccinid species live settled into the pavement turf or in crevices or under stones and emerge to hunt their prey during the night. Muricids live adhering to the pavement or surfaces of boulders where they feed on tubiculous vermitids or boring barnacles, bivalves, and sipunculids. This suite of predatory gastropods represents a significant functional group in reef-front habitats of the shelf margin oceanic reefs. On the oceanic reefs of the North West Shelf, it includes 34 or more species, the majority of which are lacking or rare in equivalent habitats of Kimberley and Pilbara fringing reefs. A selection of them is illustrated in Figures 4.46 (Conidae), 4.47 (Muricidae), and 4.48 (Buccindae).

The genus *Conus* is particularly informative. There are 17 surface-dwelling species of this genus in reef-front habitats of North West Shelf oceanic reefs (Table 4.3). Nine of them are recorded only from these reefs and may be regarded as strictly oceanic reef cones (*catus, distans, flavidus, imperialis, litteratus, marmoreus, sanguinolentus, virgo, vitulinus*). One species (*rattus*) is common at all of the oceanic reefs but otherwise the only known North West Shelf locality is the Montebellos in the West Pilbara where it is rare. Two species (*miles, vexillum*) are common on the oceanic reefs, but rare or known so far only from beach collected shells from fringing reefs in the West Pilbara and Kimberley. Only five of the reef-front cones (*coronatus, ebraeus, lividus, musicus, mustelinus*) are common in equivalent habitats in the Kimberley. The first four of these species are also common reef-front cones in the West Pilbara but not *C. mustelinus*. There are no coastal species of *Conus* that live in the reef-front habitat but not on the oceanic reefs.

The reef-front Muricidae and Buccinidae on the shelf margin reefs exhibit similar distribution patterns (Table 4.4). Of the 18 common species

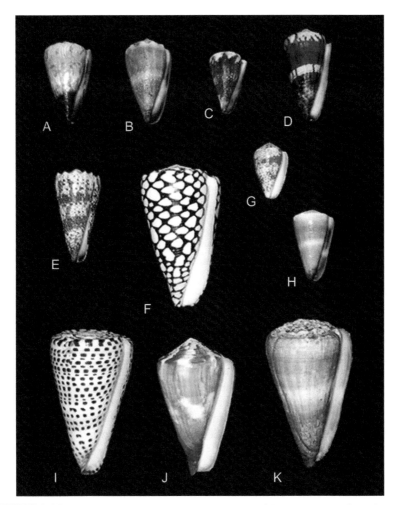

FIGURE 4.46   A selection of the predatory gastropods of the genus *Conus* from the reef-front zone of Browse Reef. These species are all widespread on Indo-West Pacific oceanic reefs, including the shelf margin reefs of the North West Shelf, but only one of them (*lividus*) is common on Kimberley and Pilbara reefs and two (*miles* and *vexillum*) are rarely recorded there (see Table 4.3). (A) *Conus miles*; (B) *Conus sanguinolentus*; (C) *Conus rattus*; (D) *Conus vitulinus*; (E) *Conus imperialis*; (F) *Conus marmoreus*; (G) *Conus catus*; (H) *Conus flavidus*; (I) *Conus litteratus*; (J) *Conus vexillum*; (K) *Conus distans*.

in this habitat, 13 are found only on the oceanic reefs, 3 are also present but not common on the coastal reefs, and only 2 are also common on the coastal reefs.

The conclusion may be drawn that the suite of surface-dwelling carnivorous gastropods in the reef-front habitat of the atoll and platform reefs of the Oceanic Shoals Bioregion is representative of this Indo-West Pacific

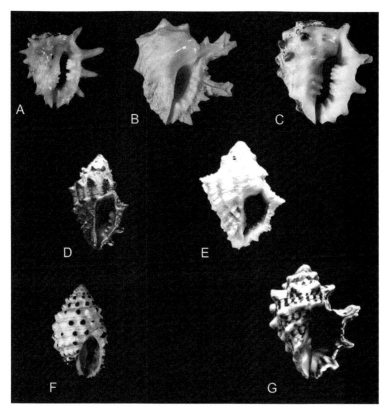

FIGURE 4.47   A selection of the predatory gastropods of the family Muricidae from the reef-front zone of Browse Reef. These species play key trophic roles in the reef-front zone of coral reefs of the North West Shelf. Of this group, *Morula spinosa* is found on reefs throughout the North West Shelf while *Drupina grossularia* is rare on Kimberley fringing reefs. The others are strictly oceanic species of the shelf margin reefs (see also Table 4.4). (A) *Drupa ricinus*; (B) *Drupina grossularia*; (C) *Drupa morum*; (D) *Morula spinosa*; (E) *Muricodrupa* cf. *fiscella*; (F) *Morula uva*; (G) *Thais tuberosa*.

functional assemblage but that it is severely restricted in equivalent habitats of the Kimberley and Pilbara.

Reichelt[145] found that space (hiding places) was the main environmental factor limiting the numbers and variety of carnivorous gastropods on mid-shelf reef-fronts in Queensland. Perhaps, the restricted variety of carnivorous cones on Kimberley reef-fronts may be because of relatively low numbers of prey, that is, burrowing worms and other small invertebrates in the pavement habitat. That, in turn, could relate to the turbid and macrotidal conditions that prevail on the Kimberley coast. If there is no satisfactory ecological explanation of that kind, there may be an historical one relating to the youth of these fringing reefs and a lack of connectivity

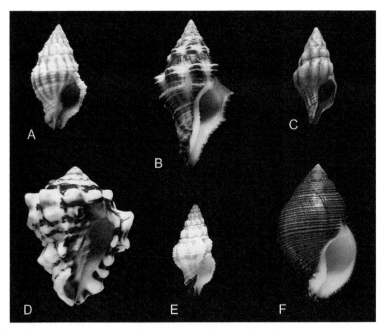

FIGURE 4.48   A selection of reef-front gastropods of the families Muricidae and Buccinidae from Browse Reef. With one exception, all of these species are key predators of reef-front communities found on all the shelf margin oceanic reefs but not on the Kimberley and Pilbara coastal reefs. The exception is *Vasum turbinellum* which is also common on Kimberley fringing reefs (see also Table 4.4). (A) *Peristernia nassatula*; (B) *Latirus polygonus*; (C) *Peristernia fastigium*; (D) *Vasum turbinellum*; (E) *Peristernia ustulata*; and (F) *Latirolagena smaragdula*.

with a source of recruits strong enough to establish and maintain breeding populations of these carnivores. If the latter were the case, it would follow that there is an underutilized resource and unfilled niches in the reef-front habitats of Kimberley fringing reefs.

4.5.3.2 Bioeroders of the Reef Crest

Coral reefs exposed to moderate to heavy wave action generally have a band of coral boulders at the top of the reef-front ramp, cast there by storm waves. Such boulder zones are a conspicuous feature of the oceanic atoll and platform reefs but more weakly developed or absent on the reefs of the coastal bioregions with less wave energy. In time the boulders are destroyed by bioeroders, including burrowing suspension-feeding molluscs, barnacles and sponges, and grazing fishes and molluscs. Many of the invertebrate burrowers also occur on boulders on mid-littoral reef flats and, on limestone shores, at the base of the upper-littoral zone where they undercut the rock face. These bioeroders are often abundant and provide a substantial food resource for specialist predators that are able to extract

TABLE 4.3   Reef-Front Species of *Conus* of North West Shelf Coral Reefs

Species		Ashmore and Cartier	Scott and Sering.	Rowley Shoals	Browse	Kimberley	Pilbara A (Dampier Arch.)	Pilbara B (Montebello and Barrow)
Conus catus	OSB	1	1	1	1	0	0	0
Conus coronatus	C	1	1	1	1	1	0	1
Conus distans	OSB	1	1	1	1	0	0	0
Conus ebraeus	C	1	1	1	1	1	0	1
Conus flavidus	OSB	1	1	1	1	0	0	0
Conus imperialis	OSB	1	1	1	1	0	0	0
Conus litteratus	OSB	1	1	1	1	0	0	0
Conus lividus	C	1	1	1	1	1	1	1
Conus marmoreus	OSB	1	1	1	1	0	0	0
Conus miles	*OSB	1	1	1	1	1*	0	1*
Conus musicus	C	1	1	1	1	1	1	1
Conus mustelinus	C	1	0	0	0	1	0	0
Conus rattus	*OSB	1	1	1	1	0	0	1*
Cobnus sanguino lentus	OSB	1	1	1	1	0	0	0
Conus vexillum	*OSB	1	1	1	1	1*	1*	1*
Conus virgo	OSB	1	0	0	0	0	0	0
Conus vitulinus	OSB	1	1	1	1	0	0	0
Total		17	15	15	15	7	3	7

C—species common on reefs of the coastal and oceanic bioregions; OSB—only found on reefs of the Oceanic Shoals Bioregion; *OSB—oceanic species that occur also but rarely on reefs in the coastal bioregions.

TABLE 4.4   Reef-Front Species of the Carnivorous Gastropod Families Muricidae and Buccinidae on North West Shelf Coral Reefs

Species		Ashmore and Carteir	Scott and Seringapatam	Rowley Shoals	Browse Reef	Bonaparte Arch.	Dampier Arch.	Ningaloo
Thais alouina	C	0	0	1	1	1	1	1
Thais armigera	OSB	1	1	1	1	0	0	1
Thais tuberosa	OSB	1	1	1	1	0	0	0
Drupa morum	OSB	1	1	1	1	0	0	1
Drupa rubusidaeus	OSB	1	1	1	1	0	0	1
Drupa ricinus	OSB	1	1	1	1	0	0	1
Drupina grossularia	OSB*	1	1	1	1	1*	0	0
Morula biconica	OSB*	1	1	1	1	1	0	0
Morula musiva	OSB	1	1	1	1	0	0	0
Morula nodicostata	OSB	1	1	0	0	0	0	0
Morula spinosa	C	1	1	1	1	1	1	0
Morula uva	OSB	1	1	1	1	0	0	1
Engina alveolata	OSB	1	1	1	0	0	0	0
Engina lineata	OSB	1	1	1	0	0	0	0
Peristernia nassatula	OSB	1	1	1	0	0	0	0
Latirolagena smaragdula	OSB	1	1	1	1	0	0	0
Latirus nodatus	OSB	1	1	1	0	0	0	0
Latirus polygonus	OSB*	1	1	1	1	1*	0	0
Vasum ceramicum	OSB*	1	1	1	1	1*	0	0
Vasum turbinellum	C	1	1	1	1	1	0	0

C, species common on reefs of the coastal and oceanic bioregions; OSB, only found on reefs of the Oceanic Shoals Bioregion; *OSB—oceanic species that occur also but rarely on reefs in the coastal bioregions.

them from their burrows. They and their predators represent a distinctive intertidal assemblage.

The bioeroders of limestone shores of the Pilbara, where they are largely responsible for the characteristic double notch, are described in Section 3.3.2.1c of Chapter 3. There are few limestone shores in the Kimberley but the bioeroder community is strongly represented there in the boulder zone of the reef crest of coral fringing reefs.

Limestone boulders of the reef crest on Pilbara and Kimberley fringing and platform reefs are usually heavily bored by bivalves, mainly the lithophagines *Lithophaga malaccana*, *Lithophaga nasuta*, *Lithophaga obesa*, *Lithophaga teres*, and *Botula fusca*, two species of *Gastrochaena*, and a species of *Petricola*. The barnacle *Lithotrya valentiana* and an unidentified sipunculid also play important bioeroding roles. All these invertebrates bore their burrows into the coral rock, opening it up to a variety of other invertebrates, especially sponges that further erode the upper layers exposing it to physical and chemical erosion.

Coral boulders of the reef crest on the oceanic reefs are also subject to rapid bioerosion (Figure 4.10). In this case, it is the barnacle *L. valentiana* that plays the principal role. The clam *Tridacna crocea* is a nestler in crevices of coral boulders but it also excavates depressions in the limestone—this is an oceanic species that is not present on the coastal reefs. The bivalves *Gastrochaena gigantea*, *L. malaccana*, and *B. fusca* are common and another lithophagine, *Lithophaga hanleyana*, may also be present. However, the common coastal lithophagines *L. nasuta*, *L. obesa*, and *L. teres* have not yet been recorded from these offshore reefs. Further study is needed to determine whether these apparent differences in the assemblages of boulder borers of the oceanic and coastal bioregions are consistent.

The surface of boulders in the reef-crest zone is usually covered by a sparse algal turf, encrusting barnacles and sometimes mats of colonial tubiculous vermetids, and sometimes by the rock-oyster *Saccostrea cucculata*. These and the boring molluscs, sipunculids, and barnacles inhabiting the boulders are prey of an assemblage of gastropod predators, mainly of the family Muricidae. There are some differences between the oceanic and coastal reefs in the species composition of this carnivorous guild (Table 4.4).

4.5.3.3 Sand Habitats of Intertidal Sand Cays and Back-Reef Pools

On most of the shelf margin coral reefs of the Oceanic Shoals Bioregion, there are intertidal sand cays on the reef platforms and extensive areas of shallow sandy back-reef pools. These habitats are extensive at Ashmore and South Scott Reefs, less so at Seringapatam Reef and the Rowley Shoals, and sparse at Hibernia, Cartier, and Browse Reefs. Where present, they support diverse infaunal and epifaunal invertebrate assemblages.

The intertidal sand of cays on reef platforms tends to be mobile (moved by wave and tide action) but more stable in back-reef and lagoon habitats. Active, suspension-feeding burrowing bivalves, and herbivorous and carnivorous gastropods are common macroinvertebrates in those situations. The majority of these species are widely distributed in coral reef habitats throughout the Indo-West Pacific realm but do not occur in comparable habitats of the North West Shelf coastal bioregions where they are replaced by other species of the same genera. To illustrate this, Tables 4.5–4.7 list sand-dwelling gastropods of the families Strombidae, Olividae, and Conidae that are common in sand habitats of the North West Shelf with some species found on intertidal reefs and benthic shelf habitats but the majority confined to one or the other.

STROMBIDAE (TABLE 4.5)

There are 25 species of the herbivorous family Strombidae in sandy habitats of the North West Shelf. Fifteen of them live in either subtidal lagoon habitats of the oceanic reefs or benthic habitats of the continental shelf. Ten species are recorded from sandy intertidal and shallow subtidal habitats. Of the latter group, six occur only on coral reefs of the Oceanic Shoals Bioregion and not on the coastal fringing reefs and may be regarded as characteristic of oceanic coral reef habitats (*aurisdianae, erythrinus, gibberulus, lentiginosus, luhuanus, microurceus*). Three (*orrae, iredalei, campbelli*) live on intertidal flats of the coastal bioregions and are not found on the oceanic reefs; all of the latter group are endemic to the continental shores of northern Australia. The tenth species of this intertidal group, *S. mutabilis*, is a very widespread Indo-West Pacific stromb that lives in sandy habitats on coral reefs and rocky shores. In Western Australia it occurs on the shelf-margin reefs, on Ningaloo Reef, on the outer islands of the Pilbara (offshore) Bioregion, and as far south as Cape Naturaliste in the temperate waters of the West Coast.

OLIVIDAE (TABLE 4.6)

The family Olividae are sand-dwelling, carnivorous gastropods. Twenty species are recorded from sandy shore, benthic shelf, and coral reef habitats on the North West Shelf. Nine of them are widely distributed Indo-West Pacific species that, in Western Australia, occur only on the oceanic coral reefs of the shelf margin (*annulata, carneola, guttata, paxillus, sidelia, tesselata, textilina, tremulina, vidua*) and these species may be regarded as oceanic species. Eight are found only on sand flats and shallow subtidal habitats of the coastal bioregions and all of these are endemic to northwestern Australia or northern Australia more generally (*Alocopspira rosea, Ancillista cingulata, A. muscae, Belloliva sp, Cupidoliva cf. nympha, Oliva cf. australis, O. caldania, O. lignea*). Three species are recorded from both the shelf margin reefs and the coastal regions. These are typically oceanic

TABLE 4.5   Species of the Herbivorous Gastropod Family Strombidae Recorded from Intertidal and Benthic Sand Habitats of the North West Shelf

Species	Depth Zone	Ashmore	Scott and Seringapatam	Kimberley	Pilbara	Ningaloo
Lambis truncata	S	1	1	0	0	0
Rimella cancellata	S	0	1	1	1	0
Strombus aurisdianae	I	1	0	0	0	0
Strombus bulla	S	1	0	0	0	0
*Strombus campbelli**	I	0	0	1	1	0
Strombus dentatus	S	0	1	0	0	0
Strombus dilatatus	S	0	0	0	1	0
Strombus epidromus	S	0	1	0	0	0
Strombus erythrinus	I	1	0	0	0	0
Strombus fragilis	S	0	1	0	0	0
Strombus gibberulus	I	1	1	0	0	1
Strombus haemastoma	S	1	0	0	0	0
Strombus labiatus	S	0	0	0	1	0
Strombus latissimus	S	0	1	0	0	0
Strombus lentiginosus	I	1	1	0	0	0
Strombus luhuanus	I	1	1	0	0	0
Strombus microurceus	I	1	1	0	0	0
Strombus mutabilis	I	1	1	0	1	1

Continued

TABLE 4.5   Species of the Herbivorous Gastropod Family Strombidae Recorded from Intertidal and Benthic Sand Habitats of the North West Shelf—cont'd

Species	Depth Zone	Ashmore	Scott and Seringapatam	Kimberley	Pilbara	Ningaloo
Strombus pipus	S	1	1	0	0	0
Strombus plicatus	S	1	1	0	1	0
Strombus sinuatus	S	1	1	0	0	0
Strombus urceus*	I	0	0	1	1	0
Strombus vittatus	S	0	0	0	1	0
Strombus vomer*	I	0	0	1	1	0
Terebellum terebellum	S	0	1	1	1	0
	Total	14	15	5	10	2

I, intertidal, sandy reef habitats; S, subtidal, sandy back-reef, lagoon, and benthic shelf habitats. Species marked with an asterisk are endemic to northern Western Australia and the Northern Territory.

species but *O. caerulea* and *O. panniculata* occur on the sandflats of the off-shore West Pilbara islands and *O. miniacea* is sometimes found on Kimberley fringing reefs. The Pilbara (10 species) has more species of these intertidal sand flat gastropods than the Kimberley (6 species).

CONIDAE (TABLE 4.7)

The very large carnivorous family Conidae includes mostly reef and rocky shore species but there are many sand-dwelling cones in intertidal and shallow benthic shelf habitats. Table 4.7 lists 11 sand-dwelling cones recorded from the North West Shelf. Two of them are found on sand cays and on back-reef pools of the shelf-margin coral reefs but do not occur on coastal reefs (*pulicarius, quercinus*). Conversely, there are three species that are found in sandy intertidal and shallow subtidal habitats of the coastal bioregions but not on the oceanic reefs and these are all endemic to the North West Shelf.

4.5.3.4 Soft Habitats of Lagoons

The back-reef slopes and shallow lagoons of the Oceanic Shoals coral reefs comprise mainly sand/rubble substrata with a mixed benthic fauna. The term "abiotic" is sometimes used for such habitats in the sense that no

TABLE 4.6 Species of the Carnivorous, Intertidal and Shallow Subtidal Sand-Dwelling Family Olividae Recorded from Sandy Habitats of the North West Shelf

	Oceanic Shoals East	Oceanic Shoals West	Kimberley	Pilbara	Ningaloo
*Alocospira rosea**	0	0	1	1	0
*Ancillista cingulata**	0	0	1	1	1
*Ancillista muscae**	0	0	0	1	1
Belloliva sp.*	0	0	0	1	0
Cupidoliva cf. *nympha**	0	0	0	1	0
Oliva annulata	1	1	0	0	0
Oliva cf. *australis**	0	0	1	1	1
Oliva caerulea	1	1	0	1	1
*Oliva caldania**	0	0	1	1	0
Oliva carneola	1	1	0	0	0
Oliva guttata	1	1	0	0	0
*Oliva lignaria**	0	0	1	1	0
Oliva miniacea	1	0	1	0	0
Oliva panniculata	0	1	0	1	0
Oliva paxillus	1	0	0	0	0
Oliva cf. *sidelia*	1	0	0	0	0
Oliva tesselata	0	1	0	0	0
Oliva textilina	0	1	0	0	0
Oliva tremulina	1	0	0	0	0
Oliva vidua	1	0	0	0	0
Total	9	7	6	10	4

All of these species live in intertidal and shallow subtidal habitats. Species marked with an asterisk are endemic to northwestern Australia or northern Australia more generally.

macroorganisms are visible. However, this is misleading. Lagoonal soft substrata support important microorganisms and often there are species-rich infaunal assemblages, sometimes indicated by extensive bioturbation. Most biological surveys of subtidal reef slope and lagoon

TABLE 4.7   Species of the Carnivorous, Intertidal and Shallow Subtidal Sand-Dwelling Family Conidae Recorded from the North West Shelf

Species	Oceanic Shoals East	Oceanic Shoals West	Kimberley	Pilbara	Ningaloo
Conus arenatus	1	1	1	1	1
*Conus dampieriensis**	0	0	0	1	0
Conus eburneus	1	1	0	1	0
Conus miliaris	1	1	1	1	0
Conus pulicarius	1	1	0	0	1
Conus quercinus	1	1	0	0	0
*Conus reductaspiralis**	0	0	0	1	0
Conus spectrum	1	0	1	1	0
Conus suturatus	0	0	0	1	0
Conus tessulatus	1	1	1	0	0
*Conus trigonis**	0	0	0	1	0
Total	7	6	4	9	2

All of these species live in intertidal and shallow subtidal habitats. Species marked with an asterisk are endemic to northwestern Australia.

habitats on the Oceanic Shoals reefs have been conducted by video transects and diving which do not reveal soft substrate infauna. They have provided information on epifaunal macroinvertebrates of soft substrata and reef species associated with patch reefs in lagoons, but little on the soft substrate infauna of the lagoon floor. This component of coral reef ecosystems of the region is poorly known.

There is significant organic material in the surface layers of lagoon sediments, indicated by an assemblage of detrital-feeding holthurians. Selected benthic invertebrate species and their habitats at Ashmore and Cartier Reefs have been surveyed on behalf of the Commonwealth management authority.[95] There were 16 species of holothurians, found in a range of habitats from the reef flat to the lagoons, with higher densities in the shallower eastern lagoon and in upper lagoon rather than deeper habitats. Species associated mostly with the back-reef slope were *Bohadschia graeffei*, *Bohadschia marmorata*, *H. edulis*, and *H. fuscopunctata*. Species associated with the upper lagoon included *Holothuria leucospilota*, *Holothuria nobilis*, and *Bohadschia argus*. More common in deeper lagoon habitats were *Holothuria fuscogilva*, *Stichopus variegatus*, *Thelenota ananus*, and *Thelenota anax*. Most of these holothurians are also recorded from the Rowley Shelf atolls[146] and are a conspicuous element of lagoon habitats throughout the Oceanic Shoals Bioregion. Several of these common oceanic reef holothurians are also recorded on coastal fringing reefs and rocky shores of the Kimberley and Pilbara[147] but only one, *H. atra*, is common there.

In the South Scott lagoon, a belt of soft substrate around the southern and south-western margin includes sand and rubble derived from the back-reef slope to depths of around 30 m, and biogenic silt-clay beyond that at depths of 30-50 m. ROV images of the silt-clay sites below 30 m in the South Scott deep lagoon revealed a heavily turbated surface with small tufts of *Halimeda* and dense colonies of spatangoid echinoids, indicating a high level of benthic secondary productivity.[138] A grab-sample survey there found a soft substrate benthic community comprising burrowing suspension-feeding, detrital-feeding, and predatory invertebrates.[148] The sand fraction of the sediment largely comprised tests of benthic foraminfera suggesting a dense living population of these microorganisms that may be a foundation of detrital-feeding trophic systems in this benthic community.

Detrital feeders of the deep South Scott infauna include a number of errant polychaete species, a sipunculid (possibly responsible for the bioturbation), the gastropods *Cerithium munditum*, *Strombus epidromus*, *Strombus dentatus*, *Strombus plicatus*, and *Terebellum terebellum*, and four species of spatangoid echinoids, *Brissopsis luzonica*, *Metalia sternalis*, *Schizaster (Ova)* sp., and *Faorina* sp. *aff. chinensis*. None of these echinoids were recorded in shallower water.[146] Suspension feeders include tubiculous polychaetes and the bivalves *Lioberus flavidus*, *Fulvia australe*, *Lyrocardium lyratum*, *Ctenocardia* sp., *Tapes* spp., and *Lioconcha polita*. Also present are the predatory gastropods *Casmaria ponderosa* (that preys on spatangoids), *Nassarius comptus*, *Natica* sp., and *Terebra* sp.

These data indicate a moderately species-rich soft substrate benthic community in the deep South Scott lagoon, perhaps with high biomass, based on secondary production and detrital-feeding. Although most of these invertebrate species of the deep lagoon occur elsewhere in benthic shelf communities of the region, no other soft substrate benthic community of this kind is yet recorded from the North West Shelf.

4.6 SECTION SUMMARY

Invertebrate assemblages of the shelf-margin reefs of the Oceanic Shoals Bioregion are complex and species rich, comparable to equivalent assemblages on oceanic reefs throughout the Indo-West Pacific realm. They comprise species, including the corals themselves, that are predominantly planktotrophic with wide Indo-West Pacific distributions and regional endemism is negligible.

Mid-littoral invertebrate assemblages of fringing and platform reefs in the coastal bioregions are diverse and comprise species that are common on rocky shores, with or without corals. Though rich in coral species and with structurally complex coral communities, reef-front assemblages of associated invertebrates are severely restricted.

Many key species in the reef faunas of the Oceanic Shoals Bioregion do not occur in the faunas of fringing and platform reefs in the coastal bioregions. This is clearly demonstrated by reef-front, lower-littoral herbivorous, and carnivorous gastropods. Key reef-front species that fill these trophic roles on the oceanic reefs are missing in the equivalent habitats of the coastal reefs and they are not replaced by continental species.

Although all the corals of the coastal reefs are widely distributed, restricted distributions, including local endemism, is common among other invertebrates.

References

1. Fox JJ, Sen S. *A study of socio-economic issues facing traditional Indonesian fishers who access the MOU box.* Unpublished report to Environment Australia; 2002. 64 pp.
2. Teichert C, Fairbridge RW. Some coral reefs of the Sahul shelf. *Geogr Rev* 1948;**38**(2):222–49.
3. Berry PF, Marsh LM. History of investigation and description of the physical environment, part I. In: Berry PF, editor. *Faunal surveys of the Rowley Shoals, Scott Reef and Seringapatam Reef. Records of the Western Australian Museum.* Perth: Western Australian Museum, Supplement No. 25; 1986. p. 1–25.
4. Berry PF. Historical background, description of the physical environments of Ashmore Reef and Cartier Island and notes on exploited species. In: Berry PF, editor. *Marine faunal surveys of Ashmore Reef and Cartier Island, North-Western Australia. Records of the Western Australian Museum.* Perth: Western Australian Museum, Supplement No. 44; 1993. p. 1–11.
5. Russell BC. The Ashmore Region: history and development. In: Russell BC, Larson CJ, Glasby RC, Willan RC, Martin J, editors. *Understanding the cultural and natural heritage values and management challenges of the Ashmore Region. Records of the museums and art galleries of the northern territory,* Supplement No. 1; 2005. p. 1–8.
6. King PP. *Narrative of a survey of the intertropical and western coasts of Australia, performed between the years 1818 and 1822,* 2 vols. London: John Murray; 1827.
7. Stokes JL. *Discoveries in Australia, etc.,* 2 vols. London: T. and W. Boone; 1846.
8. Bassett-Smith PW. On the formation of the coral-reefs on the N.W. coast of Australia. *Proc Zool Soc Lond* 1899;**1899**:157–9.
9. Darwin C. *Structure and distribution of coral reefs.* London: Smith Elder; 1842, 355 pp.
10. Darwin C. *The origin of species by means of natural selection.* London: John Murray; 1859, 502 pp.
11. Fairbridge RW. Recent and Pleistocene coral reefs of Australia. *J Geol* 1950;**58**(4): 330–401.
12. Fairbridge RW. The Sahul Shelf, northern Australia: its structure and geological relationships. *J R Soc West Aust* 1953;**37**:1–34.

13. Jones HA. Late Cenozoic sedimentary forms on the northwest Australian continental shelf. *Mar Geol* 1971;**10**(4):M20–6.

14. Jones HA. Marine geology of the Northwest Australian Continental Shelf. *Bur Miner Res Geol Geophys Bull* 1973;**136**:1–102.

15. Wilson BR. Western Australian coral reefs with preliminary notes on a study at Kendrew Island, Dampier Archipelago. *Report of the Crown-of-thorns Seminar*, Brisbane, Queensland; August 1972. p. 47–58.

16. Wilson BR, Marsh LM. Acanthaster studies on a Western Australian coral reef. In: *Proceedings of the second international coral reef symposium* 1; 1974. p. 612–630.

17. Department of Conservation and Land Management . *Indicative management plan for the proposed Montebello/Barrow Islands marine conservation reserves.* Perth: Department of Conservation and Land Management; 1994, 101 pp.

18. Wilson BR, Marsh LM. Seasonal behaviour of a "normal" population of Acanthaster in Western Australia, In: Proceedings of the Crown-of-thorns starfish Seminar, Brisbane, September 6, 1974, Canberra: Australian Government Publishing Service; 1975. p. 167–79.

19. Meagher T, Le Provost I. *Marine environment of Dampier Archipelago.* Unpublished technical report to Woodside Petroleum, North West Shelf Project; 1979. 242 pp.

20. Chittleborough RG. *Marine systems of the Dampier Archipelago: papers presented to a workshop convened by the Department of Conservation and Environment.* Perth, Western Australia: Department of Conservation and Environment; 1982 [Bulletin No. 109].

21. Semeniuk V, Chalmer PN, Le Provost I. The marine environments of the Dampier Archipelago. *J R Soc West Aust* 1982;**65**:97–144.

22. Simpson CJ. *Mass spawning of scleractinian corals in the Dampier Archipelago and the implications for management of coral reefs in Western Australia.* Perth, Western Australia: Department of Conservation and Environment; 1985 [Bulletin 244, p. 1–35].

23. Simpson CJ. *Environmental factors influencing coral growth in the Dampier Archipelago, Western Australia.* Perth, Western Australia: Department of Conservation and Environment; 1985 [Environmental Note 168, p. 1–14].

24. Simpson CJ. *Ecology of scleractinian corals in the Dampier Archipelago, Western Australia.* Technical Series No. 23. Perth, Western Australia: Environmental Protection Authority; 1988. p. 1–238.

25. Simpson CJ, Grey KA. *Survey of Crown-of-thorns starfish and coral communities in the Dampier Archipelago, Western Australia.* Technical Series No. 25. Perth, Western Australia: Environmental Protection Authority; 1988. p. 1–24.

26. Hatcher BG, Allen GR, Berry PF, Hutchins BJ, Marsh LM, Simpson CJ, et al. Australia (Western). In: Wells SM, editor. *IUCN directory of coral reefs of international importance*, vol. 2. Cambridge, UK: Conservation Monitoring Centre; 1987. p. 1–53.

27. Veron JEN, Marsh LM. Hermatypic corals of Western Australia. *Records of the Western Australian Museum*, Supplement No. 29; 136 pp.

28. Veron JEN. Reef-building corals. In: *Faunal surveys of the Rowley Shoals, Scott Reef and Seringapatam Reef, part II. Records of the Western Australian Museum.* Perth: Western Australian Museum, Supplement No. 25; 1986. p. 27–35.

29. McKinney D. A survey of scleractinian corals at Mermaid, Scott and Seringapatam Reefs, Western Australia. In: Bryce C, editor. *Marine biodiversity survey of mermaid (Rowley Shoals), Scott and Seringapatam Reefs. Records of the Western Australian Museum*, Supplement No. 77; 2009. p. 105–43.

30. Marsh LM. Scleractinian and other hard corals. In: Morgan GJ, editor. *Survey of the aquatic fauna of the Kimberley Islands and reefs, Western Australia.* Unpublished report to Western Australian Museum; 1992. p. 15–22.

31. Blakeway D. Scleractinian corals and reef development, part 9. In: Walker D, editor. *Marine biological survey of the central Kimberley coast, Western Australia.* Perth: University

of Western Australia. Unpublished report, W.A. Museum Library UR377; 1997. p. 77–85.

32. INPEX. *Biological and ecological studies of the Bonaparte Archipelago and Browse Basin.* Perth: INPEX; in press.

33. Marsh LM. Scleractinian corals of the Montebello Islands. In: Berry PF, Wells FE, editors. *Survey of the marine fauna of the Montebello Islands and Christmas Island, Indian Ocean. Records of the Western Australian Museum.* Perth: Western Australian Museum, Supplement No. 59; 2000. p. 15–9.

34. Blakeway DR, Radford BTM. Scleractinian corals of the Dampier Port and inner Mermaid Sound: species list, community composition and distributional data. *MScience Report*; 2004. 11 pp.

35. Griffith JK. Scleractinian corals collected during 1998 from the Dampier Archipelago, Western Australia. In: Jones DS, editor. *Marine biodiversity of the Dampier Archipelago, Western Australia, 1998–2002. Records of the Western Australian Museum.* Perth: Western Australian Museum, Supplement No. 66; 2004.

36. Richards Z. Hard and soft corals. Gorgon Gas Development and Janz Feed Gas Pipeline, section 6. *Coastal and marine baseline State and environmental impact report.* Chevron Document G1-NT-REPX0001838; 2010.

37. Richards ZT, Rosser NL. Abundance, distribution and new records of scleractinian corals at Barrow Island and Southern Montebello Islands, Pilbara (Offshore) Bioregion. *J R Soc West Aust* 2012;**95**:155–65.

38. Montaggioni LF. History of Indo-Pacific coral reef systems since the last glaciation: development patterns and controlling factors. *Earth Sci Rev* 2005;**71**:1–75.

39. Fagerstrom JA. *The evolution of reef communities.* New York: John Wiley and Sons; 1987, 600 pp.

40. Tomascik T, Mah AJ, Nontji A, Moosa MK. *The ecology of Indonesian seas, part 1. The ecology of Indonesia series.* **VII**. Hong Kong: Periplus Editions; 1997, 642 pp.

41. Tomascik T, Mah AJ, Nontji A, Moosa MK. *The ecology of Indonesian seas, part 11. The ecology of Indonesia series.* **VIII**. Hong Kong: Periplus Editions; 1997, 1388 pp.

42. Hopley D. Geomorphology of coral reefs with special reference to the Great Barrier Reef. In: Hutchings P, Kingsford M, Hoegh-Guldberg O, editors. *The great barrier reef—biology environment and management.* Collingwood, Victoria: CSIRO Publishing; 2008. p. 5–16.

43. Hopley D, Slocombe AM, Muir F, Grant C. Nearshore fringing reefs in north Queensland. *Coral Reefs* 1983;**1**(3):151–60.

44. Kleypas JA. Coral reef development under naturally turbid conditions: fringing reefs near Broad Sound, Australia. *Coral Reefs* 1996;**15**:153–67.

45. Smithers SG, Larcombe P. Late Holocene initiation and growth of a nearshore turbid-zone coral reef: Paluma Shoals, central Great Barrier Reef, Australia. *Coral Reefs* 2003;**22**:499–505.

46. Smithers SG, Hopley D, Parnell KE. Fringing and nearshore coral reefs of the Great Barrier Reef: episodic Holocene development and future prospects. *J Coast Res* 2006; **22**(1):175–87.

47. Davis WM. *The coral reef problem.* American Geographic Society Special Publication No. 9; 1928. p. 231–232.

48. Muscatine L. The role of symbiotic algae and energy flux in reef corals. In: Dubinsky Z, editor. *Ecosystems of the world—coral reefs.* Amsterdam: Elsevier; 1990. p. 75–87.

49. Lewis JB. Suspension feeding in Atlantic reef corals and the importance of suspended particulate matter as a food source. *Proc Third Int Coral Reef Symp* 1997;**1**:405–8.

50. Stafford-Smith MG, Ormond RFG. Sediment rejection mechanisms of 42 species of Australian scleractinian corals. *Aust J Mar Freshw Res* 1992;**43**:683–705.

51. Anthony KRN. Enhanced particle-feeding capacity in corals on turbid reefs (Great Barrier Reef, Australia). *Coral Reefs* 2000;**19**:59–67.

52. Perry CT, Larcombe P. Marginal and non-reef-building coral environments. *Coral Reefs* 2003;**22**:427–32.
53. Abbott RT. The genus *Strombus* in the Indo-Pacific. *Indo-Pacific Mollusca* 1960;**1** (2):33–146.
54. Taylor JD. Reef associated molluscan assemblages in the western Indian Ocean. *Symp Zool Soc Lond* 1971;**28**:501–34.
55. Reid DG. *The littorinid molluscs of mangrove forests in the Indo-Pacific region. The genus Littoraria.* British Museum (Natural History) Publication No. 978. London; 1986.
56. Wilson BR, Allen GA. Major components and distribution of marine fauna. In: Dyne GR, Walton DW, editors. *The fauna of Australia—general articles.* Canberra: Australian Government Publishing Service; 1987. p. 43–68 [chapter 3].
57. Vermeij GJ. The dispersal barrier in the tropical Pacific: implications for molluscan speciation and extinction. *Evolution* 1987;**41**:1046–58.
58. Williams ST, Reid DG. Speciation and diversity on tropical rocky shores: a global phylogeny of snails of the genus *Echinolittorina. Evolution* 2004;**58**(10):2227–51.
59. Hopley D. *The geomorphology of the Great Barrier Reef: Quaternary evolution of coral reefs.* New York: John Wiley Inter-science; 1982, 455 pp.
60. Scott GAJ, Rotondo GM. A model to explain the differences between Pacific plate island-Atoll types. *Coral Reefs* 1983;**1**:139–50.
61. Molengraaff GAF. The coral reefs of the East Indian Archipelago, their distribution and mode of development. *Proceedings of the fourth Pacific Science Congress, Java,* 2; 1929. p. 55–99.
62. Kuenen PH. Geology of coral reefs, part 2, geological results. In: *Snellius expedition in the eastern part of the Netherlands East Indies 1929–1930,* vol. 5. Utrecht: Kemink en Zoon N.V.; 1933. p. 1–125.
63. Umbgrove JHF. Coral reefs in the East Indies. *Bull Geogr Soc Am* 1947;**58**:729–78.
64. Wang G, Lu B, Quan S. The evolution of continental slope atolls in the South China Sea. In: Proceedings of the 1st international conference on Asian marine geology, Beijing; 1988. p. 291–306.
65. Stagg HMJ, Exon NF. Geology of the Scott Plateau and Rowley Terrace, off northwestern Australia. *BMR Bull* 1981;**213**:1–47.
66. Collins LB, Testa V, Zhao J, Qu D. Holocene growth history and evolution of the Scott Reef carbonate platform and coral reef. *J R Soc West Aust* 2011;**94**(2):239–50.
67. Hopley D, Smithers S, Parnell K. *The geomorphology of the Great Barrier Reef: development, diversity and change.* Cambridge, UK: Cambridge University Press; 2007, 532 pp.
68. Wilson BR, Blake S. Notes on the origins and biogeomorpholgy of Montgomery Reef, Kimberley, Western Australia. *J R Soc West Aust* 2011;**94**:107–19.
69. Jongsma D. Eustatic sea level changes in the Arafura Sea. *Nature* 1970;**228**:150–1.
70. Sandiford M. The tilting continent: a new constraint on the dynamic topographic field from Australia. *Earth Planet Sci Lett* 2007;**261**:152–63.
71. Collins L, Twiggs E, Tecchiato S, Stevens A. Growth history, geomorphology, surficial sediments and habitats of the Ningaloo Reef. *Ningaloo research progress report: discovering Ningaloo—latest findings and their implications for management.* Ningaloo Research Coordinating Committee, Department of Environment and Conservation, WA; 2008. p. 48–51.
72. Collins LB, Zhu ZR, Wyrwoll KH, Eisenhauer A. Late quaternary structure and development of the northern Ningaloo Reef, Australia. *Sediment Geol* 2003;**159**:81–94.
73. O'Brien GW. Some ideas on the rifting history of the Timor Sea from the integration of deep crustal seismic and other data. *PESA J* 1993;**21**:95–113.
74. Glenn K. Water properties of Ashmore Reef. In: Russell BC, Larson CJ, Glasby RC, Willan RC, Martin J, editors. *Understanding the cultural and natural heritage values and management challenges of the Ashmore Region. Records of the museums and art galleries of the northern territory.* Perth: Western Australian Museum, Supplement No. 1; 2005. p. 9–12.

75. Heyward A, Pinceratto E, Smith L, editors. *Big bank shoals of the Timor Sea: an environmental resource atlas.* Townsville: Australian Institute of Marine Science; 1997, 115 pp.

76. Lavering IH. Quaternary and modern environments of the Van Dieman Rise, Timor Sea and potential effects of additional petroleum exploration activity. *BMR J Aust Geol Geophys* 1993,**13**(4):281–92.

77. O'Brien GW, Glenn KC, Lawrence G, Williams A, Webster M, Burns S, et al. Influence of hydrocarbon migration and seepage on benthic communities in the Timor Sea, Australia. *APPEA J* 2002;**42**(1):225–40.

78. O'Brien GW, Glenn KC. Natural hydrocarbon seepage, sub-seafloor geology and eustatic sea-level variations as key determiners of the nature and distribution of carbonate build-ups and other benthic habitats in the Timor Sea. In: Russell BC, Larson CJ, Glasby RC, Willan RC, Martin J, editors. *Understanding the cultural and natural heritage values and management challenges of the Ashmore Region. Records of the museums and art galleries of the northern territory,* Supplement No. 1; 2005.

79. Orme GR, Flood PG. The geological history of the Great Barrier Reef: a reappraisal of some aspects in the light of new evidence. *Proc Third Int Coral Reef Symp* 1977;**2**:37–43.

80. Orme GR. The sedimentological importance of *Halimeda* in the development of back reef lithofacies, Northern Great Barrier Reef (Australia). *Proc Fifth Int Coral Reef Symp* 1985;**5**:31–7.

81. Davies PJ, Marshall JF. *Halimeda* bioherms—low energy reefs, northern Great Barrier Reef. *Proc Fifth Int Coral Reef Symp* 1985;**5**:1–7.

82. Drew EA, Abel KM. Biology, sedimentology and geography of the vast inter-reefal *Halimeda* meadows within the Great Barrier Reef Province. *Proc Fifth Int Coral Reef Symp* 1985;**5**:15–20.

83. Drew EA, Abel KM. Studies on Halimeda I. The distribution and species composition of *Halimeda* meadows throughout the Great Barrier Reef Province. *Coral Reefs* 1988;**6**:195–205.

84. Orme GR, Salama MS. Form and seismic stratigraphy of *Halimeda* banks in part of the northern Great Barrier Reef Province. *Coral Reefs* 1988;**6**:131–7.

85. Marshall JF, Davies PJ. *Halimeda* bioherms of the northern Great Barrier Reef. *Coral Reefs* 1988;**6**:139–48.

86. Phipps CVG, Roberts HH. Seismic characteristics and accretion history of *Halimeda* bioherms on Kalukalukuang Bank, eastern Java Sea (Indonesia). *Coral Reefs* 1988;**6**:149–59.

87. Roberts HH, Phipps CV, Effendi L. *Halimeda* bioherms of the eastern Java Sea. *Geology* 1987;**15**:371–4.

88. Roberts HH, Aharon P, Phipps CV. Morphology and sedimentology of *Halimeda* bioherms from the eastern Java Sea (Indonesia). *Coral Reefs* 1988;**6**:161–72.

89. Hine AC, Hallock P, Harris MW, Mullins HT, Belknap DF, Jaap WC. *Halimeda* bioherms along an open seaway: Miskito Channel, Nicaraguan Rise, SW Caribbean Sea. *Coral Reefs* 1988;**6**:173–8.

90. Multer HG. Growth rate, ultrastructure and sediment contribution of *Halimeda incrassate* and *Halimeda monile,* Nonsuch and Falmouth Bays, Antigua, W.I. *Coral Reefs* 1988;**6**:179–86.

91. Johns HD, Moore CH. Reef to basin sediment transport using *Halimeda* as a sediment tracer, Grand Cayman Island, West Indies. *Coral Reefs* 1988;**6**:187–93.

92. Littler MM, Littler DS, Lapointe BE. A comparison of nutrient- and light-limited photosynthesis in psammophytic versus epilithic forms of *Halimeda* (Caulerpales, Halimedaceae) from the Bahamas. *Coral Reefs* 1988;**6**:219–25.

93. Willan RC. The molluscan fauna from the emergent reefs of the northernmost Sahul Shelf, Timor Sea—Ashmore, Cartier and Hibernia Reefs; biodiversity and biogeography. In: Russell BC, Larson CJ, Glasby RC, Willan RC, Martin J, editors. *Understanding the*

cultural and natural heritage values and management challenges of the Ashmore Region. Records of the museums and art galleries of the northern territory, Supplement No. 1; 2005. p. 51–81.

94. Skewes TD, Gordon SR, McLeod IR, Taranto TJ, Dennis DM, Jacobs DR, et al. Survey and stock size estimates of the shallow reef (0–15 m deep) and shoal area (15–50 m deep) marine resources and habitat mapping within the Timor Sea MOU74 box. *Volume 2: Habitat mapping and coral dieback.* CSIRO report; 1999. p. 1–59.

95. Kospartov M, Beger M, Ceccarelli D, Richards Z. An assessment of the distribution and abundance of sea cucumbers, trochus, giant clams, coral, fish and invasive marine species at Ashmore Reef National Nature Reserve and Cartier Island Marine Reserve. *UniQuest report to the Department of Environment and Heritage;* 2006.

96. Glenn K. *Sedimentary processes during the Late Quaternary across the Kimberley Shelf, Northwest Australia.* Thesis, University of Adelaide; 2004.

97. Glenn K. *Aspects of the geological development of Ashmore Reef, North West Australia.* Unpublished thesis, Australian National University, Canberra; 1997.

98. Glenn K, Collins D. Ashmore Reef's sedimentological and morphological response to the Holocene sea level rise. In: Russell BC, Larson CJ, Glasby RC, Willan RC, Martin J, editors. *Understanding the cultural and natural heritage values and management challenges of the Ashmore Region. Records of the museums and art galleries of the northern territory,* Supplement No. 1; 2005. p. 13–28.

99. Collins DR. *Carbonate deposition model for Ashmore Reef, a shelf-edge reef, northwestern Australia.* Unpublished thesis, University of Western Australia; 2001.

100. Collins LB. Geological setting, marine geomorphology and oceanic shoals growth history of the Kimberley Region. *J R Soc West Aust* 2011;**94**(2):89–105.

101. Wilson BR. Notes on a brief visit to Seringapatam Atoll, Sahul Shelf, Western Australia. *Atoll Res Bull* 1985;**292**:83–99.

102. Bryce C, editor. *Marine biodiversity survey of Mermaid (Rowley Shoals), Scott and Seringapatam Reefs. Records of the Western Australian Museum,* Supplement No. 77; 2009.

103. Wilson BR, Blake S, Ryan D, Hacker J. Reconnaissance of species-rich coral reefs in a muddy, macro-tidal, enclosed embayment—Talbot Bay, Kimberley, Western Australia. *J R Soc West Aust* 2011;**94**:251–65.

104. Brooke BP. Geomorphology of the islands and inner shelf reefs of the central Kimberley coast, part 4. In: Walker D, editor. *Marine biological survey of the central Kimberley coast, Western Australia.* Perth: University of Western Australia. Unpublished report W.A., Museum Library No. UR377; 1997.

105. O'Connor S. A 6,700 BP date for island use in the West Kimberley, Western Australia: new evidence from High Cliffy Island. *Aust Archaeol* 1994;**39**:102–7.

106. Le Provost Environmental Consultants. Shallow marine habitats and biotic assemblages of Barrow Island. *Report to West Australian Petroleum Pty Ltd;* LEC Ref J193. Report No. R313; 1991.

107. Bowman Bishaw and Gorham. *Barrow Island intertidal survey: 1995.* Unpublished report to West Australian Petroleum Pty Ltd; 1996. 95 pp. and figures.

108. Australian Petroleum Production and Exploration Association. *Potential management arrangements for a marine management area covering the Montebello Islands to Barrow Shoals region.* Unpublished report prepared by Le Provost Dames and Moore, Murex Consultants and the Great Barrier Reef Marine Parks Authority; 1997.

109. Wells FE, Berry PF. The physical environment, marine habitats and characteristics of the marine fauna. In: Wells FE, Berry PF, editors. *Survey of the marine fauna of the Montebello Islands and Christmas Island, Indian Ocean. Records of the Western Australian Museum.* Perth: Western Australian Museum, Supplement No. 59; 2000. p. 9–13.

110. Berry PF, Wells FE. Survey of the marine fauna of the Montebello Islands and Christmas Island, Indian Ocean. In: *Records of the Western Australian Museum,* Supplement No. 59; 127 pp.

111. Wilson BR, Marsh LM, Hutchins JB. A puffer fish predator of Crown of Thorns in Australia. *Search* 1974;**5**(11–12):601–2.

112. Johnson DB, Stoddart JA. *Report on surveys of the distribution, abundance and impact of Acanthaster planci on reefs within the Dampier Archipelago (Western Australia)*. Australian Institution of Marine Science; 1988.

113. Jones DS, editor. *Marine biodiversity of the Dampier Archipelago, Western Australia 1998–2002. Records of the Western Australian Museum*. Perth: Western Australian Museum, Supplement No. 66; 2004.

114. Cassata L, Collins LB. Coral reef communities, habitats, and substrates in and near sanctuary zones of Ningaloo Marine Park. *J Coast Res* 2008;**24**(1):139–51.

115. McClanahan TR, Polunin NVC, Done TJ. Resilience of coral reefs. In: Gunderson L, Jansson B-O, Hollings CS, Folke C, editors. *Resilience and the behavior of large-scale ecosystems*. Washington, DC: Island Press; 2002. p. 111–63.

116. Bellwood DR, Hughes TP, Folke C, Nystrom M. Confronting the coral reef crisis. *Nature* 2004;**429**:827–33.

117. Marshall PA, Schuttenberg HZ. *A reef manager's guide to coral bleaching*. Townsville: Great Barrier Reef Marine Park Authority; 2006.

118. Wilson SK, Graham NAJ, Polunin NVC. Appraisal of visual assessments of habitat complexity and benthic composition on coral reefs. *Mar Biol* 2007;**151**:1069–76.

119. Hixon MA, Beets JP. Predation, prey refuges, and the structure of coral-reef fish assemblages. *Ecol Monogr* 1993;**63**:77–101.

120. Ault TR, Johnson CR. Spatially and temporally predictable fish communities on coral reefs. *Ecol Monogr* 1998;**68**:25–50.

121. Jones GP, Syms C. Disturbance, habitat structure and the ecology of fishes on coral reefs. *Aust J Ecol* 1998;**23**(3):287–97.

122. Williams D McB. Patterns in the distribution of fish communities across the central Great Barrier Reef. *Coral Reefs* 1982;**1**:35–43.

123. Done TJ. Coral zonation: its nature and significance. In: Barnes DJ, editor. *Perspectives on coral reefs*. Manuka, Australia: Brian Clouston Publisher; 1983. p. 107–47.

124. Russ GR. Distribution and abundance of herbivorous grazing fishes in the central Great Barrier Reef. I. Levels of variability across the entire continental shelf. *Mar Ecol Prog Ser* 1984;**20**:23–34.

125. Russ GR. Distribution and abundance of herbivorous grazing fishes in the central Great Barrier Reef. II. Patterns of zonation of mid-shelf and outer shelf reefs. *Mar Ecol Prog Ser* 1984;**20**:35–44.

126. Connell JH. Diversity in tropical rain forests and coral reefs. *Science* 1978;**199**:1302–10.

127. Connell JH. Disturbance and recovery of coral assemblages. *Coral Reefs* 1997;**16**:S101–13.

128. Done TJ. Coral community adaptability to environmental changes at scales of regions, reefs and reef zones. *Am Zool* 1999;**39**:66–79.

129. Mora C, Sale PF. Are populations of coral reef fishes open or closed? *Trends Ecol Evol* 2002;**17**:422–8.

130. Russell BC, Larson CJ, Glasby RC, Willan RC, Martin J, editors. *Understanding the cultural and natural heritage values and management challenges of the Ashmore Region. Records of the museums and art galleries of the northern territory*, Supplement No. 1; 2005.

131. Cairns SD. Azooxanthellate Scleractinia (Cnidaria: Anthozoa) of Western Australia. In: *Records of the Western Australian Museum* **18**:. p. 361–417.

132. Marsh LM. Corals and other cnidarians. In: Bowman Bishaw and Gorham, *Barrow Island intertidal survey*. Report No. RI5164, unpublished; 1995. p. 39–42, Tables 6–12.

133. Marsh LM. Corals and other cnidarians. In: Bowman Bishaw and Gorham, *Survey of the intertidal shores of the eastern side of Barrow Island*. Report No. RI6335, unpublished; 1997. p. 33–35, Tables 1–14.

134. Hutchins JB, Berry PF. Station lists for the diving expeditions (DA1 and DA3). In: Jones DS, editor. *Marine biodiversity of the Dampier Archipelago, Western Australia 1998–2002. Records of the Western Australian Museum.* Perth: Western Australian Museum, Supplement No. 66; 2004. p. 7–14.

135. Wilson B, Fromont J. *Addendum to habitats and invertebrate assemblages of Bouguer Passage and their regional significance: Spring Equinox (September/October 2011) low tide survey.* Report prepared for API Management Pty Ltd, Como, Western Australia; 2011.

136. Done TJ, Williams D McB, Speare PJ, Turak E, Davidson J, DeVantier LM, et al. *Surveys of coral and fish communities at Scott Reef and Rowley Shoals.* Townsville: Australian Institute of Marine Science; 1994.

137. Morrison PF. Subtidal habitats of Mermaid Reef (Rowley Shoals), Scott and Seringapatam, Western Australia. In: Bryce C, editor. *Marine biodiversity survey of Mermaid (Rowley Shoals), Scott and Seringapatam Reefs. Records of the Western Australian Museum.* Perth: Western Australian Museum, Supplement No. 77; 2009. p. 29–49.

138. AIMS. *Benthic habitat survey of Scott Reef (0–60 m).* Unpublished report for the Browse Joint Venture Partners; 2006.

139. Marsh LM. Report on the corals and some associated invertebrates of the Dampier Archipelago. In: *Report on the marine fauna and flora of the Dampier Archipelago.* Unpublished report, Western Australian Museum; 1978. p. 1–66.

140. Paling EI. *Analysis of coral community data using multivariate techniques, and their applicability to other community data.* Technical Series No. 3. Perth, Western Australia: Environmental Protection Authority; 1986. p. 1–28.

141. Huisman JM, Leliaert F, Verbruggen H, Townsend . Marine benthic plants of Western Australia's shelf-edge atolls. In: Bryce C, editor. *Marine biodiversity survey of Mermaid (Rowley Shoals), Scott and Seringapatam Reefs. Records of the Western Australian Museum.* Perth: Western Australian Museum, Supplement No. 77; 2009. p. 50–87.

142. Kohn AJ. Environmental complexity and species diversity in the gastropod genus *Conus* on Indo-West Pacific reef platforms. *Am Nat* 1967;**101**:251–9.

143. Kohn AJ, Leviten PJ. Effect of habiat complexity on population density and species richness in tropical intertidal predatory gastropod assemblages. *Oecologia* 1976;**25**:199–210.

144. Leviten PJ, Kohn AJ. Microhabitat resource use, activity patterns, and episodic catastrophe: *Conus* on tropical intertidal reef rock benches. *Ecol Monogr* 1980;**50**:55–75.

145. Reichelt RE. Space: a non-limiting resource in the niches of some abundant coral reef gastropods. *Coral Reefs* 1982;**1**:3–11.

146. Bryce C, Marsh LM. Echinodermata (Asteroidea, Echinoidea and Holthuroidea) of Mermaid (Rowley Shoals) Scott and Seringapatam Reefs, Western Australia. In: Bryce C, editor. *Marine biodiversity survey of Mermaid (Rowley Shoals), Scott and Seringapatam Reefs. Records of the Western Australian Museum.* Perth: Western Australian Museum, Supplement No. 77; 2009. p. 209–20.

147. Marsh LM, Morrison SM. Echinoderms of the Dampier Archipelago, Western Australia. In: Jones D, editor. *Marine biodiversity of the Dampier Archipelago Western Australia 1998-2002. Records of the Western Australian Museum.* Perth: Western Australian Museum, Supplement No. 66; 2004. p. 293–342.

148. URS (Australia). *Scott Reef Survey Four: ROV inspection of deep habitats in South Reef lagoon in the vicinity of potential pipeline routes and wellheads.* Unpublished report to Woodside; August 2007.

Mangrove Habitat and Associated Fauna

Following Cresswell and Semeniuk,[1] the term "mangrove" refers to individual plants defined as "woody trees and shrubs that inhabit tidal environments between mean sea level and the highest astronomical tide." The term "mangal" refers to an assemblage or community of mangroves and the habitat the mangrove plants create.

The historical biogeography of mangroves, mangal, and the associated mangal fauna is strikingly similar to that of modern coral reefs. Both mangal habitats and coral reefs are believed to have originated in the Eocene as pan-tropical ecosystems that, subsequently, were fragmented by the closure of the American and African/Europe seaways in the Miocene.[2,3] They are both characterized by a center of high species richness in the central Indo-West Pacific region and northern Australia. From this center, mangrove biodiversity decreases to the west in the western Indian Ocean and to the east in the islands of the West Pacific, with increasing latitude northward on the East Asian coast and southward on the coasts of eastern and western Australia.

Climatic, tidal, and physiographic conditions are key determinants of the variety of mangrove species and the regional development of mangals.[1,4–6] They are most developed and most species rich on shores with high rainfall and suitable geomorphology with river-fed estuaries characterized by terrigenous sediments.[7,8]

5.1 THE MANGROVE SPECIES

There is difficulty interpreting the literature on mangrove species diversity because of inconsistency in definition of the term mangrove. In this chapter, the term is used in the manner proposed by Cresswell and Semeniuk[1] as defined earlier so that samphires, herbaceous plants, ferns, and strand plants are excluded.

Most authors have cited 16 species from Western Australia following Semeniuk et al.[9] or 17 species following Semeniuk,[4] but, with several taxonomic adjustments, Cresswell and Semeniuk[1] list 15 species (Table 5.1).

5.2 MANGROVE BIOGEOGRAPHIC REGIONS IN WESTERN AUSTRALIA

On the shores of the North West Shelf, there is an attenuation in mangrove diversity from north to south (see Table 6.1) that corresponds with increasing aridity and changing habitat, but the causal factors involved in that progressive species loss are varied. Semeniuk[4] discussed the importance of regional and local freshwater seepage to mangrove species' spatial distribution. There is also a close relationship of mangrove distribution to coastal geomorphology.[1,10,11] Dispersal capacity and connectivity may also be involved in determining the geographic range of mangrove species.

Mangroves and mangal habitat are especially well developed on the humid, high rainfall Kimberley ria coast and moderately well developed on the arid to semi-arid coasts of Cambridge Gulf to the east and the Pilbara to the south. They also occur as a single mangrove species ecosystem (*Avicennia marina*) on the temperate West Coast in Shark Bay, the Abrolhos Islands, and as far south as Bunbury.

Several authors have recognized sets of mangrove "biogeographic regions" or "coastal sectors" in Western Australia,[9,11,12,13] characterized by distinctive climatic and geomorphic settings and stepped reduction in mangrove species richness from north to south (Figure 5.1). The Western Australian mangrove regions recognized by Semeniuk[11] relate closely to the IMCRA bioregions of northern Western Australia (Commonwealth of Australia 2006) and may be described as follows (with minor amendments):

1. Cambridge Gulf (Cambridge-Bonaparte Bioregion). With 12 mangrove species. Tide-dominated deltaic-estuarine environment in a gulf setting and semi-arid climate; mud component of sediment dominantly terrigenous.
2. Kimberley Coast (Kimberley Bioregion). With 15 mangrove species. Tide-dominated ria-archipelago settings with a diversity of estuarine and open sea habitats (Figures 5.2–5.4); sub-humid to humid climate; mud component of sediment dominantly terrigenous.
3. King Sound (King Sound Bioregion). With 11 mangrove species. Tide-dominated deltaic-estuarine environment and semi-arid climate; mud component of sediment dominantly terrigenous (Figure 5.5).

TABLE 5.1 Distribution of Mangrove Species on the Western Australian Coast

Species	Cambridge Gulf	Kimberley Bioregion	King Sound Bioregion	Canning Bioregion	Pilbara Coast	Rowley Shelf Province (Offshore Islands)	Carnarvon Province
Avicennia marina	√	√	√	√	√	√	√
Rhizophora stylosa	√	√	√	√	√	√	√ (Ningaloo only)
Aegialitis annulata	√	√	√	√	√	√	–
Aegiceras corniculatum	√	√	√	√	√	√	–
Ceriops tagal	√	√	√	√	√	√	–
Bruguiera exaristata	√	√	√	√	√	√	–
Osbornia octodonta	√	√	√	√	√	–	–
Excoecaria agallocha	√	√	√	√	–	–	–
Camptostemon schultzii	√	√	√	√	–	–	–
Sonneratia alba	√	√	–	√	–	–	–
Xylocarpus moluccensis	√	√	√	√	–	–	–
Lumnitzera racemosa	√	√	–	√	–	–	–

Continued

TABLE 5.1 Distribution of Mangrove Species on the Western Australian Coast—cont'd

Species	Cambridge Gulf	Kimberley Bioregion	King Sound Bioregion	Canning Bioregion	Pilbara Coast	Rowley Shelf Province (Offshore Islands)	Carnarvon Province
Bruguiera parviflora	–	√	√	–	–	–	–
Scyphiphora hydrophylacea	–	√	–	–	–	–	–
Xylocarpus granatum	–	√	–	–	–	–	–
Total	12	15	11	12	7	6	2

For additional information on the distribution and relative abundance of the species, refer to Ref. 4, Table 5 and Ref. 1.

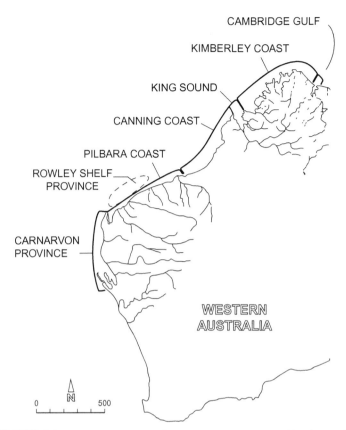

FIGURE 5.1 The Western Australian mangrove provinces. *After Semeniuk[11]*

FIGURE 5.2 A structurally complex, dendritic mangal in the northern part of Saint George Basin, an enclosed estuarine gulf on the ria Kimberley coast. Thirteen mangrove species are recorded from this location.[14] Together with a similar adjacent mangal immediately to the south, the total area of mangal habitat in the basin is estimated at 142 km², the largest in Western Australia. Creekside vegetation is comprised predominantly of *Camptostemon schulzti* interspersed with *Rhizophora stylosa* and *Avicennia marina*. *Image Courtesy of Department of Environment and Conservation.*

FIGURE 5.3   *Avicennia marina* growing on a basalt shore, the only mangrove tree on Berthier Island in the Bonaparte Archipelago. *Photo: John Huisman.*

FIGURE 5.4   Dense mangrove forest and supralittoral salt grass flats (*Sporolobus virginicus*) on Yawijaba Island, a Holocene mud island on Montgomery Reef, Collier Bay, Kimberley Bioregion. *Photo: Tim Willing.*

4. Canning Coast (Canning Bioregion). With 12 mangrove species. Barred to open tidal embayment settings and semi-arid climate; mud component of sediment dominantly carbonate (Figures 5.6 and 5.7).

5. Pilbara Coast (Pilbara (nearshore) Bioregion). With seven mangrove species. Wave-dominated settings on the shores of deltas (Figure 5.8), beach/dune coasts, limestone barrier island (Figure 3.8), and ria-archipelago rocky shores (Figure 3.6); arid climate but with many tidal creeks (Figure 5.9) and several large rivers that intermittently discharge large volumes of water into the coastal zone; mud component of sediment dominantly terrigenous.

6. Rowley Shelf Province (Pilbara (offshore) Bioregion). With six mangrove species. Embayments and lagoons of wave-dominated island shores in an arid, oceanic climate; mud component of sediment either terrigenous or carbonate.

7. Carnarvon Province (Ningaloo, Shark Bay, and Abrolhos Bioregions). With two mangrove species—only one south of Ningaloo. Embayments, lagoons, and wave-dominated deltas; mud component of sediment either terrigenous or carbonate.

FIGURE 5.5 Dendritic mangals on tidal creeks draining large supratidal mud flats at Doctors Creek on the eastern side of King Sound. The town and port of Derby on the peninsula at the bottom of the picture. *Image Courtesy of Department of Environment and Conservation.*

Thirteen species of mangrove are common to the Cambridge/Bonaparte, Kimberley, and Canning Bioregions. These 13 species may be regarded as the "regional species pool" of the Western Australian coast.[10] Eight of them comprise the mangrove flora of the Pilbara coast. The reduction of species numbers from the Cambridge/Bonaparte and Kimberley Bioregions (15) to the Canning Bioregion (13) and the Pilbara (nearshore) Bioregion (7) may be attributed largely to habitat change relating to increasing aridity.

A noteworthy difference between the mangals of the Kimberley Bioregion and those of the semi-arid and arid King Sound, Canning, and Pilbara Bioregions is the development of mangrove communities along the hinterland fringe (Semeniuk[4] Table 3). In the Kimberley Bioregion, mangrove vegetation commonly grows to the edge of the hinterland that bears terrestrial vegetation, and there is rarely mud flat habitat between. In that situation, there may be a species-rich mangrove fringe community with eight or more species present. Where there is freshwater seepage, there may be as many as 11 species in this high tidal zone. Several mangrove species are characteristic of this zone and are not found elsewhere in

FIGURE 5.6   The open sea mangal on the shores of Roebuck Bay south of Broome. *Image Courtesy of Department of Environment and Conservation.*

FIGURE 5.7   A small, mangal with *Rhizophora* bordering a tidal creek at Port Smith, Canning Bioregion. The mangal is developed in a semienclosed bay with muddy, carbonate sand flats behind limestone barrier headlands. Twelve species of mangrove are recorded from mangals of this kind in the bioregion, and there is a rich associated fauna of invertebrates. *Photo: Sue Morrison.*

FIGURE 5.8   The eastern part of the Ashburton Delta, on the arid West Pilbara coast. The mangal is a Holocene development behind a sand bar. Successive stages of its growth are evident in the series of dunal ridges between the sand bar and the Pleistocene hinterland. Six species of mangrove are recorded from this locality but two, *Avicennia marina* and *Rhizophora stylosa*, predominate. Pale green areas are primarily *Avicennia* and dark green areas are *Rhizophora*. *Photo: Barry Wilson.*

FIGURE 5.9   Hooley Creek, West Pilbara, with a narrow fringe of *Avicennia*. The pneumatophore zone is habitat of the fiddler crab *Uca flammula*. *Photo: Barry Wilson.*

the mangrove system, for example, *Pemphis acidula, Scyphiphora hydrophylacea, Excoecaria ovalis*, and *Lumnitzera racemosa*.

The rocky islands of the Kimberley archipelagos generally have restricted mangrove floras and mangal habitats. They occur most often in bays and as narrow fringes on rocky shores. Larger islands close to the coast have the best developed systems, and the outer islands in less turbid water and further from the influences of river discharge have quite limited mangrove communities. For example, on Berthier Island in the Bonaparte Archipelago, only a single *A. marina* tree was found in a survey of intertidal habitats (Figure 5.3). Surveys of the mangal invertebrates on islands in the Buccaneer Archipelago recorded the mangrove trees present, the most common being *A. marina, Sonneratia alba*, and *Rhizophora stylosa*.[15] The most diverse flora was at Kingfisher Islands where there were 10 species. Saenger[16] made a similar survey of islands and coastal sites in the northern Kimberley.

An atypical mangal occurs on Yawijaba Island, Montgomery Reef, Collier Bay (Figure 5.4). This low Holocene mud island has supralittoral salt grass flats (*Sporolobus virginicus*) and six mangrove species. The dominant mangrove association is a woodland of *Osbornia octodonta* and *A. marina*, ranging in height from 1 to 10 m. Scattered throughout the woodlands are *R. stylosa, Aegiceras corniculatum, Ceriops tagal*, and *Bruguiera exaristata*.

In the semi-arid and arid bioregions, the presence of wide high tidal mud flats above the mangrove fringe is the general rule. There the terrestrial vegetation of the hinterland is typically bordered by samphire shrubs and a mangrove fringe community is poorly developed or absent.[11] In the semi-arid Canning Bioregion, *A. marina*, *C. tagal*, *E. ovalis*, and *L. racemosa* sometimes occur in this zone where there is freshwater seepage. Where there is no freshwater seepage, only the first two of these species are present. In the arid Pilbara bioregion, this high tidal fringe zone normally has no mangroves, although *A. marina* and *C. tagal* may be present where there is freshwater seepage.

In short, climate-related biogeomorphic features, that is, high tidal mud flat development and freshwater seepage, account for the progressive latitudinal loss of species that are characteristic of the high tidal, mangrove fringe zone: *S. hydrophylacea* (restricted to the Kimberley Bioregion) and *E. ovalis* and *L. racemosa* (rare in the Canning Bioregion and absent in the Pilbara).

5.3 ECOLOGICAL ASSOCIATIONS AND DISTRIBUTION

As a result of varied adaptations to salt, immersion, desiccation, and other environmental circumstances, mangrove species generally arrange themselves spatially in patterns of ecological association and distribution (zones) that are repeated and recognizable on a regional basis.[17] In Western Australia, common zonation patterns have been described by several writers.[1,9,11]

5.4 PRIMARY AND SECONDARY PRODUCTION

Mangals are critically important primary production habitats. There are several sources of primary production in mangals, including the extensive mud and salt flats that are such features of mangals in the semi-arid and arid bioregions:

- mangrove plants produce large quantities of detrital material, derived from fallen leaves and decaying wood
- microphytobenthos (e.g., cyanobacterial layers of upper littoral mud flats produce and fix significant amounts of nitrogen in the substrate[18])
- microepiflora—on the mangrove vegetation
- planktonic microflora imported from the coastal waters by tidal flux.

The notion that mangals are net exporters of nutrients to the adjacent coastal zone may be misplaced. They are more likely net importers of nutrients from the adjacent coastal marine and terrestrial habitats. Nevertheless, it is clear that mangals are important sinks of carbon, dissolved nitrogen, phosphorous, and silicon[19] and that they play major secondary production roles in supporting coastal food webs and nutrient cycles in the coastal zone.

The substrate of mangals and associated mud flats has a high organic content and supports high microbial activity and large densities of grazing and detrital-feeding fishes and invertebrates.[20–24] While there are some predatory species and some suspensory-feeding invertebrates that live in the seaward margins of mangals, the majority of a mangal biomass (apart from the trees) comprises surface dwelling and burrowing grazers and detritivores that perform the critical role of breaking down the organic material into particulate components. The burrowers also have very important functions in the redistribution of that material in the ecosystem and maintaining favorable geochemical conditions in the substrate.

The density of biota in mangals is usually very high. The standing stock of commercial species may be many times higher than that of adjacent coastal marine habitats.[25–27] Many coastal species that have commercial importance use mangal habitats as breeding and nursery areas, taking advantage of the protection and rich food resources available there.[28,29] In these ways, the primary and secondary production of mangals have very large impacts on the biodiversity of other coastal habitats.

There have been many studies of primary and secondary production in mangals of the Australian wet tropics. The ecological roles of mangals on arid coasts are less well understood, but a significant role in primary production and nutrient recycling is evident there also. Biogeochemical processes in mangrove forests (*Avicennia* and *Rhizophora*) at localities on the arid Pilbara coast have been studied.[30] The important role of cyanobacteria in nitrogen fixation on highly saline mud flats in the Pilbara region has been examined.[18] However, the complexities of mangal ecosystems in Western Australia, and the nutrient and trophic pathways that operate, remain poorly understood.

Nevertheless, it is clear that the diverse mangal systems of the humid Kimberley coast are very different from those of the semi-arid and arid Canning and Pilbara coasts. The north-south attenuation of mangrove species is well documented, and the southern mangal systems have different, simpler floristic structures from those of the Kimberley coast. One significant difference is the development of extensive high tidal mud flats in arid southern mangals, the presence there of vast areas of nutrient-producing cyanobacteria (Figure 5.10) , and high biomass of burrowing crabs that play significant roles in secondary production and substrate rehabilitation (Figures 5.11 and 5.12). A common pattern is for a band on the mud flats immediately behind the mangrove tree line that is heavily bioturbated by fiddler and marsh crabs, indicating that this is a zone of significant secondary production (Figures 6.7 and 6.8).

The climatic conditions that have created these ecosystem differences between the wet north and arid south have been of long standing. Intuitively, it seems likely that different nutrient sources, pathways, and outputs of the mangals may have had significant influences on the patterns

FIGURE 5.10   Supralittoral mud flats with samphire shrubs on the western banks of Hooley Creek, West Pilbara. Such flats are significant as both primary and secondary producer habitat. Note the dry, salty surface with a cyanobacterial mat in the foreground and the wet area closer to the creek (in the background) that was inundated by the previous night's high spring tide. The latter is a zone of high bioturbation by sesarmid and fiddler crabs. The sesarmids also occur under the samphire shrubs right up to the spinifex margin. *Photo: Barry Wilson.*

FIGURE 5.11   Hooley Creek, Ashburton Delta. Behind the mangrove tree line on West Pilbara mud flats there is often a band, tens of meters wide, that is regularly inundated at high tide and heavily bioturbated by detritivorous crabs (species of *Uca*—mainly *U. elegans*—and sesarmids) and thereby important areas of secondary production. *Photo: Barry Wilson.*

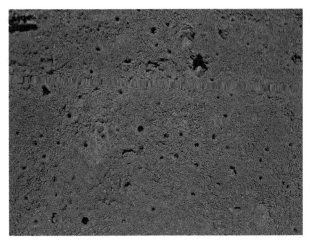

FIGURE 5.12   Hooley Creek, Ashburton Delta. Burrows of fiddler crabs, mainly *Uca elegans*, cover vast areas of mud flats behind the mangrove tree line. *Photo: Barry Wilson.*

of evolutionary development of regional biotas along the length of the North West Shelf. However, until there is much better information on the mangal ecosystems of the region, this remains largely a matter of conjecture.

5.5 ASSOCIATED MANGAL INVERTEBRATE FAUNA

There is a variety of terrestrial and marine vertebrates and invertebrates that utilize the food resources of mangals on a temporary basis.[31–33] Many marine animals spawn and develop their young in the food-rich, protected creeks and pools of mangals. Some marine animals live in mangals as well other coastal intertidal habitats. However, most of the key secondary producer species that dominate the faunas of mangal ecosystems belong to fish and invertebrate taxa that are restricted to mangal habitats—variously referred to as the "mangal dependants" or "mangal obligates."

The most conspicuous and abundant macroinvertebrates that inhabit mangals on a permanent basis are obligate species that belong to certain genera of Mollusca, Crustacea, and Polychaetae. Most of these species, or the genera to which they belong, are widespread in mangal habitats across the Indo-West Pacific Realm, but there is a significant endemic element in the mangal fauna of the North West Shelf. It is believed that the key genera coevolved with the mangal ecosystems in the late Tertiary and Quaternary periods.[3] There is no information on the polychaetes of mangals of shores of the North West Shelf.

5.6 MOLLUSCS

A number of bivalved molluscs live in mangals, including wood borers (Teredinidae) that are abundant and play a very important role in the breakdown of mangrove wood. Two species, the anomiid *Enigmonia aenigmatica* and the isognomid *Isognomon* cf. *vitrea*, live attached to mangrove prop roots and pneumatophores. However, no northern Australian bivalves have been shown to be mangal obligates.

Four families of gastropod molluscs contain species that are mangal obligates and may play key roles in the consumption and breakdown of particulate matter (plant litter and/or microbial):

* Neritidae (some species of *Nerita*)
* Littorinidae (most species of *Littoraria*)
* Potamididae (most species of *Terebralia*, *Telescopium*, and *Cerithidea*)
* Ellobiidae (some species of *Ellobium*, *Cassidula*, *Melampus*).

In studies of the resilience/recovery of a mangal fauna, where connectivity is an all-important issue, it is necessary to consider the life history of each species present as there is a large range of larval development and dispersal strategies in each of these families.

5.6.1 Family Neritidae

This family is characteristic of the intertidal zone, mostly on rocky shores. Nerites are grazers on microorganisms growing on the surface of the substrate. The females lay gelatinous egg capsules on the substrate or on the shells of neighbors. In most species, the larvae escape from the capsules as planktotrophic veligers and there is potential, at least, for wide dispersal in coastal waters, but direct development is known in the family, that is, without a pelagic larval stage.[34]

Nerita balteata is the only mangal-obligate nerite found in Australian waters. It is widely distributed in mangroves throughout the Indo-West Pacific region. In northern Australia, it is a conspicuous arboreal snail in mangroves from Moreton Bay to Shark Bay. It lives on the trunks of mangrove trees and is a common shell in middens along the shores of the North West Shelf.

5.6.2 Family Littorinidae

This family is also characteristic of the intertidal zone, primarily on rocky shores. One genus, *Littoraria*, is widespread throughout the tropical Indo-West Pacific, Eastern Pacific, and Atlantic regions. It includes many species (in the subgenera *Littorinopsis* and *Palustorina*) that are mangal obligates or at least commonly found in mangal habitats. The taxonomy and life history of *Littoraria* in the Indo-West Pacific have been reviewed.[35] In northern Australia, there are nine species that live on mangroves. Seven of them occur in mangals, the shores of the North West Shelf, and two of them endemic to the region.

Development varies within the genus. The subgenus *Littorinopsis* is ovoviviparous, the larvae being retained for a time within the mantle cavity of the female and released in gelatinous capsules. In the subgenus *Palustorina*, development is oviparous, the larvae being released as veligers. However, the development strategy does not seem to affect dispersal as both forms have late-stage planktotrophic larvae capable of wide pelagic distribution.[35]

There is little information on the feeding habits of the species of *Littoraria*. There is no evidence that they graze on the leaves, except on leaf hairs on the underside of *Avicennia* leaves.[35] Those that live on mangrove trunks and branches are believed to feed on the microflora that forms a film on the surface of the bark (Figures 5.13 and 5.14).

The species of *Littoraria* are most abundant along the seaward edges of mangals. Several species may cohabit within a mangal. In Singapore, for example, as many as 10 species may be found within one mangal. The species exhibit a degree of habitat preference, that is, preferred height or position on the trees, but there is a great deal of overlap and there is very little published information on the ecological distribution of *Littoraria* species within Western Australian mangals.

Seven species of *Littoraria* inhabitat mangals of the North West Shelf (Table 5.2). All of them are found along the entire coast of the North West Shelf as far west as Exmouth Gulf or beyond.

FIGURE 5.13   *Littoraria pallescens*, Roebuck Bay, Broome. *Photo: Peter Strain.*

FIGURE 5.14   *Littoraria filosa*, Roebuck Bay, Broome. *Photo: Peter Strain.*

TABLE 5.2   Species of Mangrove Tree Snails (Littorinidae, *Littoraria*) Found in Mangroves of North West Shelf Shores

Species	Distribution	Habitat
Littoraria (*Littorinopsis*) *scabra*	Indo-West Pacific; northern Australia from southern Queensland to Exmouth Gulf	Mangrove trunks and roots, especially Rhizophora
Littoraria (*Littorinopsis*) *pallescens*	Indo-West Pacific; northern Australia from southern Queensland to Exmouth Gulf	Mangrove branches and leaves, typically along seaward edge
Littoraria (*Littorinopsis*) *filosa*	Central Indo-West Pacific; northern Australia from southern Queensland to Exmouth Gulf	Mangrove branches and leaves, typically in the Avicennia fringe
Littoraria (*Littorinopsis*) *cingulata*	Endemic; Buccaneer Archipelago to Exmouth Gulf, with a subspecies endemic to Shark Bay (L. c. Pristissini)	Mangrove branches, pneumatophores, and strand line shrubs, especially Avicennia fringe; occasionally on rocky shores
Littoraria (*Palustorina*) *sulculosa*	Endemic; Vansittart Bay to Exmouth Gulf	Mangrove trunks, branches, and leaves, often at landward fringe of Avicennia; occasionally on rocky shores
Littoraria (*Palustorina*) *articulata*	Central Indo-West Pacific; northern Australia from southern Queensland to Exmouth Gulf	Mangrove trunk and branches, typically on seaward edge; occasionally on rocky shores and pilings
Littoraria (*Palustorina*) *undulata*	Indo-West Pacific; northern Australia from southern Queensland to Exmouth Gulf	Mangrove trunk and branches and intertidal rocky shores

5.6.3 Family Potamididae

In northern Australia, there are four genera of this family, containing eight species that are mangal obligates, seven of which are found in mangals of the North West Shelf. These are large- to small-sized snails that crawl on the mud surface among the mangrove roots (*Terebralia*, *Telescopium*, *Cerithideopsilla*—the "mud creepers") or arboreal species that climb on the trunks and branches of the trees (*Cerithidea*—the "treecreepers"). The mud creepers often occur in vast colonies and are among the most conspicuous animals seen in mangal habitats (Figure 5.15).

Mud creepers are detritivores that consume decayed mangrove leaves and other litter on the substrate surface. A great deal of research has been done on their feeding habitats and the important role these animals play in the breakdown of plant litter.[36–38] The treecreepers are grazers, feeding on surface microorganisms on the mangrove trunks and branches (Table 5.3[39]).

5.6.4 Family Ellobiidae

Ellobiids are air-breathing pulmonate snails that inhabit the intertidal and supratidal zones, especially in marshes associated with estuaries. Many species live in Indo-West Pacific mangals, but the taxonomy of the family remains poorly studied. The taxonomy of two ellobiid genera, *Ellobium* and *Cassidula*, has been studied in northern Australia (Table 5.4[40]).

FIGURE 5.15   A colony of the potamidid mud creeper *Terebralia semistriata* feeding on mud flats in front of the *Avicennia* tree line, Exmouth Gulf. *Photo: Barry Wilson.*

TABLE 5.3 Mud Creepers and Treecreepers (Family Potamididae) That Inhabit Mangals on Shores of the North West Shelf

Species	Distribution	Habitat
Telescopium telescopium Mud creeper	Central Indo-West Pacific; northern Australia, Moreton Bay to Exmouth Gulf	Gutters and puddles in the shade of mangrove trees (Ref. 36)
Terebralia palustris Mud creeper	Indo-West Pacific; northern Australia, Moreton Bay to Exmouth Gulf	Mud among mangrove roots (Refs. 36, 42, 43)
Terebralia sulcata Mud creeper	Indo-West Pacific; northern Australia, southern Queensland to the north Kimberley	Among the pneumatophore roots of *Avicennia* and on mud flats in front of the mangal
Terebralia semistriata Mud creeper	Northern Australia; central Queensland to Shark Bay; sympatric with *T. sulcata* in the North Kimberley	As for *T. sulcata*[41]
Cerithidea largillierti Mud creeper (may be misplaced generically)	Central Indo-West Pacific; northern Australia, North Queensland to Exmouth Gulf	Mud-dweller in shallow gutters among the roots of mangroves
Cerithideopsilla cingulata Mud creeper	Indo-West Pacific; northern Australia from southern Queensland to Exmouth Gulf	Mud flats at the mangrove edge
Cerithidea anticipata Arboreal tree creeper	Northern Australian; southern Queensland to Admiralty Gulf	Attaches to mangroves by mucous threads (Ref. 37)
Cerithidea reidi Arboreal tree creeper	Endemic; North Kimberley to Exmouth Gulf; sympatric with *C. anticipata* in the North Kimberley	As for *C. anticipata*. (Ref. 37)

5.7 CRUSTACEANS

There are many kinds of small crustaceans that live in litter and wood in mangals, but there is very little information on them for northern Australia.[31,32,41] There are also medium- to large-sized mangal-obligate crustaceans that are better known[44] including arboreal barnacles and burrowing mud shrimps and crabs, the latter playing key roles in mangal ecosystems.[45,46]

TABLE 5.4 Large Air-Breathing Ellobiid Snails Recorded from Mangals on the Shores of the North West Shelf

Species	Distribution	Habitat
Ellobium aurisjudae	Indo-West Pacific; northern Australia, Moreton Bay to Exmouth Gulf	Buried in mud beneath deep litter at the landward margins of mangals and among rotting timber
Ellobium semisculptum	Central Indo-West Pacific; northern Australia, Hervey Bay (Queensland) to North West Cape	Uncommon in WA, habitat unknown
Cassidula aurisfelis	Northern Australia; central Queensland to the Ashburton delta	Supratidal halophyte flats behind the Avicennia fringe
Cassidula angulifera	Northern Australia; central Queensland to Shark Bay	Under logs in the landward Avicennia fringe
Cassidula sowerbyana	Northern Australia; recorded from Kimberley and Canning coasts	Habitat unknown

5.7.1 Barnacles (Cirripedia)

In northern Australian mangals, there are several species of barnacle that live attached to the trunks, branches, leaves, or pneumatophores of the trees, especially along the seaward edges of open bay mangrove systems. These animals are not well studied on the shores of the North West Shelf, but the presence of several mangal-obligate species is known, such as

- *Fistubalanus* sp. lives attached to pneumatophores—there is at least one undescribed species on mangrove shores in the Pilbara.
- *Hexaminius popeiana* which lives on mangrove trunks and branches.
- *Hexaminius foliorum* which lives on mangrove leaves.

There are also several species that live on mangroves but are found also on wooden structures in the intertidal zone outside mangroves. These include *Chthamalus malayensis* (a northern Australian species) and *Micro-euraphia withersi* (a widespread Indo-West Pacific barnacle) (Figure 5.16).

5.7.2 Mud and Ghost Shrimps

The mud and ghost shrimps are burrowing crustaceans normally present, if not conspicuous, in northern Australian mangroves. Several families are involved.

The largest of these creatures are three species of mangrove lobster (family Thalassinidae)—*Thalassina squamifera*, *T. emerii*, and *T. anomala*, all of which have been reported from Pilbara mangroves. These are deep

FIGURE 5.16   A colony of the barnacle *Chthamalus malayensis* on the trunk of an *Avicennia* tree, Port Smith, Canning Bioregion. *Photo: Barry Wilson.*

burrowers that build conspicuous conical turrets, usually among the mangrove trees. None of these species, or their burrows, was observed within the study area. Small and inconspicuous crustaceans known as ghost shrimps are common in northern Australian mangroves but are rarely seen because of their deep and inconspicuous burrows. Two families are involved, the Callianassidae and Upogebiidae. Very little is known of these very strange creatures. They are rarely collected but are likely to play a significant role in mangrove ecosystems as secondary producers.

5.7.3 Brachyuran Crabs

Crabs are generally an important component of mangal invertebrate faunas. They include representatives of the families Portunidae and Grapsidae, although most of these crabs are not mangal obligates. However, two families of burrowing crabs, the Sesarmidae and Ocypodidiae, are adapted for life in the intertidal zone and contain mangal-obligate genera and species. As well as their ecological role in the break up and redistribution of plant material, the bioturbation they produce by their burrowing activities has important positive effects on substrate geochemistry. They also comprise a large biomass that supports mangal predators (e.g., fishes, birds, crocodiles).

5.7.3.1 Family Ocypodidae

The family includes two genera, *Ocypode* and *Uca* that live in mangals or on associated tidal mud and sand flats, the latter among the most conspicuous and ecologically important invertebrates in mangal ecosystems.

Genus Ocypode (Ghost Crabs)

These medium-sized crabs are beach burrowers. There are five species in the North West Shelf region, and one of these, *Ocypode fabricii* (Figure 5.17), is commonly found on muddy sand beaches associated with mangals on mainland shores. Like other species of the genus, it digs deep burrows that it uses as a base from which to make its foraging sorties along the beach and into the adjacent dunes.

Genus Uca (Fiddler Crabs)

Fiddler crabs live in deep burrows, often in dense colonies, and are a conspicuous feature of mangrove forests and associated mud flats where they feed on fine organic material and microphytobenthos on the surface layers of the substrate (Figures 5.18 and 5.19). In that role, they are very important secondary producers in mangrove ecosystems.

In the Canning and Pilbara Bioregions, there is commonly a zone on the mud flats behind the mangrove tree line that is heavily burrowed with very high density of fiddler crabs. The endemic North West Shelf species *Uca elegans* (Figure 5.20) is especially prominent in this situation and is clearly a key species in the mangrove ecosystem of these bioregions.

FIGURE 5.17 *Ocypode fabricii*, Ashburton Delta, West Pilbara. This species burrows in slightly muddy sand on mainland shores and is often associated with mangals. Like other species of the genus, it digs deep burrows that it uses as a base from which to make its foraging sorties along the beach and into the adjacent dunes. It is easily distinguished from *O. ceratopthalmus* that is very common burrowing on open sea beaches by its more robust body and lack of long stylets on the eye stalks. *Photo: Barry Wilson.*

FIGURE 5.18   *Uca dampieri*: a common fiddler crab that burrows on mud flats in Roebuck Bay, Broome. *Photo: Peter Strain.*

FIGURE 5.19   A colony of *Uca dampieri* living in the pneumatophore zone of an *Avicennia* mangal at Roebuck Bay, Broome. *Photo: Peter Strain.*

There are 19 species of *Uca* in northern Australia.[47,48] Ten species of *Uca* are known from Western Australian mangroves, but the number may be greater than that because the family has not been extensively studied in the diverse mangal habitats of the Kimberley coast.

In the context of biogeography, the number of *Uca* species that are endemic to the north-west coast of Australia, mostly from the Northern Territory to Exmouth Gulf or beyond, is highly significant. Davie (personal communication) has suggested that the North West Shelf has been a significant center of speciation in this genus. This may relate to the long-standing, extensive development of upper littoral and supralittoral mud flats that are associated with mangal habitats along the shores of the North West Shelf, especially in the Canning and Pilbara Bioregions (Figure 5.21; Table 5.5).

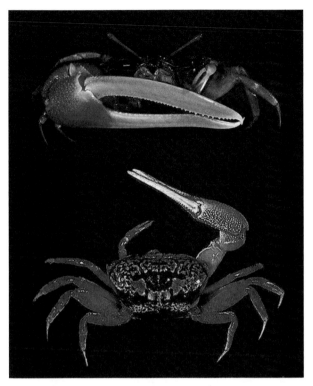

FIGURE 5.20   *Uca elegans*, an endemic North West Shelf species that is abundant on mud flats and plays a key role in ecosystem function. *Photo: Barry Wilson.*

5.7.3.2 Family Sesarmidae (Marsh Crabs)

Several genera of this family are characteristic of mangrove habitats in the Indo-West Pacific region. They are burrowing or cryptic crabs that play key ecological roles.[46,48–51] There is a large literature on the feeding habitats, reproduction, and other biological characteristics of these animals. They are small- to large-sized crabs that forage on the mud flat surface for plant material and drag it into their burrows where they shred and consume it, thereby moving fine organic material deep into the soil. The taxonomy of Western Australian sesarmids is presently under study. Three Indo-West Pacific genera known to be present in North West Shelf mangals are *Neosamartium*, *Perisesarma*, and *Parasesarma*.

Neosarmartium meinteri is a large, widespread Indo-West Pacific species that is common in mangrove habitats across northern Australia and at least as far west as Exmouth Gulf. It builds deep, hooded burrows among mangrove tree roots and in the shelter of samphire shrubs across the entire width of the high tidal mud flats (Figure 5.22).

FIGURE 5.21 *Uca flammula*, a very conspicuous, key species that lives along the banks of tidal creeks, usually making its burrows among pneumatophores. *Photo: Barry Wilson.*

TABLE 5.5 *Uca* Species Distribution and Habitats in Mangals of the North West Shelf

Species	Distribution	Habitat
Uca elegans	Endemic; Darwin to Shark Bay	Entire width of high tidal mud flats behind the mangrove tree line
Uca flammula	Endemic; Darwin to Ningaloo	Muddy banks of tidal creeks
Uca mojbergi	Northern Australia and New Guinea; Darwin to Ningaloo	Sandy mud flats close to the mangrove tree line and base of beach slopes
Uca dampieri	Endemic; Exmouth Gulf to the Northern Territory	Intertidal sandy mud flats of open bays and creeks
Uca capricornis	Northern Australia; Queensland to Exmouth Gulf	. . .
Uca polita	. . . to Ningaloo	Muddy banks of tidal creeks
Uca hirsutimanus	Northern Australia; Torres Strait to . . .	. . .

Perisesarma and *Parasesarma* are two genera of smaller marsh crabs. There are at least two species of *Perisesarmu* in North West Shelf mangals (Figure 5.23). *Perisesarma semperi* is known from Papua New Guinea, the Gulf of Carpentaria, and mangals of the Ashburton coast. In the West Pilbara, there is also an unnamed species of *Perisesarma*. *Parasesarma hartogi* is a recently described marsh crab so far known from Shark Bay and the mainland coast of the West Pilbara (Figure 5.24). All three of these crabs are abundant and active among mangrove trees and high tidal

FIGURE 5.22   The hooded burrows of the large sesarmid marsh crab *Neosarmartium meinteri* are conspicuous on mangal mud flats on the coast of the North West Shelf. *Photo: Barry Wilson.*

FIGURE 5.23   *Perisesarma holthuisi.* A widely distributed Marsh Crab in the central Indo-West Pacific region that is very common in mangals of the North West Shelf. *Photo: Barry Wilson.*

FIGURE 5.24   *Parasesarma hartogi.* This is a recently described species of Marsh Crab from Shark Bay and the West Pilbara. It is abundant among the trees and inner mud flats. *Photo: Barry Wilson.*

samphire flats in the West Pilbara and probably along the entire coast of the North West Shelf. Their burrows are a conspicuous feature of high tidal mud flats within and adjacent to the landward mangrove fringe. Like the fiddler crabs, these small marsh crabs are key species in mangal eco-systems of the region.

5.7.3.3 *Family Grapsidae*

The grapsid *Metopograpsus frontalis* is very abundant running among the trunks and roots of mangroves of the Kimberley, Canning, and Pilbara bioregions (Figure 5.25). The species is common throughout the central Indo-West Pacific Realm. It is not a burrower or a mangal obligate and is common also in rocky shore habitats.

5.7.3.4 *Family Portunidae*

There are several common portunids crabs in mangal habitats within the region. *Scylla serrata*, known as the mangrove mud crab, is a popular target of the recreational fishery in northern Australia and elsewhere in the Indo-West Pacific Realm. There are also several species of the genus *Thalamita* that are common in sand and mud flat habitats associated with mangals. None of these portunids are mangal obligates, but they are con-spicuous predators in mangal and adjacent mud flat ecosystems.

FIGURE 5.25   *Metopograpsus frontalis*. This grapsid is a common predator in mangals and rocky shores. *Photo: Barry Wilson.*

References

1. Cresswell ID, Semeniuk V. Mangroves of the Kimberley region: ecological patterns in a tropical ria coast setting. *J R Soc West Aust* 2011;**94**(2):213–37.
2. Plaziat J-C, Cavagenetto C, Koeniguer J-C, Baltzer F. History and biogeography of the mangrove ecosystem, based on a critical reassessment of the paleontological record. *Wetlands Ecol Manage* 2001;**9**:161–79.
3. Reid DG, Dyal P, Lozouet ZP, Glaubrecht M, Williams ST. Mudwhelks and mangroves: the evolutionary history of an ecological association (Gastropoda: Potamididae). *Mol Phylogenet Evol* 2008;**47**:680–99.
4. Semeniuk V. Mangrove distribution in north-western Australia in relation to regional and local freshwater seepage. *Vegetatio* 1983;**53**:11–31.
5. Woodroffe C, Grindrod J. Mangrove biogeography: the role of Quaternary environmental and sea-level change. *J Biogeogr* 1991;**18**:479–92.
6. Duke NC, Ball MC, Ellison JC. Factors influencing biodiversity and distribution gradients in mangroves. *Glob Ecol Biogeogr Lett* 1998;**7**:27–47.
7. Thom BG. Mangrove ecology—a geomorphological perspective. In: Clough BC, editor. *Mangrove systems in Australia.* Canberra: Australian Institute of Marine Science; 1982. p. 3–30.
8. Thom BG. Coastal landforms and geomorphic processes. In: Snedaker SC, Snedaker JG, editors. *The mangrove ecosystem: research methods.* UNESCO; 1984. p. 3–17.
9. Semeniuk V, Kenneally KF, Wilson PG. *Mangroves of Western Australia.* Perth: Western Australian Naturalist Club; 1978, Handbook Number 12, 92 pp.
10. Semeniuk V. Development of mangrove habitats along ria shorelines in north and north-western tropical Australia. *Vegetatio* 1985;**60**:3–23.
11. Semeniuk V. The mangrove systems of Western Australia: 1993 Presidential Address. *J R Soc West Aust* 1993;**76**:99–122.
12. Galloway RW. Distribution and physiographic patterns of Australian mangroves. In: Clough BF, editor. *Mangrove ecosystems in Australia. Structure, function and management.* Canberra: Australian Institute of Marine Science; 1982.
13. Johnstone RE. Mangroves and mangrove birds of Western Australia. *Records of the Western Australian museum* 1990;(Suppl. 32):1–120.

14. Wells AG. A survey of riverside mangrove vegetation fringing tidal river systems of Kimberley, Western Australia, Part 3. In: *Biological Survey of Mitchell Plateau and Admiralty Gulf, Kimberley, Western Australia*. Western Australian Museum: Perth; 1981. p. 95–121.

15. Hanley JR. Quantitative survey of mangrove invertebrates, part 7. In: Wells FE, Hanley JR, Walker DI, editors. *Survey of the marine biota of the southern Kimberley Islands*. Perth: Western Australian Museum. Unpublished report No. UR 286; 1995. p. 82–100

16. Saenger P. *Mangrove flora*: distribution of species and habitat descriptions, part 6. In: Walker DI, Wells FE, Hanley JR, editors. *Marine biological survey of the eastern Kimberley, Western Australia*. Perth: Western Australian Museum. Unpublished report UR353; 1996. p. 39–53.

17. Macnae W. A general account of the flora and fauna of mangrove swamps and forests in the Indo-West Pacific region. *Adv Mar Biol* 1968;**6**:73–270.

18. Paling EI, McComb AJ, Pate JS. Nitrogen fixation (acetylene reduction) in nonheterocystous Cyanobacteria mats from the Dampier Archipelago, Western Australia. *Aust J Mar Freshw Res* 1989;**40**:147–53.

19. Alongi DM. The dynamics of benthic nutrient pools and fluxes in tropical mangrove forests. *J Mar Res* 1996;**54**:123–48.

20. Alongi DM. The role of soft-bottom benthic communities in tropical mangrove and coral reef ecosystems. *Rev Aquat Sci* 1989;**1**:243–80.

21. Alongi DM. Ecology of tropical soft-bottom benthos: a review with emphasis on emerging concepts. *Rev Biol Trop* 1989;**37**:85–100.

22. Alongi DM, Sasekumar A. Benthic communities. In: Robertson AI, Alongi DM, editors. *Tropical mangrove ecosystems*. Washington, DC: American Geophysical Union; 1992. p. 137–72.

23. Odum WE, Heald EJ. Trophic analysis of an estuarine mangrove community. *Bull Mar Sci* 1972;**22**:671–738.

24. Ray S, Ulanowicz RE, Majee NC, Roy AB. Network analysis of a benthic food web model of a partly reclaimed island in the Sundarban mangrove ecosystem, India. *J Biol Syst* 2000;**8**:263–78.

25. Robertson AI, Duke NC. Mangrove fish-communities in tropical Queensland, Australia: spatial and temporal patterns in densities, biomass and community structure. *Mar Biol* 1990;**104**:369–79.

26. Morton RM. Community structure, density and standing crop of fishes in a subtropical Australian mangrove area. *Mar Biol* 1990;**105**:385–94.

27. Ronnback P, Troell M, Kautsky N, Primavera JH. Distribution pattern of shrimps and fish among *Avicennia* and *Rhizophora* microhabitats in the Pagbilao mangroves, Philippines. *Estuar Coast Shelf Sci* 1999;**48**:223–34.

28. Dall W, Hill BJ, Rothlisberg PC, Staples DJ. The biology of the Penaeidae. *Advances in marine biology*, vol. 27. London: Academic Press; 1990. p. 1–189.

29. Robertson AI, Blaber SJM. Plankton, epibenthos and fish communities. Alongi DM, editor. *Tropical mangrove ecosystems*. Washington, DC: American Geophysical Union; 1992. p. 173–324.

30. Alongi DM, Tirendi F, Clough BF. Below-ground decomposition of organic matter in forests of the mangroves *Rhizophora stylosa* and *Avicennia marina* along the arid coast of Western Australia. *Aquat Bot* 2000;**68**:97–122.

31. Hutchings P, Recher HF. The fauna of Australian mangroves. *Proc Linn Soc NSW* 1982;**106**:83–121.

32. Hutchings P, Recher HF. The faunal communities of Australian mangroves. In: Teas HJ, editor. *Biology and ecology of mangroves*, vol. 8. The Hague: W. Junk; 1983. p. 103–10.

33. Milward NE. Mangrove-dependent biota. In: Clough BF, editor. *Mangrove ecosystems in Australia*. Canberra: Australian Institute of Marine Science; 1982. p. 121–40.

34. Anderson DT. The reproduction and early life histories of the gastropods *Bembicium auratum* (Quoy and Gaimard) (Fam. Littorinidae) and *Cellana tremoserica* (Sower.) (Fam. Patellidae). *Proc Linn Soc NSW* 1962;**87**:62–8.

35. Reid DG. *The littorinid molluscs of mangrove forests in the Indo-Pacific region. The genus Littoraria*. London: British Museum (Natural History); 1986, Publication No. 978.

36. Houbrick RS. Systematic review and functional morphology of the mangrove snails *Terebralia* and *Telescopium*. *Malacologia* 1991;**33**:289–338.

37. Houbrick RS. Revision of higher taxa in the genus *Cerithidea* (Mesogastropoda: Potamididae) based on comparative morphology and biological data. *Am Malacol Bull* 1984;**2**:1–20.

38. Kamimura S, Tsuchiya M. The effect of feeding behaviour of the gastropods *Batillaria zonalis* and *Cerithiopsilla cingulata* on their ambient environment. *Mar Biol* 2004;**144**(4):705–12.

39. McGuinness KA. The climbing behaviour of *Cerithidea anticipata* (Mollusca: Gastropoda): the role of physical versus biological factors. *Aust J Ecol* 2006;**19**:283–9.

40. Wells FE. A review of the northern Australian species of the genera *Cassidula* and *Ellobium* (Gastropoda: Ellobiidae). In: Hanley JR, et al., editors. *The marine flora and fauna of Darwin Harbour, Northern Australia*. Proceedings of the 6th international marine biological workshop. Museums and art galleries of the northern territory; 1997. p. 213–29.

41. Metcalfe K. *The biological diversity, recovery from disturbance and rehabilitation of mangroves in Darwin Harbour, Northern Territory*. PhD Thesis, Charles Darwin University; 2007.

42. Wells FE. Comparative distributions of the mudwhelks *Terebralia sulcata* and *T. palustris* in a mangrove swamp in northwestern Australia. *Malacological Review* 1980;**13**:1–5.

43. Wells FE. An analysis of marine invertebrate distributions in a mangrove swamp in northwestern Australia. *Bull Mar Sci* 1983;**33**:736–44.

44. Davie PJF. The biogeography of littoral crabs (Crustacea: Decapoda; Brachyura) associated with tidal wetlands in tropical and sub-tropical Australia. In: Bardsley KN, Davie JDS, Woodroffe C, editors, *Coasts and tidal wetlands of the Australian monsoon region. Mangrove monograph No. 1*. North Australia Research Unit, Australian National University: Darwin, Northern Territory; 1985.

45. Kristensen E. Mangrove crabs as potential ecosystem engineers; with emphasis on sediment processes. In: Lee JSY, Duke NC, Connolly R, Davie PJF, Meziane T, Skilleter G, Dittman S, Burrows D, editors. Mangrove Macrobenthos meeting 11, Conference program. Abstract only, Cold Coast, Queensland; 2006, p. 36.

46. Lee SY. Ecological role of grapsid crabs in mangrove ecosystems: a review. *Mar Freshw Res* 1998;**49**:335–43.

47. George RW, Jones DS. A revision of the fiddler crabs of Australia (Ocypodidae: *Uca*). *Records of the Western Australian Museum, Supplement* 1982;**14**:1–99.

48. Von Hagen HO, Jones DS. The fiddler crabs (Ocypodidae: *Uca*) of Darwin Northern Territory, Australia. *Beagle* 1989;**6**:55–68.

49. Robertson AI. Leaf burying crabs: their influence on energy flow and export from mixed mangrove forests (*Rhizophora* sp.) in northeast Australia. *J Exp Mar Biol Ecol* 1986;**102**:237–48.

50. Robertson AI. Decomposition of mangrove litter in tropical Australia. *J Exp Mar Biol Ecol* 1988;**116**:235–47.

51. Robertson AI, Daniel PA. The influence of crabs on litter processing in mangrove forests in tropical Australia. *Oecologia* 1989;**78**:191–8.

6

Habitats and Biotic Assemblages of Intertidal Sandy and Muddy Shores

This chapter considers intertidal habitats that are characterized by soft sediment substrata. Like intertidal rocky shore habitats, there are major changes in sandy and muddy shores along the length of the North West Shelf relating to geology and climate.

For about half its length (North West Cape to Cape Leveque), the mainland coastline of the North West Shelf lies mainly over sedimentary terrain and the shores are predominantly sandy. In contrast, the northern part of the coastline (north of Cape Leveque) lies over Proterozoic igneous and metamorphic terrain and rocky shores predominate. In addition to this basic division of habitat type, "soft" substrate habitats also vary along the length of the shelf. Sediments that dominate shores overlying the major sedimentary basins (Section 2.1.4) are generally marine carbonates, while those on and adjacent to Proterozoic igneous and metamorphic shores of the Dampier Archipelago and the Kimberley are largely of terrigenous origin. Carbonate and terrestrial sediments provide soft substrate habitats of different characters. There has been no study of this factor in the geographic distribution of intertidal invertebrates in the region, but subjective observations indicate that it is a key determinant of distribution patterns for many species.

Superimposed over this geological environment, and also a modifier of it, there is a north-to-south gradient of climate along the length of the shelf from wet to arid. The north Kimberley is a region of moderately high rainfall. There are many rivers and estuaries that deliver their huge loads of fine sediment into the coastal zone. Consequently, muddy shores are prevalent and intertidal flats tend to be mud flats rather than sand flats. In the arid Pilbara, there are fewer large rivers and they flow episodically. They too deliver quantities of terrigenous sediment into the coastal zone but there is a dynamic relationship between that process and the reworking of inner shelf, mainly biogenic carbonate sediments.

This significant part of the coastal marine environment on the North West Shelf remains seriously understudied and little is known about the spatial distribution patterns of its biota. Although the taxonomy of the larger groups is moderately well established, there is no adequate inventory and very little information about invertebrate assemblages and ecology of soft substrate ecosystems. Until that situation improves, in the context of historical and ongoing environmental change, species distribution patterns within the region will remain difficult to interpret.

6.1 HISTORY OF RESEARCH

There is considerable knowledge of the biogeomorphology of intertidal flats on the mainland coast of the North West Shelf. Aspects of the geomorphology of mud flats in Cambridge Gulf and the estuary of the Ord River (Cambridge-Bonaparte Bioregion) have been described in some detail.[1-4] The stratigraphy, sedimentation, and Holocene history of tidal flats of the Kimberley coast, King Sound, and the Canning coast also have been described.[5-13]

There have been two studies of mud flat molluscs associated with mangals in the Kimberley Bioregion.[14,15] Intertidal sand and mud flat molluscan, echinoderm and larger crustacean faunas of the Pilbara are moderately well known taxonomically and have been listed in Western Australian Museum publications for the Montebello Islands[16] and the Dampier Archipelago.[17] Intertidal invertebrate taxa of soft substrate communities have been sampled during ecological studies in the Eighty Mile Beach in the Canning Bioregions (see below).

Wells[18,19] described molluscan assemblages and species' densities and biomass on a seaward mudflat in Exmouth Gulf. Semeniuk[11] described Recent and Quaternary fossil mud flats assemblages in the Canning Bioregion. The most substantial contribution to knowledge of the North West Shelf coastal soft substrate fauna is an ongoing international study, coordinated by the Royal Netherlands Institute for Sea Research (RNISR), on the vast mudflats of Roebuck Bay and Eighty Mile Beach in the Canning Bioregion.[20-27]

Across-shore and along-shore gradients in the distribution of sediments and in the composition of benthic assemblages (examined at family level) of the very wide intertidal zone at Eighty Mile Beach show that the sediments at this locality are mostly mud (less than 63 μm) and very fine sand (63-125 μm) with coarser sediments nearshore and finer sediments near low tide level.[25] Both benthic diversity and abundance are highest in the finer sediments in the lower part of the tidal range but there are significant differences between sites along the length of the beach.

Bivalve diversity and total benthic diversity are associated with sediment heterogeneity at Roebuck Bay and Eighty Mile Beach and significantly higher than at higher latitude European localities.[26] Diversity of bivalves is higher in finer-grained sediments, but this correlation does not hold for total benthic diversity. Amphipods, for example, are more diverse in coarser sediments. The RNISR study showed that bivalve and total benthic diversity is not associated with sediment heterogeneity. Instead, most species, bivalves in particular, shared a large degree of distributional overlap.[27]

6.2 SANDY SHORES

On the ria-like coasts of the Dampier Archipelago and in the Kimberley Bioregion, where the geology is igneous and metamorphic, beaches that occur between rocky headlands are mostly steep and short and often include sediments of terrestrial origin (Figure 6.1). Intertidal sand flats are common in bays but they are not as extensive as they are on the shores of the western Pilbara, Eighty Mile Beach, Canning, and King Sound

FIGURE 6.1 A short beach between lateritic headlands on the north-eastern shore of North Maret Island, Kimberley Bioregion. Note the recent high tide mark high on the beach slope and the vegetated berm between the crest of the beach and the foredunes that is turtle nesting habitat. *Photo: Barry Wilson.*

FIGURE 6.2   Eastern part of the Ashburton River Delta, Onslow coast, with a long sand barrier protecting the lagoons and mangrove forests of the delta. This is a low-energy, tide-modified shore. The beach is highly mobile, the sand transport being predominantly eastward during the summer season of westerly prevailing winds but reverses in winter. Probably partly because of this mobility, the beach fauna is sparse and species-poor. (Present mouth of the Ashburton at the top of the picture.) *Photo: Barry Wilson.*

Bioregions, where the geology is sedimentary and the sediments are predominantly marine carbonates. In those bioregions, there are long, uninterrupted stretches of beach (Figure 6.2); the adjacent seabed is often gently sloping and fronted by wide intertidal sand flats. In the Canning and Eighty Mile Beach Bioregions where the tidal range may exceed 10 m, the sand flats may be more than a kilometer wide. Many are densely populated with diverse invertebrates, fishes and feeding shore birds and forays out to the lowest tide margin is a wonderful experience, but it can be very dangerous as the flooding tide comes in so quickly.

Western Australian beaches, including those along the shores of the North West Shelf, have been described and categorized.[28–30] For the most part, North West Shelf beaches occur in areas of high to very high tidal range and are considered to be low-energy shores that are *tide-modified* (when exposed to prevailing seas) or *tide-dominated* (when protected from wave action). They vary greatly longshore in response to local and seasonal wave conditions and the effects of variable geomorphic features.

McLachlan[31] published a detailed review on the ecology of beaches. Biotic zonation of beaches and sandy shores has been intensively studied internationally. Some general principles are evident that are common throughout the world although there is little agreement in detail from one region to another. Most authors recognize distinct supralittoral beach, upper-littoral beach slope, and mid- to lower-littoral tidal flat zones and that basic division of sandy shore habitats will be followed here.

6.2.1 Supralittoral Beach

Above the high water mark on the beach slope, there is generally a flat area of beach that is sometimes washed by wave action. These berms are often vegetated by salt-tolerant terrestrial vegetation, such as spinifex and grasses (Figure 6.1). It is a terrestrial habitat but two conspicuous marine animals make use of it. Ghost crabs (Ocypodidae) that burrow on the beach slope often forage into this zone. Marine turtles make seasonal use of it for egg-laying purposes. This second activity has high conservation significance, but otherwise the supralittoral zone of beaches has little relevance to marine biogeography.

6.2.2 Beach Slope

Beaches of tide-modified and tide-dominated shores, like most of those of the North West Shelf, tend to be rather steep and there is usually a distinct change in slope where the beach and the tidal flats meet (Figure 6.3). The position of the break of slope in relation to mean sea level varies according to factors like the degree of wave action. However, it marks an abrupt change of biota and it is useful to treat it as a key level on the shore. In this study, the break of slope on sandy shores is interpreted as the boundary of the upper-littoral and mid-littoral zones.

The beach slope supports two different ecosystems. There is a modest macroinvertebrate fauna and a species-rich, interstitial meiofauna that may have significant biomass. The latter is unstudied in the region and

FIGURE 6.3   South-western shore of Pender Bay, Dampier Peninsula, Canning Bioregion at low tide. The steep beach is about 50 m wide. At the break of slope, there is an abrupt change of sediment from the coarse, smooth, white, shelly sand of the upper-littoral slope to the darker, finer, sand of the inner sand flat. The intertidal sand flats along this shore are very wide and species-rich. *Photo: Barry Wilson.*

will not be considered here. The macroinvertebrate and meiofaunas do not exchange much trophic energy; both depend on imported organic food, either as particulate matter delivered by the sea or terrestrial detritus carried to the beach by wind or water flow from streams. There is little primary production in this zone.

The strand line, high on the beach slope, with bands of decaying marine algae, is often rich in amphipod and isopod crustaceans but these have not been studied in the region. Often, the most conspicuous animals on the beach slope are ghost crabs of the genus *Ocypode*. Four species are known from the region, the most common being *O. ceratopthalmus* which occurs on beaches facing open sea.[32] A slightly more robust species *O. fabricii* (Figure 5.17) is more common on beaches in sheltered embayments, often associated with mangal habitats. These very active animals forage both up into the terrestrial habitats of the supralittoral zone and down onto the tidal flats of the mid-littoral.

Another crustacean that may be present in large numbers on the lower beach slope is the Bubbler Crab (family Dotillidae). The species *Scopimera kochi* is recorded from the region but there appears to be a second, undescribed species in the Pilbara (A. Hosie, personal communication, W.A. Museum). These small crabs are responsible for the distinctive burrows surrounded by radiating rows of sand globules that are such a feature of along the smooth, lower part of the beach slope, and also on higher parts of the inner mid-littoral sand flats.

At the base of the beach slope, in the coarse-grained sand immediately above the fine-grained sand flats, there are commonly two or three infaunal bivalves—*Donax faba*, *Paphies* (*Actodea*) *striata*, and *Paphies altenai*. On tide-dominated beaches in sheltered situations, the venerid *Placamen berryi* also may be abundant in this habitat. These animals are suspension feeders and rely on particulate matter delivered to them by wave action. (They are easily accessible for much of the daily tidal cycle even during neap tides and, where populations of them are present, they may be useful for pollution monitoring programs.)

At the very bottom of the beach slope, there may also be several predatory gastropod species. Almost ubiquitous are the naticids *Polinices* cf. *conicus* and *Natica gualtieriana*, both made conspicuous by their meandering tracks in the sand as they hunt for food (bivalves or polychaetes).

6.2.3 Mid- to Lower-Littoral Sand Flat

Intertidal sand flats are exceptionally well developed on shores of the North West Shelf, especially in the Pilbara, Eighty Mile Beach, and Canning Bioregions. They support species-rich but highly variable and patchy infaunal and epifaunal invertebrate communities made up predominantly of burrowing molluscs, polychaetes, echinoderms, and crustaceans. These faunal assemblages include a high proportion of regionally endemic species.

Intertidal sand flat habitats are heterogeneous, complex environments where benthic species diversity is associated with physical habitat structure, that is, sediment size and stability, organic content, and tidal level (exposure time). Invertebrate inhabitants may modify habitats so that biotic factors are also involved. Invertebrate communities are composed primarily of suspension and deposit feeders and their predators. There may be photosynthetic activity by seagrasses (mainly *Thalassia hemprichti*, *Halophyla ovalis*, and *Halodule uninervis*), green algae such as species of the genera *Caulerpa* and *Halimeda* and other macroalgae and microorganisms on the sediment surface. Nevertheless, sand and mud flat communities are essentially secondary production communities where trophic systems depend mainly on import of organic planktonic materials from the ocean and organic detritus from the land.

6.2.3.1 *Nearshore Sand Flats*

The sediments of sand flats of the inner mid-littoral zone, immediately below the break of slope of the beach, may be ripple-marked by tidal flow and wind (Figure 6.3) or relatively smooth except for bioturbation (Figures 6.4 and 6.5). In this zone, the sediment tends to contain a mud or silt fraction and a relatively high organic content. Some of the species

FIGURE 6.4   Mid-littoral sand flats at the entrance to Port Smith, Canning Bioregion. The carbonate sand is fine grained and well sorted and supports a diverse invertebrate infauna. Note the notched Pleistocene limestone upper-littoral bench in the background. *Photo: Sue Morrison, W.A. Museum.*

FIGURE 6.5   Nearshore muddy sand flats (inner mid-littoral level) on the lee side of Dixon Island, Pilbara (nearshore) Bioregion. This habitat has high biomass and a species-rich infauna of microorganisms and invertebrates. *Photo: Barry Wilson.*

found in the cleaner, better-sorted sand of the outer flats are also found here but there are many invertebrates that are confined to this zone, including many detrital-feeders.

Soldier Crabs (family Mictyridae) may be conspicuous in vast numbers when they emerge from their burrows at low tide and move about on the sand flats like massed armies. These are also deposit feeders. There is at least one species, *Mictyris* sp., in the region but its species identity remains uncertain.

Suspension-feeding and deposit-feeding bivalves generally feature strongly in the infauna of muddier nearshore flats. They include several species of the family Lucinidae, a family characterized by tolerance to anaerobic conditions and high sulfide levels that harbor symbiotic sulfur-oxidizing bacteria in their gills. The families Ungulinidae, Tellinidae, Cardiidae, Veneridae, and Pholadidae are also usually well represented. Some common bivalves typically found on muddier nearshore sand flats throughout the region are listed in Table 6.1.

Several predatory gastropods are usually very common on the sand flats close to shore immediately below the break of the beach slope. The naticids *Polinices* cf. *conicus* and *Natica gualteriana* are very active, both on the inner mid-littoral flat and the base of the beach slope above, buried in the sand and leaving long erratic trails as they move about in search of their prey. Three buccinids, *Cominella acutinodosa* (Figure 6.6B), *Nassarius dorsatus* (Figure 6.6A), and *N. clarus*, are also active at low tide crawling on the surface. Also commonly abundant in this habitat are detrital-feeding

TABLE 6.1   Common Bivalves of Slightly Muddy, Nearshore Sand Flats (Inner Mid-Littoral Zone) of North West Shelf Mainland and Coastal Island Shores

Anodontia edentula	*Anodontia pila*
Ctena bella	*Divaculina cumingi*
Divaricella ornata	*Felaniella* spp.
Cardium unedo	*Tellina capsoides*
Tellina staurella	*Tellina virgata*
Anomalocardia squamosa	*Gafrarium tumidum*
Pitar citrinus	*Placamen berryi*
Tapes variegatus	*Pholas* sp.

FIGURE 6.6   Predatory buccinids common to the sand flats: (A) *Nassarius dorsatus* and (B) *Cominella acutinodosa. Photos: Barry Wilson.*

gastropods including the small columbellid *Mitrella essingtonensis* and a tiny trochid, *Isanda coronata,* two large detrital-feeding ceriths, *Pseudovertagus aluco* and *Rhinoclavus vertagus,* and a scaphopod. About half of the gastropods inhabiting this zone are endemic to the North West Shelf and the Gulf of Carpentaria.

Infaunal polychaetes are diverse and abundant in this intertidal sand flat habitat but they have not been studied in the region. One very common errant polychaete, active on the wet surface of the sand at low tide, is a bright green carnivorous phyllodocid, one of many of its class that remain to be identified. Tubiculous polychaetes may be conspicuous also with the tops of their tubes emergent from the sand. Echinoderms are not prevalent on the inner sand flats although the small sand-dollar *Peronella orbicularis* may be abundant and the asteroid *Astropecten granulatum* may also be present preying on small bivalves.

6.2.3.2 Mid- and Lower-Littoral Sand Flats

On sand flats of the outer mid- and lower-littoral zones, a little further from shore, the substrate tends to be medium to coarse grained, well aerated, and mobile. They are often strongly rippled by tidal flows sometimes forming large sand waves, especially along the margins of drainage channels (Figures 6.7 and 6.8). These flats tend to have moderately high species diversity but a low infaunal biomass. They are usually structurally complex communities comprising secondary production assemblages based on suspension and deposit feeding.

Molluscs, especially bivalves, are a significant part of the biodiversity of the macroinvertebrate fauna of mid- and lower-littoral sand flats (Table 6.2). Many of the species are also found on soft substrates of the sublittoral zone although there are some that are strictly intertidal. The majority of sand flat

FIGURE 6.7   Sand waves on an outer (lower-littoral) sand flat at Cape Borda, Dampier Peninsula, Canning Bioregion. The sand is moderately coarse grained and mobile. These habitats support diverse infaunal assemblages of invertebrates but biomass is generally low. *Photo: Barry Wilson.*

FIGURE 6.8   Sand waves of coarse sand exposed at low tide on the side of a fast-flowing drainage channel; lee side of Dixon Island, Nickol Bay, Pilbara (Nearshore) Bioregion. In this situation, the sand is highly mobile and reworked with each tide cycle. *Photo: Barry Wilson.*

bivalves are active burrowers capable of maintaining their position in an unstable substrate. Bivalve families that are particularly well represented on sand flats of the region include suspension and detrital-feeding shallow burrowers with short siphons (e.g., Cardiidae and Veneridae) and deep burrowers with long, divided siphons (e.g., Mactridae, Tellinidae, and Psammobiidae) (Figure 6.9A and B).

Byssate bivalves are not a conspicuous element in these habitats, except where there is a significant rubble component in the substrate or patches of rock. Exceptions are razor clams and hammer oysters of the families Pinnidae (*Atrina vexillum*, *Pinna bicolor*, *P. deltodes*) and Malleidae (*Malleus alba*, *M. malleus*) that live fixed lives attached to objects buried deep in the sediment. A noteworthy species is the regionally endemic clavagellid bivalve *Brechites australis*. This remarkable animal has a thick tubular shell with a colander-like cap on the base and lives vertically in the sand attached to rubble.

Most gastropods that live in this habitat are active predators that prey on polychaetes or bivalves but there are four common detrital-feeding species on sand flats of the region are the strombs *Strombus campbelli*, *S. urceus*, and *S. vomer iredalei* and the cerith *Rhinoclavis fasciatus*. Common gastropod predators include representatives of the families Naticidae, Olividae, Muricidae, Nassaridae Turbinellidae, Volutidae, Conidae, and Terebridae. The prey of most of these species is unknown. More than half of them are endemic to north-western Australia (Table 6.3). A selection of gastropod

TABLE 6.2 Common Bivalves on Mid- and Lower-Littoral Sand Flats in the Pilbara (Nearshore) Bioregion

Species	Lower-Littoral, Clean Sand	Mid-Littoral, Muddy Sand
Anadara antiquata	1	1
Anadara crebricostata	1	–
Trisidos semitorta	1	–
Trisidos tortuosa	1	–
Glycymeris persimilis	1	–
Atrina vexillum	1	1
Pinna bicolor	1	1
Pinna deltodes	1	1
Malleus alba	1	1
Malleus malleus	1	1
Anodontia edentula	–	1
Anodontia pila	–	1
Ctena bella	–	1
Divaculina cumingi	1	1
Divaricella ornata	1	1
Felaniella sp.	–	1
Fimbria sowerbyi	1	–
Acrosterigma dupuchense	1	–
Acrosterigma reeveanum	1	–
Acrosterigma fultoni	1	–
Fragum erugatum	1	–
Fragum unedo	1	1
Lunulicardium hemicardium	1	–
Lunulicardium retusum	1	–
Fulvia aperta	1	–
Hemidonax arafurensis	1	–
Hemidonax donaciformis	1	–
Lutraria australis	1	–
Mactra antecedens	1	–
Mactra explanata	1	–
Mactra grandis	1	–
Mactra incarnata	1	–
Mactra luzonica	1	–
Mactra olorina	1	–
Mactra sericea	1	–
Mactra westralis	1	–
Spisula coppingeri	1	–
Spisula triangularis	1	–
Meropesta nicobarica	1	–
Anomalocardia squamosa	–	1
Antigona chemnitzii	1	–
Callista impar	–	1
Callista planatella	1	–
Circe scripta	1	–
Dosinia deshayesi	1	–
Dosinia scalaris	1	–
Gafrarium tumidum	1	1
Gomphina undulosa	1	–
Pitar citrinus	1	–
Placamen berryi	–	1
Placamen tiara	1	–
Tapes literatus	1	1
Tapes variegatus	–	1
Sunetta contempta	1	–
Exotica assimilis	1	–
Tellina capsoides	–	1
Tellina inflata	1	–
Tellina radians	1	–
Tellina rostrata	1	–
Tellina piratica	–	1
Tellina staurella	1	1
Tellina virgata	–	1
Brechites australis	–	1
Pholas sp.	–	1
Laternula creccina	–	1

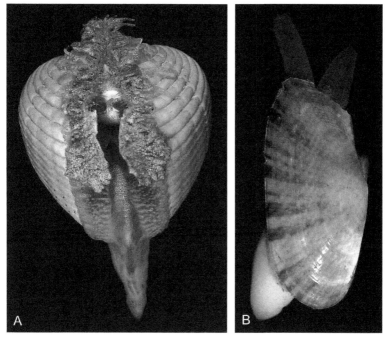

FIGURE 6.9   (A, B) Two examples of suspension-feeding, sand flat bivalves: (A) *Fragum unedo* (Cardiidae) and (B) *Gari amyethystus* (Psammobiidae). The cardiids are shallow burrowers with short siphons and they draw water and suspended food particles into the mantle cavity through a wide incurrent gap. The psammobiids are deep burrows with long, divided siphons. *Photos: Barry Wilson.*

TABLE 6.3   Common Gastropods of Intertidal Sand Flat Habitats That Are Endemic to the North West Shelf Bioregions

		Kimberley	Canning	Pilbara (Offshore)	Pilbara (Nearshore)
Strombus campbelli	D	1	1	1	1
Strombus urceus	D	1	–	–	–
Strombus vomer iredalei	D	–	1	1	1
Ficus eospila	P	1	1	1	1
Alocospira rosea	P	1	1	1	1
Ancillista cingulata	P	–	–	1	1
Ancillista muscae	P	–	–	1	1
Murex acanthostephes	P	–	1	1	1

Continued

TABLE 6.3 Common Gastropods of Intertidal Sand Flat Habitats That Are Endemic to the North West Shelf Bioregions—cont'd

		Kimberley	Canning	Pilbara (Offshore)	Pilbara (Nearshore)
Murex macgillivrayi	P	1	1	1	–
Pterynotus acanthopterus	P	–	–	–	–
Syrinx aruanus	P	–	–	–	–
Ancillista muscae	P	–	–	–	–
A. cingulata	P	–	–	–	–
Oliva australis	P	1	1	1	1
Oliva caldania	P	1	1	1	1
Oliva lignaria	P	–	–	–	–
Amoria damoni	P	–	1	1	1
Amoria dampieria	P	–	–	1	1
Amoria ellioti	P	–	–	–	1
Amoria grayi	P	–	1	1	1
Amoria jamrachi	P	–	1	1	1
Amoria macandrewi	P	–	–	1	–
Amoria praetexta	P	1	1	1	1
Amoria turneri	P	1	–	–	–
Cymbiola oblita	P	–	1	1	1
Melo amphora	P	1	1	1	1
Conus reductaspiralis	P	–	1	1	1
Conus thevenardensis	P	–	–	1	1
Duplicaria australis	P	1	1	1	1
Duplicaria crakei	P	–	1	1	1
Duplicaria duplicata	P	1	1	1	1
Duplicaria jukesi	P	–	–	1	1
Hastula rufopunctata	P	–	–	1	1

D, detrital-feeder; P, predator.

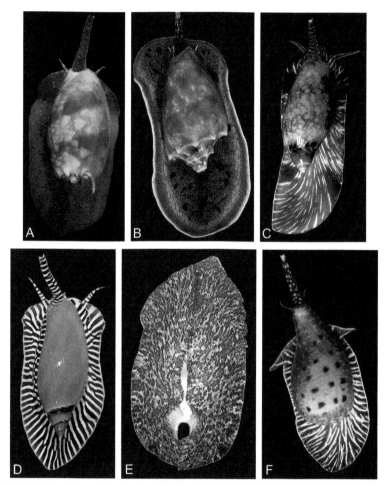

FIGURE 6.10   (A–F). Some Predatory Gastropods of Intertidal Sand Ecosystems on the North West Shelf. (A) *Melo amphora* (Volutidae), endemic to northern Australia. (B) *Cymbiola oblita* (Volutidae), endemic to the North West Shelf. (C) *Amoria damoni* (Volutidae), endemic to the North West Shelf and Gulf of Carpentaria. (D) *Amoria grayi* (Volutidae), endemic to the North West Shelf and the West Coast. (E) *Ancillista cingulata* (Olividae), endemic to the North West Shelf and the West Coast. (F) *Ficus eospila* (Ficidae); endemic to the North West Shelf. The baler shell, *Melo amphora* preys on other molluscs. Ficids generally prey on echinoderms. Nothing is known of the prey of the other species illustrated. *Photos: Barry Wilson.*

predators found in sand flat habitats on the mainland and coastal island shores of the North West Shelf are illustrated in Figure 6.10A–F.

A diverse assemblage of crustaceans lives in these sandy habitats. The taxonomy of the abundant small crustaceans that are an important element in the sand flat fauna remains poorly studied. Larger species, especially brachyuran crabs, are diverse and their taxonomy is moderately

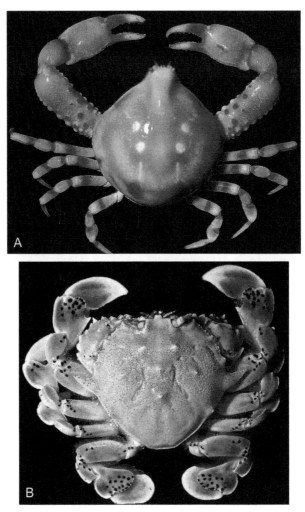

FIGURE 6.11   (A, B). Two of the common sand flat crabs of mainland and coastal island shores of the North West Shelf. (A) *Ashtoret lunaris* and (B) a pebble crab, *Leucosia* sp. *Photos: Barry Wilson.*

well known, but the trophic roles they play are not recorded. Two common crabs of North West Shelf mainland and coastal sand flat habitats are illustrated (Figure 6.11).

Errant polychaetes are an important element in this habitat but they remain unstudied within the region.

There are many echinoderm species living on the sand flats, mainly deposit feeders and predators. The asteroids *Luidia maculata*, *Astropecten granulatus*, *A. polyacanthus*, *A. sumbawanus*, and *A. vappa* are predators that crawl on the sand surface and feed on small bivalves and perhaps other invertebrates. It is evident in the field that these sea stars live in different parts of the sand flats but their individual habitat preferences are as yet undescribed. Detritivorous sand dollars *Peronella lesueuri, P.* cf. *macroproctes, P. orbicularis* and *P. tuberculata* and *Echinolampus* live on the sand surface. The irregular echinoids *Lovenia elongata, Breynia desori, Schizaster compactus,* and *Nacospatangus interruptus* are also detritivorous but burrow and are seen at the surface only when they emerge at low tide. There are also banana-shaped sand-burrowing holothurians that live in U-shaped burrows (e.g., *Paracaudina* cf. *chilensis*) and some burrowing ophiuroids such as *Amphiura tenuis.* A selection of sand flat echinoderms is illustrated in Figure 6.12.

6.3 MUDFLATS

Upper-littoral mud flats behind the mangrove tree line are a feature of the arid and semi-arid coastline of the North West Shelf's southern bioregions. Typically, they support huge populations of burrowing crabs (Chapter 5) but very little other invertebrate fauna.[14] However, seaward mudflat habitats, that is, in the mid- and lower-littoral zones in front of the mangrove line (Figure 6.13), generally support high biomass and diverse infaunal assemblages. These habitats are a feature of sheltered bays and estuaries in the Kimberley and northern Canning Bioregions but much less significant south of Cape Bossut. In this section, what little is known of invertebrate assemblages of mudflats seaward of the mangrove boundaries is considered.

As part of a study of a mangal, Wells[18,19] described the molluscan assemblages and species' densities and biomass, on a seaward mudflat in Exmouth Gulf. The assemblages included species also found in sandflat habitats lower in the tidal zone or in the mangrove tree zone above. Nevertheless, in this format, they comprise a recognizable community of species that is repeated in similar situations throughout the Pilbara (nearshore) and Canning Bioregions. The majority of the species are deposit-feeders but there are some suspension-feeding bivalves and predatory gastropods. There is little freshwater surface runoff at this locality and hence little organic material of terrestrial origin enters this ecosystem. The moderate biomass of the seaward mudflat may be dependent on organic material derived mainly from the adjacent mangrove and the sea.

The most intensive study of intertidal mud flat invertebrate assemblages in the region is the ongoing international project coordinated by the Royal Netherlands Institute for Sea Research (RNISR), on the vast mudflats of Roebuck Bay (Figure 5.6) and Eighty Mile Beach in the Canning Bioregion. The benthic invertebrate fauna of these mudflats comprise a diverse ecosystem of immense scientific interest as it provides the rich

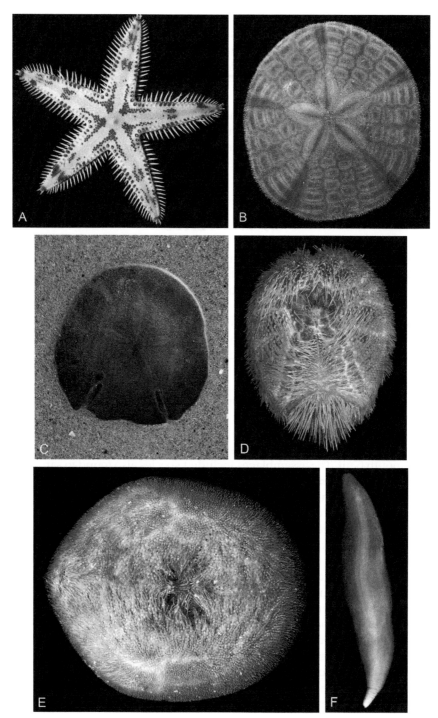

FIGURE 6.12   (A–F) A selection of sand flat echinoderms: (A) *Astropecten* cf. *vappa*; (B) a biscuit urchin, *Peronella orbicularis*; (C) *Echinodiscus auritus*; (D) *Schizaster compactus*; (E) *Echinolampus ovata*; (F) *Paracaudina* cf. *chilensis*. Photos: Barry Wilson.

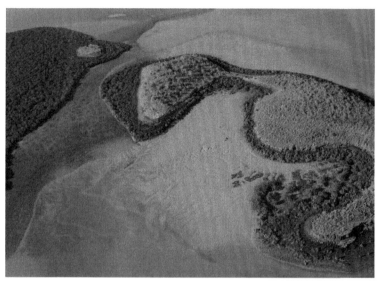

FIGURE 6.13   An intertidal mud flat in front of a fringing mangal on the shore of a small rocky island in Talbot Bay on the Yampi Peninsula, Kimberley Bioregion. The fine surface sediment of mud flats of this kind is subject to resuspension during periods of strong tides but may support high biomass of microorganisms and invertebrates. Note the high rock platform on the shore of the adjacent island (left side). Photo: Barry Wilson.

food resource that supports the migratory wader habitat for which the area was designated a Ramsar Wetland.

The infauna of the flats in Roebuck Bay is diverse and seasonally abundant with high densities of bird-food species such as polychaetes and very small molluscs and crustaceans.[24] For example, the tiny gastropods *Isanda coronata* and *Salinator burmana* may be present at densities up to 2500/m^2. As this is the food resource that sustains the extraordinary annual migrations of the birds between Australian and northern Asia, these sites and their ecology have international significance of the highest degree.

The wide tidal flats of Eighty Mile Beach and Roebuck Bay, characterized by high tidal range, low wave action, and mud to very fine-grained sand sediments, are representative of open tidal flat systems with high mollusc densities.[27] It is yet to be established whether these habitats and the species-rich invertebrate assemblages they support are representative of North West Shelf shores, or peculiar to the Canning and Eighty Mile Beach Bioregions. Preliminary indications are that the invertebrate communities of these intertidal mudflats are probably typical of the region faunistically but exceptional in terms of key species' densities and biomass.

At Eighty Mile Beach, predominantly a mud and muddy sand flat, the RNISR found that the four invertebrate taxa most numerically abundant were bivalves, polychaetes, ophiuroids, and crustaceans but the relative importance of these groups varied between the five stations studied.[24] One species, the ophiuroid *Amphiura tenuis*, was so consistently abundant in the lower-littoral, thus this was referred to as an "ophiuroid zone." Also in areas where the largest fraction in the sediment was in the 63–125 μm range, bivalves were most abundant, especially *Siliqua pulchella* and *Tellina amboynensis*.

Mollusc and other invertebrate species and the range of sediment size observed in their habitats at the RNISR Roebuck Bay and Eighty Mile Beach study sites have been listed.[23,25,27] Tables 6.4 and 6.5 of this report are compiled from these sources. This is unlikely to represent the entire molluscan fauna of these localities but is an excellent indication of the assemblages present.

TABLE 6.4   Echinoderm Species Collected at the RNISR Study Sites of Ref. 24 on the Muddy Tidal Flats of Roebuck Bay (Identifications by L.M. Marsh, Western Australian Museum)

Ophiroidea

   Amphiura catephes

   Amphiura (Ophiopeltis) tenuis

   Amphiura sp.

   Amphioplus (Lymanella) depressus

   Amphioplus sp.

   Ophiocentrus verticillatus

   Dictenophiura stellata

   Phiocnemis marmorata

Asteroidea

   Astropecten granulatus

   Astropecten monacanthus

Echinoidea

   Peronella tuberculata

Holothuroidea

   Protankyra verrelli

   Paracaudina chilensis

   Leptopentacta grisea

TABLE 6.5 Mollusc Species Collected at the RNISR Study Sites of Refs. 24 and 28 on the Muddy Tidal Flats of Roebuck Bay and Eighty Mile Beach

Species	Roebuck Bay	Eighty Mile Beach
Bivalves		
Ledella sp.	1	1
Nucula cf. *astricta*	1	–
Solemya cf. *terraereginae*	1	–
Anadara granosa	1	–
Modiolis micropteris	1	–
Anodontia bullula	1	1
Anadontia omissa	1	–
Divaricella irpex	1	1
Cardiolucina pisiformis	–	1
Ctena sp.	1	–
Heterocardia gibbosula	1	1
Pseudopythina macrophthmensis	–	1
Galeomna spp.	1	–
Scintilla sp.	1	–
Mactra abbreviata	1	–
Mactra grandis	1	–
Corbula sp.	1	–
Siliqua pulchella	1	1
Ensis sp.	1	–
Cultellus cultellus	1	–
Solen sp.	–	1
Donax cuneatus	1	1
Paphies cf. *altenai*	–	1
Tellina amboynensis	1	1
Tellina capsoides	1	–
Tellina cf. *exotica*	1	–
Tellina inflata	–	1
Tellina iridescens	1	–
Tellina piratica	1	–
Tellina cf. *serricostata*	1	–
Tellina aff. *tenuilirata*	1	1
Tellina sp. A (oval)	1	–
Tellina sp. B (mud tellin)	1	–
Tellina sp. C	1	–
Theora fragilis	–	1
Gari lessoni	1	–
Anomalocardia squamosa	1	–
Placamen calophyllum	1	–

Continued

TABLE 6.5   Mollusc Species Collected at the RNISR Study Sites of Refs. 24 and 28 on the Muddy Tidal Flats of Roebuck Bay and Eighty Mile Beach—cont'd

Species	Roebuck Bay	Eighty Mile Beach
Placamen gravescens	1	–
Placamen gilva	1	–
Gafrarium dispar	1	–
Tapes sp.	1	–
Corbula sp.	1	–
Laternula creccina	1	–
Gastropods		
Clanculus sp.	1	–
Isanda coronatus	1	–
Cerithidea cingulata	1	–
Polinices conicus	1	–
Natica sp.	1	–
Nitidella essingtonensis	1	–
Zafra sp.	1	–
Nassarius bicallosum	1	–
Nassarius dorsatus	1	–
Oliva australis	1	–
Vexillum radix	1	–
Haminoae sp.	1	–
Acteon sp.	1	–
Tornatina sp.	1	–
Salinator burmana	1	–
Leucotina sp.	1	–
Syrnola sp.	1	–
Scaphopods		
Laevidentalium cf. *lubricatum*	1	–
Dentalium cf. *bartonae*	1	–

Other invertebrate taxa from the Roebuck Bay sites included 14 species of echinoderm—eight ophiuroids, two asteroids, one echinoid, and three holothurians. These animals were identified by echinoderm specialist Loisette Marsh and the list is reproduced here (Table 6.4). This collection is probably the only account of intertidal mud flat echinoderms on the North West Shelf and some of these species clearly play key roles in the ecosystem. Brittlestars "were encountered at over half the sampling stations ... the short-arm brittlestar *Dictenophiura stellata* living on the sediment surface, unlike the *Amphiura* species that live deeply buried in the sediment with their long brittle arms stretching to the surface to catch food particles." *Dictenophiura* were found only on the lower flats. *Amphiura tenuis* and *A. catephes* were found on the higher flats and were much more wide-spread around the bay.[23]

The polychaetes of this study were identified only to family level. Quantitative core-sample data showed densities of polychaetes of the family Owenidae to be as high as 4000 individuals/m^2—rich pickings for wader birds. The crustaceans were also mostly identified to family. Although these data are not adequate for any biogeographic conclusions (because there are no comparable data for other locations), they do demonstrate the richness of the Roebuck Bay benthic fauna.

Some invertebrates found on the mud flats of Roebuck Bay are especially noteworthy, because they have not been recorded previously on intertidal flats of the North West Shelf. A small sipunculid worm was found widespread on the lower flats with densities in the core samples as high as 400-500/m^2. These animals live in burrows in the mud and feed on surface organic material. Other species of sipunculid were also found in other microhabitats on the mud flats. Also common were the mud-burrowing brachiopod *Lingula* sp., carpets of surface dwelling "solitary" tunicates, three kinds of anemones, nemertines, phoroinids, the enteropneustan *Balanoglossus* sp., some marine oligochaete worms, some terrestrial choronomid larvae, a pycnogonid, and the agnathan "lancelet fish" *Amphioxus* sp.

The RNISR study is the first and so far only account of an intertidal mud flat community of this kind in northern Western Australia. Similar tidal flat systems occur in the open gulfs of the north Kimberley, for example, near the W.A./N.T. border at the head of Joseph Bonaparte Gulf, but their mid- and lower-littoral mud flat faunas remain unstudied. A challenge for future biologists will be to determine whether the rich mud flat assemblage of Roebuck Bay is unique or widespread on muddy shores of the North West Shelf.

References

1. Wright LD, Coleman JM, Thom BG. Emerged tidal flats in the Ord River Estuary, Western Australia. *Search* 1972;**3**:339–41.
2. Wright LD, Coleman JM, Thom BG. Processes of channel development in a high tide-range environment: Cambridge Gulf-Ord River delta, Western Australia. *J Geol* 1973;**81**:15–41.
3. Thom BG, Wright LD, Coleman JM. Mangrove ecology and deltaic-estuarine geomorphology; Cambridge Gulf-Ord River, Western Australia. *J Ecol* 1975;**63**:203–32.
4. Wolanski E, Moore K, Spagnol S, D'Adamo N, Pattiaratchi C. Rapid, human-induced siltation of the Macro-Tidal Ord River Estuary, Western Australia. *Estuar Coast Shelf Sci* 2001;**53**(5):717–32.
5. Cresswell ID, Semeniuk V. Mangroves of the Kimberley region: ecological patterns in a tropical ria coast setting. *J R Soc West Aust* 2011;**94**(2):213–37.
6. Mathews D, Semeniuk V, Semeniuk CA. Freshwater seepage along the coast of the western Dampier Peninsula, Kimberley region, Western Australia. *J R Soc West Aust* 2011;**94**(2):207–12.

7. Semeniuk V. Quaternary stratigraphy of the tidal flats, King Sound, Western Australia. *J R Soc West Aust* 1980;**63**(3):65–78.

8. Semeniuk V. Sedimentology and the stratigraphic sequence of a tropical tidal flat, north-western Australia. *Sediment Geol* 1981;**29**:195–221.

9. Semeniuk V. Geomorphology and Holocene history of the tidal flats, King Sound, north-western Australia. *J R Soc West Aust* 1982;**65**(2):47–68.

10. Semeniuk V. Development of mangrove habitats along ria shorelines in north and north-western tropical Australia. *Vegetatio* 1985;**60**:3–23.

11. Semeniuk V. Holocene sedimentation, stratigraphy, biostratigraphy, and history of the Canning Coast, north-western Australia. *J R Soc West Aust* 2008;**91**(1):53–148.

12. Semeniuk V. Stratigraphic patterns in coastal sediment sequences in the Kimberley region: products of coastal form, oceanographic setting, sedimentary suites, sediment supply, and biogenesis. *J R Soc West Aust* 2011;**94**(2):133–50.

13. Semeniuk V, Brocx M. King Sound and the tide-dominated delta of the Fitzroy River: their geoheritage values. *J R Soc West Aust* 2011;**94**(2):151–60.

14. Wells FE, Slack-Smith SM. Zonation of molluscs in a mangrove swamp in the Kimberley, Western Australia, In: *Biological survey of Mitchell Plateau and Admiralty Gulf, Kimberley, Western Australia, Western Australian Museum, Perth*; 1981. p. 265–74.

15. Hanley JR. Quantitative survey of mangrove invertebrates, Part 7. In: Wells FE, Hanley JR, Walker DI, editors. *Survey of the marine biota of the Southern Kimberley Islands, Western Australian Museum, Perth*. Unpublished report no. UR 286; 1995. p. 82–100.

16. Berry PF, Wells FE. Survey of the marine fauna of the Montebello Islands and Christmas Island, Indian Ocean. *Records of the Western Australian Museum Supplement No. 59*; 2000. 127 pp.

17. Jones DS, editor. Marine biodiversity of the Dampier Archipelago, Western Australia 1998–2002. *Records of the Western Australian Museum Supplement No. 66*; 2004.

18. Wells FE. An analysis of marine invertebrate distributions in a mangrove swamp in northwestern Australia. *Bull Mar Sci* 1983;**33**:736–44.

19. Wells FE. Distribution of molluscs across a pneumatophore boundary in a small bay in northwestern Australia. *J Molluscan Stud* 1986;**52**:83–90.

20. Piersma T, Pearson GB, Hickey R, Lavaleye M, editors. The long mud: benthos and shorebirds of the foreshore of Eighty-Mile Beach, Western Australia. Report 2005/2. Texel, Netherlands: Royal Netherlands Institute for Sea Research; 2005. p. 1–218.

21. Piersma T, Rogers DI, Gonzalez PM, Zwarts L, Niles LJ, de Lima S, et al. Fuel storage rates before northward flights in red knots worldwide: facing the severest ecological constraint in tropical intertidal environments? In: Greenberg R, Marra PP, editors. *Birds of two worlds: ecology and evolution of migration*. Baltimore: Johns Hopkins University Press; 2005. p. 262–73.

22. Piersma T, Pearson G, Hickey B, Lavaleye M, Rogers D. Preliminary research report. *Southern Roebuck Bay invertebrate and bird mapping 2002*. Unpublished report. Broome Bird Observatory; 2002. 47 pp.

23. Piersma T, Pearson G, Hickey B, Dittmann S, Rogers D, Folmer E, et al. *Roebuck Bay invertebrate and bird mapping 2006*. Unpublished report. Broome Bird Observatory; 2006. 47 pp.

24. Goeij P, Lavaleye M, Pearson GB, Piersma T. Seasonal changes in the macro-zoobenthos of a tropical mudflat. Report 2003/4. Texel, Netherlands: Netherlands Institute for Sea Research; 2003. 49 pp.

25. Honkoop PJC, Pearson GB, Lavaleye MSS, Piersma T. Spatial variation of the intertidal sediments and macrozoo-benthic assemblages along Eighty-Mile Beach, north-western Australia. *J Sea Res* 2006;**55**:278–91.

26. Compton TJ, Rijkenberg MJA, Drent J, Piersma T. Thermal tolerance ranges and climate variability: a comparison between bivalves from differing climates. *J Exp Mar Biol Ecol* 2007;**352**:200–11.

27. Compton TJ, Troost TA, Van der Meere J, Krann C, Honkoop PJ, Rogers D, et al. Distributional overlap rather than habitat differentiation characterises co-occurrence of bivalves in intertidal soft sediment systems. *Mar Ecol Prog Ser* 2008;**373**:25–35.

28. Short AD. *Beaches of the Western Australian coast: Eucla to Roebuck Bay.* Sydney: Sydney University Press; 2006 432 pp.

29. Short AD. *Beaches of the Northern Australian coast: the Kimberley, Northern Territory and Cape York.* Sydney: Sydney University Press; 2006, 463 pp.

30. Short AD. Kimberley beach and barrier systems: an overview. *J R Soc West Aust* 2011;**94**:121–32.

31. McLachlan A. Sandy beach ecology—a review. In: McLachlan A, Erasmus T, editors. *Sandy beaches as ecosystems. Developments in hydrobiology*, vol. 19. The Hague: Junk Publishers; 1983. p. 321–80.

32. George RW, Knott ME. The *Ocypode* ghost crabs of Western Australia (Crustacea: Brachyura). *J R Soc West Aust* 1965;**48**:15–21.

Benthic Shelf and Slope Habitats

7.1 THE CONTINENTAL SHELF

On the inner shelf to depths of around 50 m, communities of seagrasses, thaloid algae, and surface-dwelling microorganisms may be an important source of primary production. There may also be particulate organic food delivered from the land via river discharge. But on the middle and outer shelf, below the photic zone and beyond much influence of terrestrial ecosystems, primary producers are largely absent, and benthic communities are essentially assemblages of secondary producers and their predators, relying on delivery of organic material from the photic zone of the ocean. In middle and outer shelf benthic habitats, suspension feeding and detritus gathering predominate. Nevertheless, biodiversity in deeper benthic shelf habitats may be very high, especially in areas like the outer shelf margins of the North West Shelf where there is upwelling from the deep sea.

Soft substrate communities vary enormously in their composition according to the nature of the substrate. Mud, grades of sand, and gravel offer diverse opportunities and constraints for infaunal living. Suspension-feeding and deposit-feeding polychaetes, bivalved molluscs, echinoderms, and small crustaceans dominate these communities, along with their predators. In keeping with their diverse adaptations to varied sediment type and sea floor water movements, soft substrate infaunal communities are characteristically extremely patchy in their spatial distribution.

Hard substrate fauna are predominantly sessile invertebrates such as sponges, alcyonarians, and bryozoans, all filter-feeders that live attached to the firm substrate. These assemblages support diverse assemblages of predatory and commensal invertebrates and provide important habitat for deep-sea demersal fishes.

The demersal fish fauna of the North West Shelf and its continental slope is moderately well known as a result of fishery stock surveys[1–6] and studies of fish species distribution and stock variability.[7–9] Benthic invertebrates of the continental shelf in the region are less well known. The Swedish

Mjöberg Expedition to Australia (1910–1913)[10] and I I.L. Clark (1929 and 1932)[11] made collections of benthic invertebrates dredged on the inner shelf of the Canning Bioregion. Since then staff of the Western Australian Museum have participated in various trawling operations, for example, CSIRO scampi resource surveys aboard the vessels *Courageous* and *Soela*.[1,2] As a result of these collecting ventures, museum collections have provided source material for taxonomic studies on invertebrates that have revealed the high biodiversity of the benthic shelf fauna, but this large component of the North West Shelf marine fauna remains imperfectly known.

An indication of the poor state of knowledge about the North West Shelf benthic shelf invertebrate fauna may be seen in a summary of available information on the ecology of the Browse Basin by Geoscience Australia which searched the online Ocean Biogeographic Information System database and found records of just 53 species of benthic molluscs (R. Przeslawski, personal communication, December 2009). In a review of key ecological features of the northwest marine region, a Geoscience report[12] noted that "one of the challenges in determining the ecological importance of the proposed KEFs is the absence of comprehensive biological datasets."

There have been just a few systematic studies of benthic invertebrate communities on the North West Shelf. An intensive survey of benthic molluscs of the inner shelf in the Dampier Archipelago has been carried out.[13] The report includes a taxonomic list of the molluscs sampled, information on their diverse habitats, and a discussion of species assemblages and biogeographic affinities. This is the only published account of this kind.

Fry *et al.*[14] carried out a dredging and video transect survey of inner shelf benthic habitats in the Canning Bioregion[15] aimed at defining the distribution and extent of the major benthic habitat types in that part of the Kimberley rather than species distribution patterns.

The most comprehensive benthic survey in the region has been done by Geoscience Australia in the eastern part of Joseph Bonaparte Gulf that has characterized the diverse benthic habitats of the Sahul Shelf and produced large collections of invertebrates that await taxonomic studies.[15,16]

7.1.1 Soft Substrate Communities

The sediments of the North West Shelf are described in Chapter 2.1.4. With the exception of areas of terrestrial sediment in the nearshore zone of the inner shelf in the Kimberley and central Pilbara bioregions, shelf sediments of the region are primarily biogenic carbonates, generally conducive to development of diverse infaunal invertebrate communities. Across the width of the shelf, there are distinct zones of sediment type (Figure 2.9), and great variation in the species composition of soft substrate communities may be expected.

7.1.2 Hard Substrate Benthic Shelf Communities

In Section 2.1.4.1, note was made of the prevalence of exposed rock pavement on the inner shelf of the Pilbara Bioregion. Also, in the Kimberley Bioregion and on the Sahul Shelf, there are extensive areas of base rock exposed on topographic highs and in scour channels on the sea floor of the inner shelf. It is known that there are significant areas of hard substrate along terraces of the middle and outer shelf, most notably the LGM shoreline terrace along the 120 m contour that stretches intermittently from at least the Montebello Islands to the Sahul Banks (Section 2.1.6). Industry video surveys in these areas indicated the presence of dense, species-rich epifaunal assemblages, predominantly sponges and alcyonarians. Divers searching for specimen shells at depths to 60 m in the vicinity of the Muiron Islands also report dense "sponge gardens" that dominate the sea floor (Figure 7.1).

Existing information on the sponges of northern Australia has been collated, revealing a species-rich sponge fauna in the region, although most of the records are from coral reef communities, and the epifaunal species of benthic shelf habitats have received little attention to date.

The widespread epifaunal, filter-feeding community in benthic shelf habitats of the North West Shelf remains almost undescribed. Surveys to determine its taxonomic composition and spatial distribution are needed. This community appears to be especially significant along the

FIGURE 7.1 "Sponge garden" habitat: an example of the epifaunal filter-feeding community, dominated by sponges and alcyonarians, that is, prolific on rocky substrata on the inner shelf of northern Western Australia; 21 m, off Point Murat, Exmouth Gulf, Ningaloo Marine Park. *Photo: Peter Clarkson.*

shelf margin, supported by the shelf-edge upwelling and mixing of ITF and coastal water, representing an important, extensive, and diverse element in the biodiversity of the Australian marine fauna.

7.2 CONTINENTAL SLOPE (BATHYAL)

The slope, terraces, and plateaux from depths of 200 to 2000 m are referred to as the bathyal or deep-sea zone. The bathyal zone and the abyssal and hadal zones below it are referred to as the deep sea. In much of the hydrocarbon resource area of the North West Shelf, the sea bed lies in this depth zone.

Sediments of the North West Shelf slope are carbonates, generally silty sands composed of skeletal remains of pelagic foraminifera. Ripple marks indicate bottom currents, but apart from these, the topography of the sea-bed is regular and featureless except for the high biohermic banks such as the Rowley Shoals and Scott and Seringapatam Reefs on the slope terraces. Typically, the physical conditions of these benthic bathyal habitats are relatively uniform and constant compared to the benthic shelf. However, reviews of the world's deep-sea benthos[17] show that this ecosystem is patchy but with surprisingly high biomass in places and with high levels of species-richness that surpass that of many terrestrial and other marine systems that are commonly regarded as supporting high biodiversity.

The most abundant and species-rich component of the bathyal fauna is small infaunal invertebrates, predominantly polychaetes, nematodes, foraminifers, crustaceans, and bivalved molluscs.[18] These may be in dense communities, feeding on detrital organic material sinking to the sea floor from the water column. There are also infaunal pogonophoran and sipunculid worms and many epifaunal species including echinoderms, sponges, and coelenterates. Supported by this trophic base of secondary producers are assemblages of predatory crustaceans, gastropod molluscs, and fishes.

The deep-sea benthic bathyal fauna of Australia, especially that of the North West Shelf, is little known.[19] A review of key ecological features in the Northwest Marine Region[12] provides a summary of knowledge of the deep-sea environment and identified databases of relevant information and specimens from the region. Two systematic accounts of benthic slope invertebrates in the region deal with ophiuroids[20] and azooxanthellate corals[21] and include references to species that inhabit this depth zone.

There is some information on demersal fishes of the continental slope including the results of trawling conducted by CSIRO. In the 1970s, a scampi trawling industry operated on the Rowley Terrace. There have been several inspections of the sea floor, using ROV technology, and some grab sampling on behalf of the petroleum and gas industry operating in the region. For the most part, the reports of these surveys are not publicly accessible. Nevertheless, some fish and invertebrate specimens from these surveys and from the commercial scampi trawlers have been deposited in the collections of the Western Australian Museum. In particular, there are samples of benthic molluscs, notably from the scampi grounds in the vicinity of the Rowley Shoals at depths from 300 to 450 m and from grab samples taken at the Pluto gas field off the Montebello Islands.[22]

Of particular note are specimens of gastropods (e.g., the pleurotomarid *Perotrochus westralis* and unidentified species of the trochid genus *Calliostoma*) that feed on organic detritus gathered from the sediment surface, deposit-feeding bivalves of the genera *Verticordia*, *Poromya*, and *Amygdalum*, and many species of large predatory gastropods of the families Olividae, Volutidae, Muricidae, Conidae, and Turridae. These gastropods are mostly vermivores. Their variety and numbers indicate the presence in these bathyal habitats of a dense and diverse infaunal community of small detrital-feeding invertebrates. A selection of the shells of these predatory gastropods is illustrated in Figure 7.2.

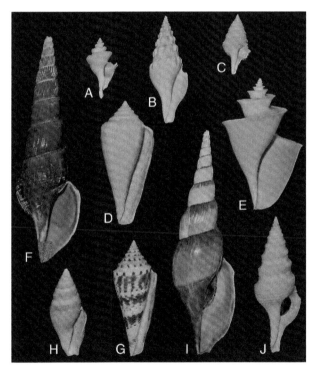

FIGURE 7.2 A selection of large predatory gastropods of the bathyal zone trawled at depths of 300-660 m on the Rowley Terrace, on the continental slope of the North West Shelf. (A) *Mipus vicdani* (Muricidae), (B) *Comitas* sp. (Turridae), (C) *Pinguigemmula philippensis* (Turridae), (D) *Conus teramachii* (Conidae); (E) *Thatcheria mirabilis* (Turridae), (F) *Teramachia dalli* (Volutidae), (G) *Conus ichinoseana* (Conidae), (H) *Bathytoma atractoides* (Turridae), (I) *Teramachia johnsoni* (Volutidae), (J) *Gemmula unedo* (Turridae).

From the biogeographic perspective, these molluscs appear to be representative of taxa that are widespread bathyal species in the Indo-West Pacific realm, although this may be a simplistic assumption based on little information. Some of the large gastropod species are described from comparable habitats in the Western Pacific. They have no biogeographic affinity with the benthic molluscan fauna of the adjacent shelf and represent a deep-sea fauna that probably has had a quite different evolutionary history to that of the shelf fauna. They are likely to respond to disturbance in different ways. The bathyal fauna along the margins of the North West Shelf is a significant element of the Australian marine biota that appears likely to be rich in species and patchily dense in biomass but which is virtually undescribed.

References

1. Sainsbury KJ, Kailola PJ, Leyland GG. *Continental shelf fishes of Northern and North Western Australia: an illustrated guide.* London: John Wiley and Sons; 1985.
2. Sainsbury KJ. Assessment and management of the demersal fishery on the continental shelf of north western Australia. In: Polovina JJ, Ralston S, editors. *Tropical snappers and groupers, biology and fisheries management.* Boulder, USA: Westview Press; 1987. p. 465–503.
3. Nowara G, Newman S. A history of foreign fishing activities and fishery-independent surveys of the demersal finfish resources in the Kimberley region of Western Australia. Department of Fisheries, Report no. 125; 2001.
4. Newman SJ, Dunk IJ. Growth, age validation, mortality, and other population characteristics of the red emporer snapper, *Lutjanus sebae* (Cuvier, 1828) off the Kimberley coast of north-western Australia. *Estuarine Coastal Shelf Sci* 2002;**55**:67–80.
5. Newman S, Evans D. Demersal finfish resource assessment survey of North West Slope of Western Australia. FRDC Report, no. 1998/152, Department of fisheries, North Beach, Western Australia; 2002.
6. Last P, Yearsley G, Gledhill D, Gomon M, Rees T, White, W. Validation of national demersal fish datasets for the regionalisation of the Australian continental slope and outer shelf. Report to the National Oceans Office, Sept., 2004, CSIRO Marine Research, Hobart; 2005.
7. Holliday D, Beckley LE, Weller E, Sutton AL. Natural variability of macro-plankton and larval fishes off the Kimberley, north-western Australia: Preliminary findings. *J R Soc West Aust* 2011;**94**(2):181–94.
8. Hutchins JB. Biogeography of the nearshore marine fish fauna of the Kimberley, Western Australia, In: Séret B, Sire J-Y, editors. Proceedings of the Fifth Indo-Pacific Fish Conference, Noumea 1997; 1999. p. 99–108.
9. Travers MJ, Potter IC, Clarke KR, Newman SJ, Hutchins JB. The inshore fish faunas over soft substrates and reefs on the tropical west coast of Australia differ and change with latitude and bioregion. *J Biogeogr* 2010;**37**(1):148–69.
10. Mortensen TH, editors. Results of Dr. E. Mjöberg Swedish Expedition to Australia, 1910–1913; 1918.
11. Clark HL. The echinoderm fauna of Australia. *Publ Carnegie Inst* 1946;**566**:1–567.
12. Falkner I, Whiteway T, Przeslawski R, Heap AD. Review of ten key ecological features (KEFs) in the northwest marine region. Geoscience Australia Record 2009/13. Geoscience Australia, Canberra; 2009.

13. Taylor JD, Glover EA. Diversity and distribution of subtidal benthic molluscs from the Dampier Archipelago, Western Australia; results of the 1999 dredge survey (DA2/99). *Rec West Aust Museum* 2004;(Suppl. 66):247–91.

14. Fry G, Heyward A, Wassenberg T, Colquhoun J, Pitcher R, Smith G, et al. Benthic habitat surveys of potential LNG hub locations in the Kimberley region. A joint CSIRO and AIMS report for the Western Australian Marine Science Institution; 2008. 131 pp.

15. Heap AD, Przeslawski R, Radke LC, Trafford J, Battershill C. Seabed environments of the Eastern Joseph Bonaparte Gulf, Northern Australia: SOL4934—post survey report. Geoscience Australia Record 2010/09; 2010.

16. Anderson TJ, Nichol S, Radke LC, Heap AD, Battershill C, Hughes MG, et al. Seabed environments of the Eastern Joseph Bonaparte Gulf, Northern Australia: GA0325/SOL5117—post survey report. Geoscience Australia Record 2011/08; 2011.

17. Thistle D. The deep-sea floor: an overview. In: Tyler PA, editor. *Ecosystems of the deep oceans*. Elsevier; 2003 [chapter 2].

18. Gage JD, Tyler PA. *Deep-sea biology: a natural history of organisms at the deep-sea floor.* Cambridge, UK: Cambridge University Press; 1991, 504 pp.

19. Ponder WF, Hutchings P, Chapman R. Overview of the conservation of Australian marine invertebrates. A report for Environment Australia, Sydney; 2002.

20. O'Hara T. Bioregionalisation of Australian waters using brittle stars (Echinoderrmata: Ohiuroidea), a major group of marine benthic invertebrates. Department of the environment, water, heritage and arts (Australia); 2008.

21. Cairns SD. Azooxanthellate Scleractinia (Cnidaria: Anthozoa) of Western Australia. *Rec West Aust Museum* 1998;**18**:361–417.

22. Fromont J, Moore G, Titelius MA, Jones DS, Slack-Smith SM. Report on the processing and identification of deep water marine specimens from Pluto field. Western Australian Museum; 2006.

8

Patterns of Life and the Processes That Produce Them

The distribution patterns of plants and animals on the surface of the earth are consequences of evolutionary history, means and mode of dispersal, and the distribution of habitats, the latter the subject of Chapters 2–7. As we have seen, distribution patterns of habitats are themselves outcomes of geological history shaped by geomorphic processes and they, in turn, are determined by physical forces like wind, ocean currents, and tides. Geological history is generally long term, measured in eons. Geomorphological processes that shape what is given at geological timescales are changeable, often short-term, seasonal, even daily, and either episodic or cyclic, but they too are responsive to longer term climate change.

A wonderful thing about life is the way it so quickly and so effectively responds to environmental change. Species adjust their physiology and behavior, rearrange themselves spatially, and adapt. When the degree or rate of change is so great that species or communities of species fail to do this, they become locally extinct. Better adapted kinds may arrive from elsewhere and take over the vacated ecological roles, the evolutionary process may produce new kinds that are better fitted to cope with the new conditions, or ecosystems may collapse and be replaced by assemblages of different species. These life processes of adaptation superimpose themselves over the changeable physical environment, and dynamic patterns of spatial distribution of life are the result.

However, both the physical and life processes are subject to limitations and the rules of nature. Patterns of species' ecological, spatial, and evolutionary change are not random but determined by a set of key principles, and it is an objective of natural science to learn enough about these principles so that patterns of distribution and the responses of life to environmental change may be understood, predicted, and perhaps managed. To this end, knowledge and understanding of certain key processes are required. In the present context of marine species' distribution patterns on the North West Shelf, key subjects include evolutionary history of habitats and their biota, past and present levels of biodiversity, past and present connectivity with the biota of adjacent bioregions, endemism, and the speciation process. These are the subjects of this chapter.

8.1 BIOGEOGRAPHIC AFFINITIES

A biogeographic affinity of one place with another refers to similarities of their flora and fauna and is an outcome of their common evolutionary histories and historical and contemporary connectivity. The existence of it and the level of it are commonly expressed by a hierarchical classification of areas into biogeographic regions such as that produced by the Integrated Marine and Coastal Regionalisation of Australia (IMCRA—see Section 1.3.3 of Chapter 1).

8.1.1 Affinities of the North West Shelf Marine Fauna

The primary biogeographic affinity of the North West Shelf lies with the Indo-West Pacific realm. The affinity derives from common ancestry in the pantropical Sea of Tethys initiated when the Australian plate came in contact with the Eurasian and Pacific plates during the mid-Tertiary 25-30 million years ago.[1,2] It is sustained in contemporary times, at least for species with pelagic larvae, by means of pelagic connectivity enabled by flow of water from the Central Indo-West Pacific region through the passages between the eastern islands of Indonesia—referred to as the Indonesian Throughflow or ITF (Section 2.2.2 of Chapter 2). In this sense, the North West Shelf biota is, historically, a sink supplied by and through the intensely species-rich Central Indo-West Pacific realm. But it also has affinity with the tropical east coast of Australia and the warm temperate coasts of higher latitudes in Western Australia, with which it shares elements of common biogeographic history.

8.1.1.1 Affinities with the Central Indo-West Pacific Region (East Indies Triangle)

The South East Asian or Sunda Shelf (bearing the western islands of Indonesia—Sumatra, Java, and Kalimantan) is separated from the continental shelf of Australasia (bearing New Guinea and Australia) by a transition zone that biogeographers call Wallacea (Figure 9.1C of Chapter 9). In this zone lie many of the eastern islands of Indonesia including Sulawesi, the Maluku group (Moluccas), and Nusa Tenggara (Lesser Sunda Islands). It is bounded by Wallace's Line and the Weber/Lydekker Line, biogeographic boundaries that have engaged the attention of terrestrial biogeographers for almost two centuries. Wallacea has such high significance because it occupies the region of sudden transition between the grossly different terrestrial floras and faunas of the Asian and Australasian continents. As you might expect, such a difference is not so apparent in the marine floras and faunas of the two continents' coastal waters because of the greater connectivity of marine species by means of pelagic dispersal. There are significant differences none the less, especially among marine animals without pelagic dispersal capacity (Section 8.4).

Wallacea occupies much of the Central Indo-West Pacific biogeographic region. It is especially important to marine biogeographers because it is part of the region revered as the world's region of greatest marine biodiversity (next section). Figure 8.1A shows the present proximity of the northern Australasian shelf to the Banda Sea and the series of Indonesian islands that form its eastern boundary. Australasia and these eastern Indonesian regions are separated by the very narrow, deep-sea Timor Trough and its northern extensions. At the present time, there are vast areas of shelf habitat (Arafura and Timor Seas) on that northwestern part of Australasia's continental margin. However, Figure 8.1B and C shows that shelf habitats on the Australasian side were very much narrower during most of the Holocene and even more so at the peak of the Last Glacial Maximum (LGM) when sea level was 125 m or more below where it is at present. Through that long period of low sea level in the Holocene, the Australasian coastline and its coastal habitats were very close to the easternmost islands of Wallacea. Such circumstances probably prevailed through much of the Quaternary. Close affinity, as well as proximity, of the marine biota of the Australian northwestern margin with that of the East Indies Triangle may be presumed. It could be said that such close proximity justifies viewing the Arafaura and Sahul Shelves as part of the East Indies Triangle, or at least an extension of it.

There is strong fossil evidence of a direct Tertiary connection of northern Australasian and Indonesian reef and benthic shelf faunas. Although there are few exposures of Tertiary fossil-bearing sediments in Northern Australia (because of subsidence),[3] the Tertiary marine fossil record in Indonesia and Irian Jaya is rich in both reef and benthic shelf species and has been extensively studied.[4-7] Many modern species of tropical Australian marine molluscs may be found there, and the ancestors of some.

After the Australasian and Eurasian first encroached upon each other during the Miocene, there would have been easy passage between the two adjacent continental shelves for species with pelagic larvae and this circumstance continues. And for a brief time, there was a shallow sea connection between the two[1,2] that would have allowed exchange of Tethyan marine species between the Sunda and Australasian shelves whatever their dispersal mode. However, nonpelagic dispersal across this transition zone ceased or became severely impeded later in the Miocene and Pliocene when the Australasian shelf began to fold beneath the Eurasian shelf and the deep Timor Trough was formed. Elements of the modern Australasian benthic shelf fauna that lack pelagic larval dispersal possess affinity with their modern congeners in the Central Indo-West Pacific region, but it would be an ancient one that is not refreshed by modern connectivity.

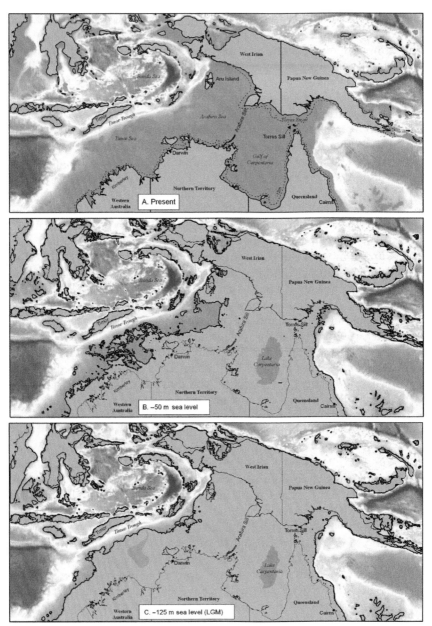

FIGURE 8.1   Changes in the area of the land connection between Australia and New Guinea since the Last Interglacial, and corresponding changes in the position of the Australasian coastline. (A) The present; (B) during most of the Holocene with sea level averaging around −50 m; (C) at the nadir Last Glacial Maximum sea level during at −125 m. The −12 m sill between Cape York and New Guinea was breached only briefly at the end of the Holocene transgression. The Arafura sill between the Wessel Islands and New Guinea, with a depth of around −50 m, left the Gulf of Carpentaria and much of the Arafura shelf exposed for most of the Holocene. During those prolonged periods, the coastline of Australasia was proximal to the islands of Maluku and could be regarded as part of the Central Indo-West Pacific biogeographic region.

There is another significant aspect relating to reef faunas that are predominantly planktotrophic. Connectivity between the North West Shelf and the Central Indo-West Pacific region is discussed in more detail in a Section 8.4, but we need to note here that it involves passage of western Pacific water, the ITF, between the islands of eastern Indonesia onto the Arafura and North West Shelves. This establishes potential for ongoing affinity not only with the species-rich reef fauna of the East Indies Triangle but also with the oceanic reef faunas of the western Pacific.

From these considerations, we may conclude that, since the Miocene, the North West Shelf has been a sink with the Central Indo-West Pacific its primary source. But the relationship may not have been always one-way traffic as it appears to be at the present time. During periods of low sea level and closer proximity of the coastlines of the two regions, there may have been two-way dispersal by means of local currents.

8.1.1.2 Affinity with the Coast of Eastern Australia

The marine fauna of Australia's northwestern margin also has an affinity with that of the tropical eastern coast of Australia. Affinity of the North West Shelf marine fauna with that of eastern Queensland derives from common ancestry in the Tethyan/Indo-West Pacific region. At the present time, while the Torres Strait is open, there is potential for direct connectivity between the Arafura Sea and Coral Sea, but this circumstance has occurred only during brief phases of high sea level. The marine fauna of the North West Shelf has closer evolutionary affinities with that of Indonesia than it does with eastern Australia. This matter is discussed in more detail in the following discussion on connectivity.

8.1.1.3 Affinity with the Southern Coast of Australia

Southeastern Australia has been a center of endemicity with very different evolutionary origins, derived partly from a shelf fauna that inhabited the margins of Gondwanaland in the Mesozoic, termed the Austral Province[7] or the Weddelian Province.[8] Many lineages of Tertiary molluscs that had their origins in that southern, high latitude province are referred to as Palaeoaustral elements in the southern Australian and New Zealand marine faunas.[9] Palaeoaustral elements have dominated the molluscan benthic fauna of southeastern Australia since the Eocene and a distinct Tertiary Southeastern Australian Province may be recognized.[10] However, during the period of global warming in the Miocene when the world's tropical zone expanded into higher latitudes, Tethyan elements invaded higher latitudes of southern Australia establishing the distinctive contemporary southern Australian Realm, a temperate biogeographic region with a very high level of regional endemism and a mixed marine fauna of Palaeoaustral, Tethyan, and cosmopolitan elements.[10] In this sense, the North West Shelf also has an affinity with the southern

Australian Realm having been an avenue and source for the southern migration of Tethyan marine fauna.

The tropical Indo-West Pacific and the temperate southern Australian faunas overlap on the lower West Coast (Shark Bay to Cape Leeuwin), a region which, embracing nearly 10° of latitude and a thousand kilometers of continental shelf, is regarded as a biogeographic overlap or transition zone between the two primary biogeographic realms of northern and southern Australia (Figure 1.3 of Chapter 1). Since the mid-Tertiary, apparently driven by the pulses of global climate change, the boundaries between these tropical and temperate faunas have shifted back and forth along the coast. The transition zone has moved into higher latitudes in periods of global warming and retracted to lower latitudes in periods of global cooling. That process was responsible for the evolution of many endemic species that now characterize the West Coast. The derivative species in the shelf faunas of the southwestern corner of the continent have mainly either North West Shelf (Tethyan-Indo-West Pacific) or southeastern Australian (Palaeoaustral) ancestors. (There is also a temperate circum-polar element.)

8.1.2 Two Faunal Elements

Within the contemporary Indo-West Pacific fauna two ecological categories stand apart:

1. Species adapted to living in oligotrophic conditions that occur on oceanic coral reefs throughout the realm. Oceanic reefs being often widely separated, connectivity between them may be achieved only by species with pelagic larvae. "Oceanic species" are generally geographically widespread and are characterized by planktotrophic larval development.
2. Benthic shelf and shore species confined to nutrient-rich continental shelves or the shores of large islands. "Continental species" have a broad variety of dispersal strategies and often have limited geographic range and a high frequency of polytypy.

Although there is much overlap, very many marine species inhabit either one or the other of these environments. Regarding the gastropod family Strombidae, it has been observed that there are "species living mainly in clear, oceanic waters" and "species limited to the rich, nitrogenous shores of continents or well-vegetated volcanic islands."[11] The same phenomenon has been noted for other Indo-West Pacific marine molluscs.[12–16] A similar distinction has also been made regarding reef and continental echinoderms on the Queensland coast[17] and the North West Shelf.[18,19] It may apply to most groups of tropical benthic animals.

This distinction is pertinent to the benthic fauna of the North West Shelf. The reef species inhabiting the shelf-edge and outer shelf reefs of the Oceanic Shoals Bioregion are predominantly widespread Indo-West Pacific, planktotrophic, oceanic species, capable of long-distance dispersal and colonization of remote localities (Section 8.4). Most benthic shelf fauna are continental species and also planktotrophic but commonly with short-lived pelagic larvae capable only of relatively short-range dispersal. Many benthic shelf invertebrates lack pelagic larvae altogether. In this latter category, philopatry is the rule (the larvae hatching and growing up at the natal site, often nurtured by their parents) and regional endemism is very high.

8.2 BIODIVERSITY

On the North West Shelf, there is a discrepancy in patterns of biodiversity of three major invertebrate taxa (corals, echinoderms, and molluscs) that is an important biogeographic anomaly needing further elaboration and explanation. Nevertheless, the provisional evidence so far available clearly indicates that the North West Shelf is a bioregion with very high diversity of benthic shelf and reef invertebrates, derived from its close proximity and biogeographic affinity with the species-rich Central Indo-West Pacific region.

The scleractinian coral fauna of the North West Shelf (as recorded at November 2012) is shown to be rich in species and genera, and the region may be regarded as a major center of coral diversity at the southern margin of the East Indies Coral Triangle. Even though collecting is far from complete, the Kimberley coast is shown to be the richest bioregion in terms of coral diversity and heterogeneity, with the western Pilbara Bioregions not far behind. Many coral species found in the nutrient-rich waters of either or both the Kimberley and Pilbara Bioregions are not found in the oligotrophic conditions of the Oceanic Shoals Bioregion along the shelf margin. These coastal corals may be regarded as a "continental shelf" element in the coral fauna of the region, adapted to the macrotidal, nutrient-rich and turbid conditions of the coastal zone.

The patterns of diversity of molluscs and echinoderms on the North West Shelf are very different to that of scleractinian corals. Available data indicate low diversity in the Kimberley reef faunas of both molluscs and echinoderms but high diversity in the Pilbara. With its diverse mix of reef, benthic shelf, and shore habitats, the Pilbara is clearly a "hot spot" of marine diversity in these two phyla. In both phyla, there are important differences between the reef faunas of the oceanic shelf margin and the coastal zone, with conspicuous "oceanic" and "continental" elements in their respective intertidal assemblages.

8.2.1 Patterns of Biodiversity on the North West Shelf

Biodiversity may be measured as numbers of species (species-richness) genetic variability within a species, or the total genetic variability embraced by the biota or selected higher taxa that comprise it. *Species diversity*, meaning species richness, may be differentiated from *heterogeneity* meaning the range of higher taxa.[20] The distinction is important in groups like corals where some higher taxa, for example, the genus *Acropora*, have vastly more species than other genera so that a place rich in species of *Acropora* but with few other genera may have high species diversity but low genetic heterogeneity. The following discussion refers mainly to species-richness data, based on presence records, but generic diversity and other factors are considered where appropriate.

Species-richness data are given in Tables 8.1–8.4 for three groups of marine animals—corals, molluscs, and echinoderms—recorded from surveyed areas on the North West Shelf. The collecting/taxonomic phase of research is yet far from complete and collecting effort has not been the same at all localities. Also, there is wide variation in the size of the areas sampled, and in the kinds of habitat at the surveyed localities, so that different assemblages of species are being compared. Nevertheless, these available data are informative when these limitations are taken into account.

8.2.1.1 Corals

The coral fauna of Western Australia was first described as a biogeographic unit by Veron and Marsh.[21] Since then, rapid progress has been made in documenting the corals of the region and their ecological and geographic distributions. Published and unpublished lists of coral species are now available for the following localities:

- oceanic reefs of the Rowley Shelf margins (Scott, Seringapatam, and Rowley Shoals)[22,23];
- oceanic reefs of the Sahul Shelf margin (Ashmore, Hibernia, and Cartier)[24–26];
- coastal reefs of the Kimberley[27–29];
- coastal reefs of the western Pilbara (Montebello Islands,[30] Dampier Archipelago,[31] and Barrow Island[32,33]).

Taxonomic work on these regional coral faunas is ongoing, and there will be amendments to these lists. Field collecting is also ongoing, and

TABLE 8.1 Numbers of Recorded Coral Taxa on the North West Shelf (at November 2012)

Region	Genera	Species
Ashmore-Cartier[24,26]	56	255
Scott-Seringapatam Reefs[23]	56	300
Kimberley coast[29]	68	318
Pilbara-Ningaloo[30–33]	65	300

the lists are probably underestimates, especially for the Kimberley and Pilbara. Nevertheless, when taken as they stand and merged together, the total known fauna of scleractinian reef corals of the North West Shelf now stands at around 400 species of 69 genera. (This figure does not include the 87 species that live in soft substrate benthic shelf habitats.[34])

Almost all of the reef coral species are widely distributed in the Indo-West Pacific region. One species, *Montigyra kenti*, appears to be endemic to the North West Shelf. To date, it is only known from the Lacepede Islands in the Canning Bioregion but probably it will be found elsewhere in coastal waters of the North West.

Survey work at Scott, Seringapatam, and the Rowley Shoals has been extensive, and the 300 species (56 genera) recorded from those localities is probably close to the real number present. Of this species total, 82 (27%) belong to the genus *Acropora*, dominance of that genus being typical of oceanic reef faunas. At species level, this fauna is characteristic of oceanic reefs of the Indo-West Pacific region. However, only one genus, *Anacropora*, with two species, is not represented on the coastal reefs. (The two species of *Anacropora* have been found in the deep lagoon of Scott Reef, a habitat not present on the coastal reefs.) Around 40 species found at the Rowley Shelf slope and outer shelf reefs (ca. 13%) have not been not found on the coastal reefs and may be regarded as an "oceanic" element in the regional coral fauna. It would be interesting to compare this well-described coral assemblage with that of shelf and slope atolls in the Banda Sea which share many biogeomorphic similarities and lie directly "upstream" in the Indonesian Through Flow.

Earlier notions that the turbid, macrotidal reefs of the Kimberley coast were not rich in coral species are no longer tenable. Survey work there has only begun but with a provisional coral list standing at 318 species and 68 genera, the Kimberley fauna is already established as diverse. That figure will probably increase as the bioregion is fully explored. In spite of having received such recent and limited attention, the Kimberley coast has the richest coral fauna of any of the North West Shelf Bioregions, in both species and genera. Of the species total, 62 (19%) belong to the genus *Acropora*, indicating that, with its 68 genera, the coral fauna of the Kimberley coast is significantly more heterogeneous as well as more species rich than the offshore reefs of the Oceanic Shoals Bioregion. What is more, not only are there many coral species that are well adapted to the turbid conditions that prevail, but also reef growth and reef development are extensive in the Kimberley as illustrated in Chapter 4.

The diversity of the scleractinian corals in the western Pilbara is also now established as high. Environmental surveys in the Dampier Archipelago and the Montebello-Barrow Islands reef complexes have recorded more than 300 species of 65 genera.[31–33] In global terms, the western Pilbara, more than 1000 km southwest of the Kimberley, is a high diversity coral reef province in its own right. Of the total number of species, 51 (18%) belong to the genus *Acropora*. It is evident that the coral fauna of the Pilbara coast is about as species rich as that of the Oceanic Shoals Bioregion but, like the Kimberley, significantly more heterogeneous.

On the present figures, the Kimberley and Pilbara reefs combined have about 350 reef coral species, of which 256 are also found on the shelf-edge reefs (73% in common). However, there are about 60 species found on the coastal reefs (ca. 17% of the fauna) that are not found on the shelf-edge reefs and they constitute a "continental" element in the fauna.

There is a high degree of similarity between the Kimberley and Pilbara coral faunas. Of the species recorded in the Pilbara region, 276 (88%) are also found in the Kimberley. This is a significantly higher level of similarity than either region has with the coral fauna of the Oceanic Shoals Bioregion. A simple cluster analysis demonstrates the high significance of this result (Figure 8.2).

The continental group of coral species in the Kimberley and Pilbara Bioregions includes a suite of monotypic and small genera that are known to be characteristic of continental shores (*Barabottoia, Moseleya, Oulastrea,*

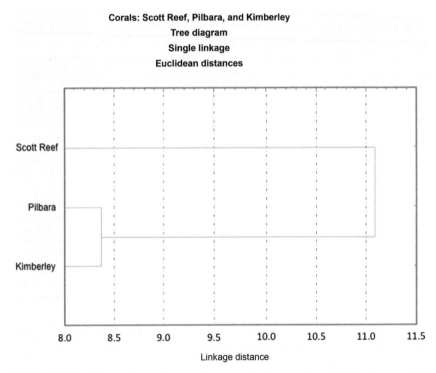

FIGURE 8.2   Relationships of the coral faunas of the Oceanic Shoals, Kimberley, and Pilbara Bioregions of the North West Shelf. *See Table 8.1 for sources of the data.*

Trachyphyllia, Blastomussa, Cyanarina, Paulastrea, Catalaphyllia, Montigyra, Pseudosiderastrea, Duncanopsammia, Diaseris, Scolymia).[30] The species of these genera are adapted to the turbid, sometimes muddy, conditions of coastal waters. Most of them are regarded as rare or uncommon, but this may be because their habitats are less commonly collected than the clear waters of open sea reefs.

There is some variation in the representation of the 15 families of scleractinian corals in the Oceanic Shoals, Kimberley and Pilbara Bioregions. This is illustrated in Figure 8.3.

The Central Indo-West Pacific region, including the Philippines, with well over 400 coral species is substantially richer in coral species than any other region.[20] It is referred to as the Coral Triangle (the coral taxonomists' version of the East Indies Triangle) and is the world's greatest center of marine biodiversity. The central Great Barrier Reef with 343 coral species[20] is a lesser coral diversity "hot spot." The known coral fauna of the North West Shelf, now standing at around 400 species of 69 genera and likely to increase, indicates that the Oceanic Shoals, Kimberley, and Pilbara Bioregions may be regarded as coral biodiversity centers at the margin of the central Indo-West Pacific Coral Triangle, with distinctive reef geomorphologies.

A similarity analysis is needed comparing the corals of the North West Shelf, Queensland, and the Indonesian Coral Triangle to examine

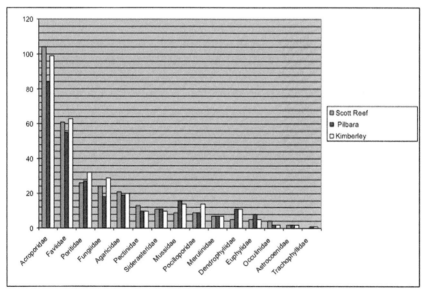

FIGURE 8.3   Representation of the families of scleractinian corals in the Oceanic Sholas, Kimberley, and Pilbara Bioregions. *See Table 8.1 for sources of the data.*

whether the Western Australian coral fauna exhibits more affinity with that of Indonesia than it does with the coral fauna of eastern Australia. That this may be so is suggested by the proximity and likely continuous connectivity of the Timor and Banda Seas and the eustatic history of the region that indicates only brief periods during the Quaternary when the Timor Sea has had connectivity with the Coral Sea (see Sections 8.1 and 8.4).

Comparison of the reef faunas of the North West Shelf slope atolls with the oceanic Cocos-Keeling atoll shows that Cocos is far less rich in corals (99 species)[35] than Scott Reef (300 species).[23] It is unclear whether this is because, as oceanic atolls go, Cocos is unusually poor in corals, or Scott is unusually rich. Further comparison of the coral fauna and ecological assemblages of Scott Reef and the other slope atolls of the Rowley Shelf with oceanic atolls of the western Pacific and Indian Oceans, and the shelf and slope atolls of Indonesia and the South China Sea would be informative.

8.2.1.2 Echinoderms

The data used in this section are derived almost entirely from publications and reports produced by Loisette Marsh and her coauthors, and the discussion is based largely on publications by Clark[36,37] and Marsh and Marshall.[18] As a result of the work of those authors, echinoderms are perhaps the best known marine invertebrates of the region although even with this group the inventory of the reef and benthic shelf fauna of the North West Shelf is still incomplete.

At the present time, the known reef and benthic shelf echinoderm fauna of the North West Shelf is exceptionally rich, comprising 480 species of 185 genera. The breakdown of these taxa between the five classes of the phylum is given in Table 8.2. Easily the most speciose class is the Ophiuroidea, followed by the Holothuroidea, Asteroidea, Echinoidea, and Crinoidea, in that order.

TABLE 8.2   Number of Genera and Species of the Classes of Echinoderm in the Reef and Benthic Shelf Fauna of the North West Shelf

Class	Genera	Species
Crinoidea	29	67
Asteroidea	42	85
Ophiuroidea	40	148
Echinoidea	39	70
Holothuroidea	35	110
Totals	185	480

TABLE 8.3   Numbers of Reef and Benthic Shelf Species in Each of the Five Echinoderm Classes at Three North West Shelf Bioregions[18]

Bioregion	Crinoids	Asteroids	Ophiuroids	Echinoids	Holothurians	Total Spp.
Oceanic	39	38	73	34	57	241
Kimberley	12	8	28	12	22	82
Pilbara	39	47	112	62	80	340

Data from Marsh and Marshall.[18]

Table 8.3 shows the numbers of species of each of the classes in three of the North West Shelf Bioregions—Oceanic Shoals (east and west combined), Kimberley, and Pilbara—and these data are expressed graphically in Figure 8.4. The Pilbara, with its mixed reef and benthic shelf habitats, has by far the richest echinoderm fauna (340 species). With regard to this phylum, the Pilbara clearly warrants recognition as a biodiversity "hot spot." The echinoderm fauna of the Kimberley Bioregion is, as yet, poorly collected, and most effort has been targeted on the intertidal reefs and

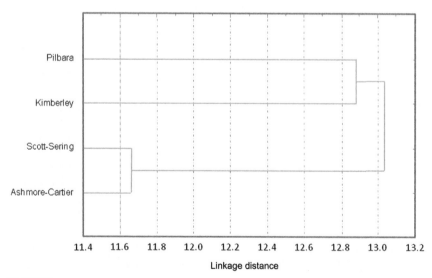

FIGURE 8.4   Relationships of the echinoderm faunas of the Oceanic Shoals (East and West combined), Kimberley and Pilbara Bioregions of the North West Shelf. *See Table 8.3 for sources of the data.*

shallow sublittoral zone. The known fauna of that region is certain to increase substantially although it is already evident that it is not rich in echinoderm species.

There are significant faunistic differences between the reef echinoderm faunas of the "offshore reefs" (which we now equate with the slope atolls and platform reefs of the Oceanic Shoals Bioregion) and the "inshore" which relates primarily to the fringing and patch reefs and the benthic shelf habitats of the Canning and Pilbara Bioregions.[18] Distribution patterns of echinoderms within the North West Shelf region are quite similar to those described on the Queensland coast.[17] Of the 109 reef species of echinoderm recorded from the offshore North West Shelf Reefs, 17 (16%), though widespread in the Indo-West Pacific region, have not been found in the inshore zone.[18] With the advantage of considerably larger species numbers since recorded, the similarities between the echinoderm faunas of Oceanic Shoals and Pilbara Bioregions are reevaluated and the substance of this observation is confirmed.

Of the 376 species (not including holothurians) recorded from the northwest of Western Australia in 1983, 49 species (13%) are endemic to the region.[18] There is a close relationship of the fauna to that of Indonesia with 66% of the northwest species also found there. In contrast, 42% of the northwest echinoderms were also found in the Great Barrier Reef region, 36% in the West Pacific, and 33% in the western Indian Ocean. This result may be interpreted as an outcome of a high level of historical and contemporary connectivity between the North West Shelf and Indonesia and infrequent phases of connectivity with eastern Australia.

8.2.1.3 Molluscs

Information on the molluscan fauna of the North West Shelf is derived from a collation of published and unpublished area survey reports produced by the Western Australian Museum, The Museums and Art Galleries of the Northern Territory, and by this author, at localities between the Ashmore Reef complex on the Sahul Shelf and the Muiron Islands in the West Pilbara.[38–50] The resulting taxonomic list includes records of 1909 molluscan species from the North West Shelf localities that have been surveyed. Table 8.4 shows the numbers of species of each of the classes in three of the North West Shelf Bioregions—Oceanic Shoals (east and west), Kimberley, and Pilbara—and these data are expressed graphically in Figure 8.5A–C for all the molluscs taxa combined and for bivalves and gastropods separately.

This result is far from complete because most of the surveys involved collecting by scuba divers or reef walks with a strong bias toward species inhabiting rocky shores and coral reefs. Soft substrate communities of benthic shelf habitats are severely undercollected, and the continental slope deep-sea fauna is hardly represented at all. There is also strong taxonomic

TABLE 8.4 Species of Mollusc Recorded by Surveys Within Four of the IMCRA Bioregions on the North West Shelf

Taxon	East Oceanic Shoals[39,40]	West Oceanic Shoals[41,42]	Kimberley[43–45]	Pilbara[46–50]	Total Species
Bivalvia	150	122	136	493	590
Gastropoda (totals)	655	458	447	783	1271
Prosobranchia	534	409	376	600	989
Opisthobranchia	118	49	58	168	262
Pulmonata	3	0	13	15	20
Scaphopoda	1	0	1	5	7

The Oceanic Shoals Bioregion is divided into east and west sections.

bias in the collections with some speciose groups virtually excluded, especially taxa characterized by many species with small (<5 mm) shells or bodies such as the rissooid and turrid gastropods and the galeommatoid bivalves. No attempt has been made to extract additional data from museum collections or the taxonomic literature. Nevertheless, taken as they stand, the data considered in this study indicate a very large, species-rich, and heterogeneous molluscan fauna on the North West Shelf and allow comparisons to be made between areas.

As expected, prosobranch gastropods, bivalves, and opisthobranch gastropods, in that order, dominate the fauna. Pulmonate gastropods live intertidally in coastal mangal, marsh, and rocky shore habitats and are very poorly represented at the offshore reefs. (Cephalopods, polyplacophorans, and scaphopods are poorly collected, and the taxonomy of the representatives of these classes is virtually unstudied in the region so that the numbers of species recorded in this study are not a fair account and they are excluded from the analysis.)

The disproportionately high number of bivalves in the Pilbara, compared to the other three bioregions, is largely due to the prevalence of sandy shores and soft substrate subtidal habitats. Most of the bivalves recorded from the offshore reefs and fringing reefs of the Kimberley coast are byssal-attached reef and rocky shore pteriomorphs. Sandy habitats are less common in those bioregions, and speciose bivalve families, like the Veneridae, Mactridae, and Tellinidae that are typically infaunal burrowers, are not well represented there.

A conspicuous feature of the oceanic reef faunas is the prevalence of predatory prosobranch gastropods, especially the families

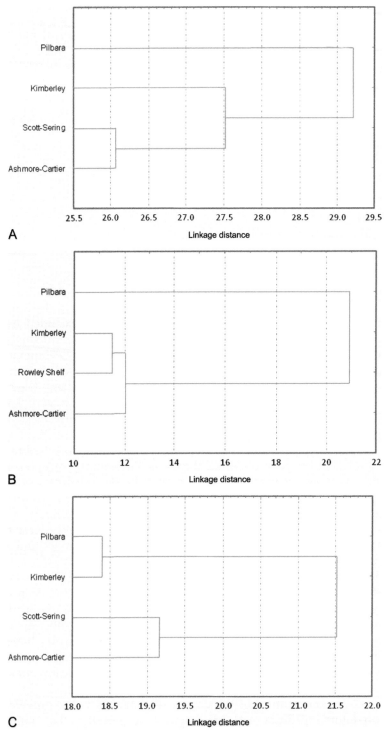

FIGURE 8.5   Tree diagrams with single linkage, Euclidean distances, showing the relationships of the mollusc faunas of the Oceanic Shoals (East and West combined), Kimberley and Pilbara Bioregions. *See Table 8.4 for sources of the data.* (A) Gastropods and Bivalves combined; (B) Bivalves; (C) Prosobranch gastropods.

Muricidae, Buccinidae, Fasciolariidae, Mitridae, Costellariidae, and Conidae. A suite of predatory gastropods that characterize the reef-front zone of the oceanic reefs[38] but poorly represented on the coastal reefs is discussed in Chapter 4.

The conid fauna of the region is particularly revealing. There are 86 species of cone recorded so far from the North West Shelf, 76 of which are widespread Indo-West Pacific species that are found on the outer oceanic reefs (37 at the Rowley Shelf atolls and 74 at the Ashmore-Cartier-Hibernia Reef complex). There are 46 cones found on the fringing reefs or rocky shores of either the Kimberley (31 species) or Pilbara (42 species), and of these, the majority, 37 species, are also found on the offshore oceanic reefs and are part of the Indo-West Pacific reef fauna. Nine species found in the coastal bioregions do not occur on the oceanic reefs, and of these, seven are endemic to the North West Shelf and one is endemic to the North West Shelf and West coast. The assemblages of cones found on the outer reefs of the North West Shelf are typical of coral oceanic reef assemblages of the Indo-West Pacific realm. Of the eight endemic coastal species, five are rocky shore inhabitants (with or without the presence of corals) and three are sand-dwellers, none being coral reef cones, and these may be regarded as continental species.

The main herbivorous, reef, and rocky shore family Cypraeidae (cowries) is also speciose in the region with 65 species recorded from the North West Shelf. Of these, 40 species are common to the coastal bioregions and the outer reefs, while 17 are so far recorded only from the coastal bioregions and 8 are only recorded from the outer reefs. The latter group includes species that are known to be oceanic coral reef cowries (*Cypraea depressa, childreni, globulus, margarita, mariae, microdon, minoridens*). Of the cowries so far found only in coastal waters, one (*C. miliaris*) lives on sponges on muddy sand flats. Four nominal species of the subgenus *Zoila* (*decipiens, eludens, marielae, perlae*) are direct-developing endemics that are sponge eaters living in benthic shelf habitats. All the North West Shelf cowries, except the four species of the subgenus *Zoila*, are widespread in the Indo-West Pacific, but regional polytypy (with the morphs sometimes recognized at subspecies rank) is evident in many of them.

The carnivorous, sand-dwelling gastropod family Volutidae tells a different story. All volutes are direct developers. There are 21 species on the North West Shelf. Except for two deep-sea species of *Teramachia* that live on the continental slope, all these volutes are endemic to the North West Shelf or the continental shelves of Northern Australia and southern New Guinea and represent an autochthonous element in the fauna. Three species (*Amoria spenceriana, Cymbiola baili*, and *Melo ashmorensis*) are short-range endemics confined to the isolated Ashmore platform, apparently derived from congeners of the adjacent continental shelf.

Taken overall, these figures indicate that the Pilbara Bioregion is substantially more species rich in molluscs than the other bioregions of the North West Shelf. The Kimberley coastline (south to Cape Leveque) is predominantly rocky with rocky shore and fringing reef habitats, and extensive mangals in bays and gulfs. Compared to the Pilbara and Ashmore areas, the nearshore molluscan fauna of the Kimberley Bioregion is depauperate. However, inner shelf benthic habitats in the Kimberley have not been sampled and it is certain that the known molluscan fauna of that bioregion will increase significantly when that has been done.

At Ashmore-Cartier-Hibernia Reefs, referred to in Table 8.4 as the East Oceanic Shoals Bioregion (Sahul Shelf), 814 mollusc species are recorded, substantially more than the number recorded at Scott-Seringapatam Reefs and the Rowley Shoals (West Oceanic Shoals Bioregion) which have 586 species. There appear to be several factors in the difference. At the East Oceanic Shoals Reefs, there is a significant continental element in its fauna, perhaps a consequence of its much closer proximity to the continental shelf. At the present time, the platform on which Ashmore Reef stands is separated from the main continental shelf by a channel 200-300 m deep and several kilometers wide. During the LGM low sea level phase, benthic shelf invertebrates may have been able to cross between these two habitats. Also, there are extensive areas of reef platform, back-reef, and lagoon sand habitats at Ashmore that support a moderately diverse infauna fauna with species of burrowing bivalves and gastropods (especially Terebridae, Olividae, and Costellariidae) that are absent at the Rowley Shelf slope atolls.

The diversity of molluscs in the Canning Bioregion remains to be assessed. With its very extensive coastal and inner shelf sand habitats, it is likely to be rich in infaunal molluscs. But as there are few rocky shores and no coral reefs in this bioregion, it is also likely to be relatively deficient in gastropods other than infaunal species. As a result, the total molluscan biodiversity of the Canning Bioregion may be found to be significantly less than that of the Pilbara.

8.3 LATITUDINAL GRADIENTS IN BIODIVERSITY

A conspicuous feature of global biogeography is much greater biodiversity in the tropics than in temperate climates of higher latitudes, a transition known as the latitudinal biodiversity gradient. It applies in both marine and terrestrial environments and across taxa and habitats.[51,52] The shelf areas surrounding the Indo-Malay Archipelago, known as the East Indies Triangle or the Coral Triangle in the case of coral biologists, are the world's most biodiverse marine biogeographic regions (Section 8.2). From this center, three axes of distribution radiate into higher latitudes—north to Japan, southeast down the eastern Australian coast, and southwest down the Western Australian coast. In each case, from low to high latitude, diversity of marine species declines. It is an important biogeographic principle, and there have been many attempts to explain it.[20,53-55]

In the context of the biogeography of Australia's western margin, there is no doubting the existence of a strong species-richness gradient of tropical marine fauna down the axis from the high diversity center of the Central Indo-West Pacific region to the temperate southwestern corner of Australia. In respect of the North West Shelf, it is sometimes said, usually in relation to coral reef fauna, that there is a progressive attenuation in marine species richness from Ashmore Reef to Ningaloo. This concept has tended to take on a life of its own, and the causes of it, if it exists, are often not explored. Sometimes, it is used in a way that implicitly affords causality to the correlated latitudinal temperature gradient, but there is no strong evidence that temperature has anything to do with it. Other factors are involved as well as important regional environmental affects. The presence of a diversity gradient along the North West Shelf from north to south needs to be examined in relation to species that possess long-lived, planktotrophic larvae and, separately, in relation to benthic shelf species where pelagic larvae are often short lived and various forms of philopatry are commonplace. The matter has relevance to management in the context of connectivity and the resilience of faunal assemblages to change.

The issues considered here are whether there is such a diversity gradient down the length of the North West Shelf, does it apply to all taxa, to what extent it is continuous, and what regional influences distort it.

8.3.1 Causes of the Global Latitudinal Diversity Gradient

One much discussed but controversial concept is the mid-domain effect (MDE) model.[56,57] This null model holds that when any assembly of objects is placed in a bounded domain and "shaken up" so that they become randomly distributed, there will be a concentration of them at the center, based on range overlap counts. Considering the world as a bounded domain, this effect will result in maximum diversity of overlapping species and higher taxa in equatorial latitudes, even when biotic, environmental, or historical factors are excluded. Proponents of this model do not deny the importance of such modifying factors but argue that the MDE model is a useful tool that allows them to be evaluated.

Various authors claim that the diversity gradient arises from latitudinal variation in rates of cladogenesis (evolution of new species and their radiation into higher taxa), extinction, immigration, and/or emigration of taxa. It is claimed that rates of cladogensis, including speciation rates, are higher at low latitudes because of higher energy levels and higher temperature and are highest in the Central Indo-West Pacific region because of greater shelf area, greater topographic complexity, the frequency of vicariant events, and other historical factors relating to the overlap of the Indian and Pacific Oceans. Some argue that the primary causes of the present-day latitudinal diversity gradients away from the Central Indo-West Pacific are higher rates of cladogenesis and subsequent range expansion of new species from low to higher latitudes.[58–60] Jablonski and coauthors coined the phrase "out of the tropics" referring to genera of marine bivalves whose lineages originate in the tropics and persist there while expanding their geographic ranges poleward.[58] In this concept, expansion of range is an essential part of the process and that begs the question of dispersal and its mode and means. Rate of expansion would be high for species with larval capacity for long-distance dispersal when there is an effective oceanographic means of delivery. Conversely, it would be low for species lacking pelagic larvae and impeded if there were bigeographical barriers to dispersal.

One overlooked but relevant issue in the Australian context is the presence of another, though lesser, center of cladogenesis, with a different evolutionary origin, in the temperate waters of southern Australia (Section 8.1.1.3). While the tropical marine fauna of Northern Australia is derived from the fauna that inhabited the pantropical sea of Tethys, the temperate element in the marine fauna of the contemporary southern Australian region is derived mainly from the fauna of the Southeastern Australian Province, itself a descendant of the Palaeoaustral marine fauna of southern high latitudes, but with a strong Tethyan element introduced during the Miocene global warming.[10,12] There are distinct latitudinal gradients away from this center but in the opposite direction to that from the tropics, i.e., from south to north and west. Dispersal streams away from the southern Australian region (northward) and away from the Central Indo-West Pacific region (southward) interact on the west and east coasts of Australia, giving rise to the overlap zones of the tropical and temperate marine faunas on the east and west coasts (Figure 1.3 of Chapter 1). This does not negate the Joblonski hypothesis that latitudinal diversity gradients are caused by outward dispersal from a center of cladogenesis but shows that dispersal from such centers may be from high to low latitudes as well. It is true that the Central Indo-West Pacific region is significantly richer in most taxa than the southern Australian region, and it may be true that it is the result of higher rates of cladogenesis in the former as an outcome of higher temperatures, productivity, habitat complexity, and vicariant history. However, dispersal away from a center of high diversity and the establishment of latitudinal gradients (or longitudinal ones for that matter) are not necessarily related to a low latitude origin.

In reviewing the generality of marine species' latitudinal diversity gradients, Hillebrand noted that, in reality, diversity patterns are more complex than simple trends with latitude.[52] There are "hot spots" and "cold spots" of diversity for different taxa and habitats[61–63] and strong regional influences.[64,65]

8.3.2 Latitudinal Diversity Gradient on the North West Shelf

8.3.2.1 Long-Distance Pelagic Dispersal

There is a clear means of delivery for pelagic dispersal poleward along the margin of the North West Shelf from an EIT source, i.e., the ITF-Holloway-Leeuwin Current that is likely to be effective for "oceanic" coral reef species that are typically planktotrophic. The distance from the historical source (eastern Indonesia) to the end point (Ningaloo near the margin of the tropical zone) is more than 2500 km, encompassing around 18° of latitude. One could expect there to be a diversity gradient of reef species over that distance in accordance with the latitudinal attenuation theory. Such an attenuation could be simply a consequence of variation in species' dispersal capacity, that is, ability to travel long distances and perhaps to prolong the period of competence. The distances across open ocean between the major reef systems suggest that there could be dropout of species along the way.

8.3.2.2 Short-Lived Pelagic Dispersal and Philopatry

A latitudinal diversity gradient in the continental benthic shelf and shore fauna, if it exists, would be more difficult to explain. There is no distinct north-south current along the inner shelf but complex tidal flows and wind-driven currents prone to reversals and multiple directionality (Section 2.2.2 of Chapter 2). Along-shelf spread of species that have short lived or no pelagic larvae is likely to be a slow, incremental process, and an attenuation diversity gradient away from the source of origin is likely to be less well developed and developed less quickly than that observed in planktotrophic species along the shelf margin.

There is also an historical complication in relation to species with direct development or other forms of philopatry (e.g., gastropods of the family Volutidae and the cypraeid genus *Zoila*). An Indonesian ancestral source for them does seem to have been the case but that occurred in the Miocene, and there has been no possibility of further, ongoing expansion of Indonesian philopatric species onto the North West Shelf since then because of the tectonic event that created the Java Trough and its northern extensions. Subsequent cladogenesis and speciation of these taxa must have been *in situ* on the Australian continental shelf. The North West Shelf could be regarded as a diversity hot spot in regard to benthic shelf species, generating its own diversity and disrupting any general latitudinal diversity gradient. It is noteworthy that the North West Shelf has a very large area of benthic habitat at times of high sea level like the present and during the last interglacial. It is topographically variable, subject to significant climate variation from north to south, and profoundly affected by repeated Quaternary transgressions and regressions. It is likely that the benthic shelf and shore marine faunas have sustained a high vicariant speciation rate throughout the late Tertiary and Quaternary, and in that context, it might be expected that the North West Shelf could be a hot spot of biodiversity and regional endemism for benthic shelf fauna. That appears to be the case and there is little evidence of a diversity gradient.

8.3.2.3 *Corals*

None of the coral provinces on the North West Shelf has a history of coral cladogenesis independent of the Indo-West Pacific realm. With an assumed ancestral Central Indo-West Pacific origin and an effective delivery current, the Jablonksi hypothesis of dispersal from a low latitude center of cladogenesis would lead us to expect a latitudinal gradient in coral diversity down the length of the North West Shelf. Indeed, there is evidence of such a gradient in corals, but it is confounded by strong environmental factors.

The numbers of coral species and genera recorded for the coral bioregions provide scant support for the notion of an overall attenuation of coral diversity along the length of the North West Shelf when the shelf margin and coastal reefs are considered together (Table 8.1). However, these regions represent two different oceanic and continental reef assemblages with different connectivity regimes. Noting that the strongest latitudinal delivery mechanism (ITF-Holloway-Leeuwin Currents) is likely to most strongly affect oceanic species with pelagic dispersal capacities, we might expect the oceanic components of the coral fauna to exhibit a more conspicuous latitudinal gradient than the inner shelf species. This is based on the premise that the coastal species are likely to have shorter-lived pelagic dispersal—a notion that needs to be tested—and a multidirectional current system.

It is not possible to distinguish the oceanic and continental elements of these faunas. Their mixing, especially in the Pilbara, may well be obscuring evidence of a latitudinal gradient of the former. However, we may address the issue by considering the known species-richness figures for the oceanic reefs separately. The sequence from north to south goes: Ashmore 256 coral species, Scott 300, Rowley Shoals 216, Ningaloo Reef 217. These figures are not up to date. Taken as they are, they do suggest a north-south diversity gradient, a total loss of 40 species over 10° of latitude, although the data are compromised by the likely mixing with continental elements at both ends of the sequence.

There are confirmed records of around 318 coral species (68 genera) in the Kimberley and 301 species (63 genera) in the West Pilbara/Ningaloo, indicating a small disparity in species richness and heterogeneity between those two regions (Section 8.2.1.1). Taxonomically, there is a high level of correspondence between the coral faunas of the Kimberley and Pilbara Bioregions, the latter being a restricted version of the former. There does not seem to be a strong physical support mechanism for rapid dispersal between these two coral provinces (Section 2.2.2 of Chapter 2). If connectivity exists between them, it must be an incremental process over long time periods.

Weak connectivity and greater distance from the ancestral source (i.e., the Jablonski hypothesis) could explain the smaller coral fauna in the inshore patch reef assemblages of the more southerly region. However, it could be explained also by the very different biogeomorphic reef structures, the difference between a monsoonal climate and an arid one, and possible differences in the primary production of the two regions. In other words, known coral distribution patterns on the North West Shelf support a weak latitudinal gradient but regional environmental circumstances confound it.

8.3.2.4 Echinoderms

The data (Tables 8.2 and 8.3) clearly indicate moderate species richness in echinoderms in the Oceanic Shoals Bioregion (241 species), low diversity in the Kimberley (82 species), and a high diversity hot spot in the Pilbara (340 species). These figures for echinoderms do not support an overall north-south diversity gradient along the North West Shelf. On the contrary, the highest diversity of species in this group is in the Pilbara at the southern end of the sequence.

8.3.2.5 Molluscs

There is no overall latitudinal trend in the mollusc species richness from north to south on the North West Shelf (Table 8.4). When taxa are pooled (total 1909 species), differences in biodiversity between the bioregions may be accounted for by significant differences in habitat, the highest diversity occurring at the southern end of the shelf in the western Pilbara Bioregion (1289 species) where there is a very broad range of mangal, mud and sand flat, rocky shore, coral reef, and benthic shelf habitats.

However, evidence of a latitudinal diversity decline does appear with certain functional groups of the outer oceanic coral reef assemblages when they are considered separately from assemblages of benthic shelf and shore habitats. The species involved are predominantly planktotrophs with widespread geographic distributions elsewhere within the Indo-West Pacific realm. Two functional groups of reef gastropods are considered to illustrate the matter: reef-front herbivorous and predatory gastropods and reef flat sand-dwelling assemblages.

(a) *Reef-Front Herbivorous and Predatory Gastropods*. Reef-front functional groups of herbivorous and predatory gastropods on oceanic reefs are particularly informative. Common reef-front species found on the oceanic reefs of the North West Shelf are listed in Table 8.5 (see also Section 4.6.3 in Chapter 4). They are all planktotrophs that are widespread in reef-front habitats of oceanic reefs of the Indo-West Pacific realm.

Throughout the Oceanic Shoals Bioregion of the North West Shelf, these functional groups are almost identical in terms of their species numbers and identities—Ashmore-Cartier (37 species), Scott-Seringapatam (36 species), Rowley Shoals (36 species), and Browse Reef (31 species) with no evidence of a latitudinal diversity decline. Strong along-shelf pelagic connectivity within that bioregion is indicated. The next reefs in this latitudinal sequence are at the Montebellos (12 species) and Ningaloo (10 species), representing a species loss of around 75% and 80%, respectively. However, the term latitudinal "gradient" of species diversity does not describe this situation very well. The functional groups are virtually intact as far south as the Rowley Shoals (latitude 18°), and there is an abrupt change (species loss) beyond that. This is an abrupt step, not a gradient. The low numbers at the Montebellos (latitude 20.5°) and Ningaloo (latitude 22°) may be due to either distance-related reduction in the effectiveness of the dispersal means (Holloway/Leeuwin Current) beyond the Rowley Shoals or a much less suitable environment at the southern localities.

The Montebellos and outer reefs of the Dampier Archipelago are not, strictly speaking, oceanic reefs. The Montebello seaward reefs lie close to the edge of the Leeuwin Current, and it is reasonable to suppose that oceanic gastropods of reef assemblages are delivered there by oceanic water mixing with shelf water along the outer shelf margin (Hatcher's "Mixing Influence Hypothesis"—see Section 8.4.3.1). The Dampier Archipelago Reefs are much further away from the shelf edge and, as expected, there is very small representation of the oceanic reef-front gastropods in their reef faunas. In that location, the equivalent functional group of predatory gastropods is dominated instead by continental species, many of which are direct-developing regional endemics.

Notably, about a quarter of the Oceanic Shoals species (10) in this functional group of oceanic reef-front predatory gastropods also occur on the fringing reefs of the Kimberley, but several of these are recorded from single live specimens. The Kimberley coastal fringing reefs may be unsuitable habitat for these oceanic species, or the low representation of the reef-front gastropods assemblage may relate to weak connectivity with the shelf-edge and outer shelf reefs with the immigrants arriving in such low numbers that they are unable to support breeding populations.

TABLE 8.5 Herbivorous and Predatory Gastropods Characteristic of Reef-Front Assemblages of Oceanic Coral Reefs on the North West Shelf

Species	Ashmore	Scott	Rowley Shoals	Browse Reef	Kimberley	Dampier Arch.	Montebellos	Ningaloo
Turbo argyrostomus	1	1	1	1	0	1	1	1
Turbo chrysostomus	1	1	1	1	0	0	0	0
Trochus maculatus	1	1	1	1	0	0	1	1
Lambis chiragra	1	1	1	1	0	0	1	1
Cypraea depressa	1	1	1	1	0	0	0	0
Thais armigera	1	1	1	1	0	0	0	1
Thais tuberosa	1	1	1	1	0	0	0	0
Drupa morum	1	1	1	1	0	0	1	1
Drupa rubusidaeus	1	1	1	1	0	0	0	1
Drupa ricinus	1	1	1	1	0	0	1	1
Drupina grossularia	1	1	1	1	1	0	0	0
Morula biconica	1	1	1	1	1	0	1	0
Morula musiva	1	1	1	1	0	0	0	0
Morula nodicostata	1	0	1	0	0	0	0	0
Morula spinosa	1	1	1	1	1	1	1	0
Morula uva	1	1	1	1	0	0	1	1
Engina alveolata	1	1	1	0	0	0	0	0
Engina lineata	1	1	1	0	0	0	0	0
Peristernia nassatula	1	1	1	0	0	0	0	0
Latirolagena smaragdula	1	1	1	1	1	0	0	0

Continued

TABLE 8.5 Herbivorous and Predatory Gastropods Characteristic of Reef-Front Assemblages of Oceanic Coral Reefs on the North West Shelf—cont'd

Species	Ashmore	Scott	Rowley Shoals	Browse Reef	Kimberley	Dampier Arch.	Montebellos	Ningaloo
Latirus nodatus	1	1	1	0	0	0	0	0
Latirus polygonus	1	1	1	1	1	0	0	0
Vasum ceramicum	1	1	1	1	1	0	0	0
Conus catus	1	1	1	1	0	0	0	0
Conus coronatus	1	1	1	1	1	0	1	0
Conus distans	1	1	1	1	0	0	0	0
Conus ebraeus	1	1	1	1	0	0	1	0
Conus flavidus	1	1	1	1	0	0	0	0
Conus imperialis	1	1	1	1	0	0	0	0
Conus litteratus	1	1	1	1	0	0	0	0
Conus lividus	1	1	1	1	1	1	1	1
Conus marmoreus	1	1	1	1	0	0	0	0
Conus miles	1	1	1	1	1	0	1	1
Conus musicus	1	1	1	1	1	1	0	0
Conus rattus	1	1	1	1	0	0	1	0
Conus virgo	1	0	1	1	0	0	0	1
Conus vitulinus	1	1	1	1	0	0	0	0
Total	37	36	36	31	10	4	12	10

Kimberley and Dampier reefs are not oceanic reefs but are included for comparison.

(b) *Reef Flat Sand-Dwelling Gastropod Assemblages.* Functional trophic groups of sand-dwelling reef flat molluscs include microphagous and predatory gastropods that crawl on the sand surface or just below it on sand cays and in shallow sandy pools of reef flat and back-reef habitats of oceanic coral reefs. The families Strombidae, Olividae, Conidae, and Terebridae are chosen here to illustrate latitudinal distribution patterns exhibited by such species (Table 8.6). Sand-dwelling functional groups are significantly more species rich on reefs at the northern end of the North West Shelf (Ashmore—39 species in the chosen families) than at the southern end (Ningaloo—16 species), while the numbers at intermediate reefs are variable (South Scott, 27 species; North Scott, 9 species; Rowley Shoals, 9 species; Browse Reef, no species). While there are significant end differences (Ashmore vs. Ningaloo), any latitudinal trend is masked by major differences in the area of sand habitats at the various reefs in the Oceanic Shoals Bioregion. Ashmore has extensive intertidal and shallow subtidal sand habitats, South Scott has much less while at North Scott, and the Rowley Shoals sand habitats are limited. Browse Reef has virtually no sand habitats and is excluded from the table. Ningaloo is an outpost of this kind of oceanic reef habitat, with extensive sills of shallow sand in the back-reef, but 12 of the 16 sand-dwelling species present are terebrids, a family characterized by long-lived pelagic larvae.

The Montebellos have extensive areas of sandy lagoon, but much of it is silty and unsuitable habitat for these coral reef sand-dwelling gastropods and only seven of the oceanic species are present there. The oceanic sand-dwelling gastropod assemblage at the Montebellos is outnumbered by other species of the same genera that are continental species common on sandy habitats of fringing reefs and rock platforms of the Pilbara Bioregion inner shelf (Table 8.6).

8.3.2.6 *Mangroves and Mangals*

Worldwide, there is a strong diversity gradient in mangroves away from the Central Indo-West Pacific region of highest species richness, closely paralleling the case of coral reefs. Along the shores of the North West Shelf, there is a distinct decline in mangrove species and associated biota from north to south, and onwards down the West Coast (Section 5.2). The decline is stepped, and the steps relate closely to major climate change from the wet tropics of the Kimberley coast to the arid coast of the Pilbara. In fact, there is also a lesser decline in mangrove diversity in the reverse direction, i.e., from the north Kimberley eastwards onto the semiarid coast of Bonaparte Gulf. Species richness in mangroves correlates strongly with wet climate and the presence of large rivers and estuaries. A latitudinal species gradient is present on the North West Shelf but it correlates with stepped changes in climate, and an argument that it is causally related

TABLE 8.6 Species of Four Representative Families of Gastropods of Sandy Intertidal and Shallow Subtidal Habitats of Coral Reefs in the Oceanic Shoals, Pilbara (Offshore), and Ningaloo Bioregions

	Ashmore	Scott	Mermaid	Montebello	Ningaloo
Lambis truncata	1	1	1	0	0
Strombus aurisdianae	1	0	0	0	0
Strombus erythrinus	1	0	0	0	0
Strombus gibberulus	1	1	1	0	1
Strombus haemastoma	1	0	0	0	0
Strombus lentiginosus	1	1	0	0	0
Strombus luhuanus	1	1	0	0	0
Oliva annulata	1	1	1	0	0
Oliva caerulea	1	1	0	1	1
Oliva carneola	1	1	0	0	0
Oliva guttata	1	1	0	0	0
Oliva miniacea	1	0	0	0	0
Oliva tesselata	0	1	0	0	0
Oliva textilina	0	1	0	0	0
Oliva tremulina	1	0	0	0	0
Oliva vidua	1	0	0	0	0
Conus arenatus	1	1	0	1	1
Conus eburneus	1	1	0	1	0
Conus miliaris	1	1	1	1	0
Conus pulicarius	1	1	0	0	1
Conus quercinus	1	1	0	0	0
Conus tessulatus	1	1	1	0	0
Hastula albula	1	1	0	0	0
Hastula lanceata	1	1	0	0	0
Terebra affinis	1	1	0	1	1
Terebra amanda	0	0	0	1	0
Terebra areolata	1	1	0	1	1
Terebra argus	1	0	0	0	0
Terebra babylonia	1	0	0	0	1
Terebra cerithina	1	0	0	0	1

TABLE 8.6 Species of Four Representative Families of Gastropods of Sandy Intertidal and Shallow Subtidal Habitats of Coral Reefs in the Oceanic Shoals, Pilbara (Offshore), and Ningaloo Bioregions—cont'd

	Ashmore	Scott	Mermaid	Montebello	Ningaloo
Terebra chlorata	1	0	0	0	1
Terebra columellaris	1	0	0	0	0
Terebra conspera	1	0	0	0	0
Terebra crenulata	1	1	1	0	1
Terebra dimidiata	1	1	0	0	1
Terebra felina	1	1	1	0	1
Terebra funiculata	1	0	0	0	1
Terebra guttata	1	1	0	0	0
Terebra maculata	1	1	1	0	1
Terebra nebulosa	1	1	1	0	1
Terebra undulata	1	1	0	0	0
Terebra subulata	1	1	0	0	1
Total	39	27	9	7	16

to distance from a source is hard to sustain against an equally plausible argument that it is due to change of habitat.

8.3.3 Section Summary and Conclusions

Down the West Coast, south of Ningaloo Reef, there is a rapid attenuation of tropical species. It is the defining feature of the transition zone between the tropical and southern temperate shelf faunas. However, evidence for a north-south diversity gradient along the 1000 km of the North West Shelf is equivocal and varies between the taxa and functional groups of marine organisms.

Available data suggest that there may be a decline in species richness of planktotrophic oceanic reef species, that is, of oceanic corals and associated reef assemblages, from the Sahul Shelf to Ningaloo Reef. However, for gastropod molluscs at least, there is no gradient and the decline is abrupt beyond the Rowley Shoals. This is interpreted as most likely a result of the greater distance and loss of connectivity between the Rowley Shoals Reefs and the reefs further south, but less suitable habitat in the more southerly reefs may also play a part.

There is also ambivalent evidence of a latitudinal diversity gradient in the coral fauna of inner shelf fringing and patch reefs. A smaller coastal coral fauna in the Pilbara, compared to that of the Kimberley Bioregion, may be thought of as a latitudinal reduction in diversity, although there are significant environmental differences between these two coral provinces that are not related to latitude or distance from the assumed ancestral source.

There are similar differences between a low diversity mangal province in the Pilbara and a moderate-high diversity mangal province in the Kimberley, but again, there are major environmental differences between these two regions that have nothing to do with latitude or distance from a source.

Excluding coral reef and mangal assemblages, available evidence does not support the notion of a general latitudinal diversity gradient along the length of the North West Shelf. On the contrary, data on overall diversity patterns of benthic shelf and shore molluscs and echinoderms indicate a diversity low spot in the Kimberley and a hot spot in the West Pilbara a thousand kilometers further south. The following hypotheses are suggested to explain the pronounced regional differences in benthic and reef biodiversity of the inner shelf:

- higher nutrient levels in the West Pilbara due to a narrow shelf and greater input of nutrients from the deep sea as a result of shelf break upwelling;
- weak connectivity that is networked rather than driven linearly;
- historical factors that have led to the development of a secondary center of speciation and endemism on the shelf;
- enhancement of biodiversity in the shelf fauna of the West Pilbara as a result of introduction of nutrient tolerant oceanic reef species from the Holloway Current by means of mixing of ITF and shelf water long the narrow outer shelf.

The conclusion could be reached that, on the North West Shelf, strong regional environmental factors override any latitudinal diversity gradient effect that results from dispersal from a source. There is evidence of a latitudinal decline in the oceanic shelf-edge reef fauna, driven by the ITF-Holloway-Ningaloo Currents, but it does not apply to the benthic and shore faunas of the shelf itself.

A center of diversity of benthic shelf species in the Pilbara may have derived from the Central Indo-West Pacific region in ancient times, but its high diversity does not depend on regular recruitment or genetic input from that or any other source. Rather, the North West Shelf is a modern center of diversity and evolution of continental shelf fauna in its own right but with ongoing, episodic input of oceanic reef species along its margin, especially at the narrow western end of the shelf, delivered by mixing of ITF-Holloway-Leeuwin Current and coastal water.

8.4 CONNECTIVITY

Connectivity here refers to exchange of recruits between breeding populations of plants and animals that are spatially separated from each other. In the marine environment, connectivity is commonly, but not always, achieved at the larval development stage.

Wise and effective management of any species, or community of species, are critically dependent on an understanding of connectivity. It is a key ecological process that fosters resilience, recovery from localized disaster, and the ability to spread when times are good. In the longer term and at broad spatial scales, connectivity, or the severance of it, is also a key evolutionary process that fosters adaptation to a changing environment and the expansion of species' territories.

It is often said that a major ecological difference between terrestrial and marine life is the much greater interconnectedness of marine animal populations by virtue of their planktotrophic larvae and pelagic dispersal. However, there is increasing evidence that local and regional populations of widespread species that apparently have a capacity for long-distance pelagic dispersal do not always utilize that potential. Self-recruitment may be the norm that sustains populations through their breeding cycles. For most marine species, wide dispersal occurs as rare events that serve a different purpose. The general paradigm that most marine populations are broadly "open" with wide larval dispersal[66-69] may be misplaced or at least overstated.

In the context of the North West Shelf, much attention has been given to along-shelf connectivity between the shelf margin coral reefs that lie in the path of the Holloway-Leeuwin Currents. Coral reef invertebrates are predominantly planktotrophic, and with such an effective means of dispersal, connectivity between the shelf margin reefs is potentially possible. As an along-shore transport mechanism, the Holloway-Leeuwin Current is variable and probably intermittent at geological timescales. With this limitation, there is no doubt that flow of ITF water onto the Sahul Shelf and southward along the shelf margin has been a dominating influence in the region.

However, linear connectivity of this kind is unlikely to be a general rule that applies generally on the North West Shelf. Benthic shelf communities are not aligned linearly, but in mosaic patterns and coastal currents are mostly tide-driven and multidirectional. Many benthic invertebrates do not have long-lived pelagic larvae, and some do not have pelagic larvae at all. Reproductive connectivity among these benthic communities is likely to be very complex, regionally constrained and in a constant state of flux.

Nevertheless, biological connectivity with the superdiverse East Indies Triangle, established historically and maintained in modern times by the ITF (at least for planktotrophic species), has been a principal driver in the biogeographic history of the North West Shelf.

8.4.1 Levels of Connectivity

Connectivity operates at many levels. At one extreme, the local ecological level among adjacent breeding populations, it means a regular process operating at relatively short timescales. It is important to conservation and sustainable resource management on a year-to-year basis. At the other extreme, connectivity operates regionally at longer timescales, where distances are great and rates of propagule exchange are low. This level of connectivity has little to do with population ecology and demography but is responsible for expansion of species' geographic range and the establishment or reestablishment of populations distant to the natal one. From a distant source, with connectivity that is a low rate event, recolonization is likely to be a long-term, uncommon, and erratic process. Every possible intermediate level of connectivity exists, tempered by varying pelagic and local dispersal mechanisms.

Benthic shelf and shore habitats, including coral reef habitats, are characteristically patchy and discontinuous. Breeding populations of the species that inhabit them are spatially isolated from each other to varied extent. In sessile species, connectivity between separated breeding populations is commonly achieved by means of pelagic larval dispersal. This is generally true of vagile animals like reef fishes as well although they may also have a capacity to migrate as adults between adjacent populations.[70,71] The capacity to exchange larval recruits between populations depends on the duration of pelagic larval life and the velocity, direction, and seasonality of current flows. In the case of reef fishes, it also depends on the ability of the larvae to control where they are and where they go. Late-stage larvae of most reef fishes are not passive drifters—they are efficient swimmers with sensory and behavioral adaptations that enable them to seek and locate suitable habitat for settlement.[72,73] Marine invertebrates have less ability in this regard, but they are able to choose to settle or not at the sites to which they are carried and to defer settlement for a time if the recognition cues of site suitability are not met.[74]

Local dynamics of tropical reef fish populations are sensitive to the rates of exchange of individuals among populations, and this is probably true of most benthic shelf, shore, and reef animals. Three demographic states may be recognized as follows (Figure 8.6):

- when exchange is slight the local populations function independently;
- when it is high they function as a single large, but spatially subdivided population;
- when it is intermediate the exchange buffers fluctuations in abundance of the otherwise separate local populations.

Local clusters of spatially separated breeding populations that are demographically interacting may be thought of as a "metapopulation"

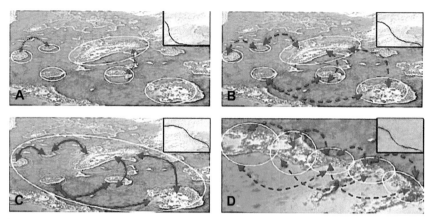

FIGURE 8.6   A hypothetical case of a fish species' "metapopulation" in a complex of coral reef patch reefs (after Sale and Kritzer[78], fig. 1). Each panel shows a patchy array of reef habitat, some of which is occupied (ovals = local aggregations of fish). Dispersal (chiefly by larvae) among sites is shown by arrows, graded to show slight (A), moderate (B, D), or extensive exchange (C). Mean scale of dispersal is shown as a graph of proportion of larvae (y-axis) against a distance from source (x-axis) in the upper right corner of each panel—mean dispersal distance is least in (A), intermediate and identical in (B) and (D), and greatest in (C). Cases (A)–(C) differ only in the scale of dispersal relative to the scale of patchiness of habitat, yet yield: (1) essentially independent local populations (A), (2) a metapopulation (B) in which local populations are sufficiently connected by dispersal for some interaction, and (3) a single but subdivided population (C) occupying a number of patches of habitat. Case (D) is typical of regions where coral reef habitat is more contiguous, yet the spatially explicit mating pattern and scale of larval dispersal still provide a functional metapopulation even though patch structure is primarily an analytical construct.

or a "population of populations".[75–78] Metapopulations have particular relevance in regard to networks of patchy marine benthic shelf and reef habitats. The limits of a metapopulation are determined, on a species by species basis, by the dispersal capacity of the organisms in question and by their distances apart and the effectiveness of circulation.

This has clear relevance to biogeography and management. At local demographic levels, connectivity within metapopulations is a key factor in their resilience and response to environmental variability. When local population extinction occurs as a result of environmental stress, reestablishment by means of propagules from an external source may be essential. When propagules cannot be sourced locally, through normal metapopulation exchange processes, the recovery and regional presence of a species may depend on rare, episodic, or low-level connectivity with remote populations at longer timescales.

Low-level connectivity, at longer timescales, may be especially important to populations near the margin of species' geographic range, especially in times of climate change. At this scale, connectivity is a dynamic, perhaps unpredictable process. Ocean currents, tidal flux, storm frequency, and seabed topography, all of which affect larval dispersal, change drastically over time. Past climate change events affecting dispersal have reshaped the distribution patterns of marine organisms and will do so again. The biogeographic distribution patterns we see today have an historical causal element and cannot be interpreted solely in terms of contemporary dispersal. This is especially the case in the transition between the tropics and temperate zones, such as the west coast of Western Australia.

The marine environment is often perceived as "open" with broad scale pelagic larval dispersal. This is only more-or-less true and can be regarded only as a common circumstance. It applies only to planktotrophic species, and even within that group, there are many exceptions. As a working principle for marine conservation or fishery resource management purposes, it must always be qualified. It cannot be assumed that there is panmixa between geographically widespread coastal species. On the contrary, there is increasing evidence that self-populating events are more frequent than long-distance connectivity.[79,80]

8.4.2 The Importance of Life History

8.4.2.1 Larval Development

An organism's mode of dispersal and the physical (environmental) means by which it is given effect are both causally linked to connectivity potential. Many nektonic marine fishes and other vertebrates regularly migrate as adults between feeding and breeding areas. Marine benthic invertebrates have diverse means of dispersal, and while pelagic larval dispersal by means of currents is the most common and the most rapid, especially among tropical reef species, many benthic species do not practice it. Direct-developing gastropods, for example, that have no pelagic larval stage, may disperse as adults by crawling across the seabed—at a snail's pace—or by episodic transport of adults or juveniles displaced by means storm surge or exceptional tidal currents. Accordingly, consideration of connectivity requires knowledge about reproductive seasons, spawning/mating processes, larval development strategies and dispersal capacities, and the relationship of these things to local and regional oceanographic circumstances, especially seasonality and directions and rates of flow of currents that are the prime means of pelagic dispersal.

Dispersal mode (and thereby connectivity potential) is closely linked to reproduction and larval development strategies that are infinitely varied. There are two general categories:

1. Pelagic larval development, either:
   - lecithotrophic—where the larvae do not feed themselves until settlement but rely on food provided within their bodies and tend to have pelagic lives of short duration;

- planktotrophic—where later stage larvae feed themselves in the plankton and have long-distance pelagic larval dispersal capacity (although that long-distance potential may not always be utilized).
2. Philopatry, i.e., reproductive strategies by which the larvae are retained at their birthplace (viviparity, oviparity, and ovoviviparity) and there is no pelagic larval stage so that dispersal must be at the adult stage, or as posthatchling juveniles.

There is a vast array of dispersal modes between these extremes although variants of pelagic larval dispersal are the most common in tropical benthic shelf communities. Total philopatric reproductive strategies, where there is no pelagic larval stage at all, are most common in the cooler waters of higher latitudes and in the deep sea but also occur frequently among benthic shelf species of the tropics. The significance of connectivity is very different for oceanic reef communities where planktotrophy predominates and benthic shelf and shore communities where short pelagic larval life is the norm and total philopatry common (Figures 8.7 and 8.8).

8.4.2.2 The Duration of Pelagic Larval Life

Among the species with pelagic larvae, there is great variation in the duration of normal planktonic life. There is also variation in the period of competency, that is, the period after the larvae become mature enough to metamorphose and settle into the adult habitat, should they find one.

Reproductive and larval development strategies of scleractinian corals are extremely varied involving both asexual and sexual processes.[81]

FIGURE 8.7   A female *Cymbiola oblita* nurturing a gelatinous egg mass she has laid on a stone on the sea bed—Cape Preston, Pilbara Bioregion. Like all volutid gastropods, there is no pelagic larval stage in this species; the young hatch directly from the capsules as shelled snails and begin their benthic life in the place of their birth. Each capsule contains a single fertilized egg, the fecundity of the females is very low compared to gastropods that spawn and produce pelagic larvae, and dispersal is very restricted. *Photo: Barry Wilson.*

FIGURE 8.8   A female *Cypraea talpa* tending her egg mass (pustulose, bilobed mantle spread over the shell): North Maret Island, Kimberley Bioregion. Female cowries cover the egg mass until the larvae hatch. In this species, like most tropical cowries, there are several hundred eggs in each capsule. Each one develops and the larvae hatch as advanced pelagic veligers that are capable of swimming and feeding themselves in the water mass. There is a pelagic dispersal phase lasting one or more weeks. *Photo: Barry Wilson.*

Sexual reproduction may involve brooding after internal fertilization or mass spawning with external fertilization, but in both cases, the end products are pelagic planula larvae. However, the period of competence of the pelagic planulae varies considerably. Planulae released by brooding corals may settle virtually immediately.[81,82] Some species may settle within 48 h.[83] Most coral mass spawners have been shown to settle after 4-6 days[84] although some may be competent for up to 105 days.[85]

There is another important factor when the connectivity of coral populations is considered. These colonial sessile animals are capable of asexual reproduction, and once established, a coral population may not need regular recruitment to sustain a stable demographic state over many years. Rates of connectivity to sustain a coral population may be very different to that required for reef invertebrates and fishes that have short generations.

There is similar variability in the duration of pelagic larval life among benthic molluscs. Chitons and bivalves include some brooders but most release pelagic larvae that are of short duration (i.e., several days) and either lecithotrophic or planktotrophic. Gastropods of lower phylogenetic levels are generally mass spawners with either short-lived lecithotrophic or planktotrophic larvae. In gastropods of higher phylogenetic levels, capsular development is the general rule whereby the larvae are protected and sustained within gelatinous strings or capsules at the laying site and released as swimming veligers only at a late stage of development. In such species, early larval development is nourished at the natal site, but advantage is also taken of the rich food resources available in the plankton near the end of larval development, and there is opportunity for pelagic dispersal as well. For example, in most tropical cowries, females incubate their eggs and early larvae in gelatinous capsules for periods between 11 and 18 days followed by a pelagic veliger phase lasting several days.[86,87] Upper littoral, rocky shore species of the tropical genus *Echinolittorina* have pelagic veligers lasting up to 4 weeks.[88,89] Tonnoidean gastropods have long-lived pelagic veligers, sometimes lasting many months.[90] There are many examples of gastropods that brood eggs or larvae within body cavities of the mother or lay eggs externally but nurture them until hatching. In extreme cases, there is no pelagic larval stage at all, the hatchlings emerging from the capsules as miniature crawling snails, e.g., all species of the family Volutidae[91] and cowries of the genus *Zoila*.[87] Brooders and direct developers are commonplace among benthic shelf and shore gastropods but rare in oceanic coral reef communities.

Echinoderms generally have moderately long-lived, planktotrophic, and pelagic larvae ranging between 10 and 30 days duration, but for some species, this period is less than 9 days, and for others, it is up to 50 days.[92–94]

Nearly all bony reef fishes have a pelagic larval stage, whether the eggs are pelagic, demersal, or brooded.[73] A rare exception has been described in the pomacentrid genus *Altrichthys* where larvae hatch from demersal eggs and are tended by their parents without there being a pelagic phase.[95] The pelagic larval duration of reef fishes averages about 1 month but varies enormously.[96,97] In some damselfishes, the pelagic stage may last only a week,[98,99] while at the other extreme, in some porcupine fishes, it may last more than 64 weeks.[100]

Marine species with long-duration larvae tend to have wide distributions while species with short-duration larvae, or lack a pelagic larval stage, have more restricted distributions.[101–103] However, duration of larval life is not the only factor that determines the breadth of species' geographic distributions. The asteroid *Astropecten polycanthus*, a common species on the North West Shelf, has a short larval life of 3-4 days, yet it is a widespread species in the Indo-West Pacific.[104]

As a broad generalization, it might be said that, in the tropics, most benthic shelf, shore, and reef invertebrates and fishes have pelagic larvae with a duration of a few days to several weeks, but there is enormous variation from zero to many months. Because of this variation, it is difficult to make credible generalizations about connectivity in the sea, unless the life cycles of the organisms in question are known.

Even when the potential duration of pelagic larval life is known, attempts to determine connectivity potential by relating that to distance and ocean current velocity or storm frequency are fraught. Species' potential for long-distance dispersal may often not be realized because of mechanisms by which competent larvae find their way back to their natal site.

Retention of recruitment to natal sites and short-distance dispersal are prevalent among reef fishes[73,105–109] and may be the norm in corals and other invertebrates of reef communities.[110–112]

Accordingly, long-distance pelagic dispersal potential may have little relevance to the demographic dynamics of populations, especially in coral reef communities. Species management programs should not assume wide dispersal and genetic panmixa. However, long-distance dispersal capacity may be crucial to the expansion of species' geographic range and the recovery of remote populations that have suffered local extinction.

8.4.3 The Ecological and Evolutionary Significance of Life History

8.4.3.1 *Apparently Conflicting Strategies*

Reproductive strategies of all species serve two seemingly conflicting ecological and evolutionary requirements, and the balance struck between them is as varied as the strategies themselves. Larval retention (at the natal site) is an adaptation for individual and population survival. The larval birthplace, *ipso facto*, is a suitable place for a larva to establish itself for adult life, and there is a strong evolutionary imperative to ensure its own survival by returning to the place it was born. There is also selection pressure, at population level, for the natal population to be replenished. Reef fishes, with their relatively advanced sensory and locomotory capacities, have evolved diverse adaptations in response to these selection pressures. In contrast, long-distance larval dispersal is an adaptation for expansion of the species' geographic range and is important in circumstances when local populations are stressed by environmental change and recolonization is necessary.

The evolutionary and ecological outcomes of these diverse strategies are profoundly different. It is known from the fossil record that marine mollusc taxa with long-distance larval dispersal capacity have lower rates of extinction than taxa that have short-lived veliger larvae.[58] On the other hand, taxa that have short-duration pelagic larvae and limited dispersal capacity, or that practice philopatry of one form or another, have greater propensity for population fragmentation and vicariant speciation.

In the biogeographic context, marine communities like coral reefs that are characterized by species with pelagic larvae and a capacity for wide dispersal, even if that potential is rarely exercised, are the most likely to respond to change by adjustments to geographic distribution. Benthic shelf communities, with a greater prevalence of philopatry and short-range pelagic dispersal, are the most likely to suffer local disturbance, restrictions of range, vicariant speciation, and local population extinction.

8.4.3.2 Sources and Sinks

While a local population may be more-or-less self-recruiting, it may be also a "source" that provides recruits to remote areas or a "sink" that receives recruits from other areas (or both). Identification of source populations is vital to conservation and resource management. They may play crucial roles at either local ecological connectivity levels or regional levels on geological timescales. At the margins of a species' geographic range, populations may be sinks that are not capable of replenishing themselves and depend on ongoing external recruitment from a remote source. Referring to maintenance of coral reefs on the subtropical West Coast, Hatcher[113] proposed a recruitment hypothesis defined as: "advective delivery of the larvae of reef-building organisms ... [that] replenishes populations after local extinctions, and maintains populations of reef organisms where they are not reproductively viable."

Climate change with changes of sea level and transgressions and regressions of the sea over coastal land result in the alternation of terrestrial and marine habitats of continental margins. Source populations with pools of appropriate emigrants and the means of their delivery are essential to the total reconstruction of marine benthic communities in these circumstances.

Clearly, transformation of a grassy plain or eucalypt woodland to inner shelf benthic habitat for sessile invertebrates and the construction by them of habitat for demersal fishes are a cataclysmic process. (Just as it was for the terrestrial inhabitants that were displaced.) This scenario has played out several times during the Quaternary with very large tracts of coastal land subjected to it (the last time during the period of human occupation). The processes of ecological succession must come into play, and connectivity with a source population is pivotal. Species with long-distance dispersal capacity may play important colonizing roles, and chance must be a big factor. Species without pelagic larvae are likely to be late arrivals.

8.4.4 Connectivity on the North West Shelf

8.4.4.1 Connectivity Between Oceanic Coral Reefs of the Shelf Margin

Recent discussion of this matter has focused most attention on along-shelf connectivity between the shelf margin coral reefs that lie in the path of the Holloway-Leeuwin Currents[114–116] (Figure 8.9). The notion has arisen of these reef communities having a linear, source-sink relationship characterized as "island hopping" or "stepping stone" with each reef being a sink for recruits from up-current reefs and a source of recruits for reefs downcurrent. This concept applies to planktotrophic species. It has little relevance to species that lack a pelagic larval stage but is important in regard to the oceanic reefs aligned along the margin of the North

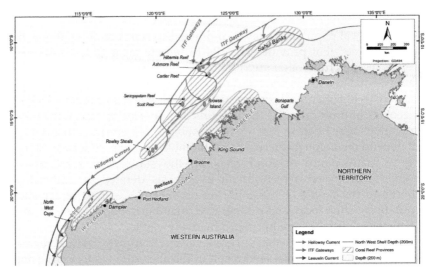

FIGURE 8.9   Flow of the Indonesian Through Flow, Holloway Current, and Leeuwin Current along the North West Shelf margin in autumn and winter, with potential connectivity between the Oceanic Shoals and Pilbara coral reef provinces. It is proposed that the shelf margin mixing zone includes the offshore part of the Pilbara but that mixing may not extend into the Kimberley zone and that connectivity of the oceanic reef biota with that of the Kimberley coral reef province is weak. The distance of the mixing zone at the shelf margin from the shore of the Canning coast precludes connectivity of the oceanic ecosystems with coastal waters in that region. *Drawn by URS Australia.*

West Shelf whose reef communities comprise mainly planktotrophic species.

Although larval retention to natal sites may be prevalent as a demographic process in coral reef communities, capacity for long-distance pelagic dispersal is also probably common. Duration of larval life is not the only factor involved, but estimates of the maximum duration of pelagic life are a good place to begin consideration of the potential of a reef species for long-distance dispersal and recolonization.

As a starting point for investigation of the potential for connectivity of reef organisms of the North West Shelf marginal reefs by means of transport by the Holloway Current, Condie, and Andrewartha[114] took a 14-day larval duration as a working standard. They found that connectivity among the West Pilbara reefs of the Montebellos, Dampier, Barrow, and Ningaloo may be common over 1-2 week timescales under favorable seasonal conditions, well within the probable dispersal capacity of most reef species. However, their modeling indicated that there was significant interannual variability in that connectivity potential, related to variability in strength of current flow. Connectivity potential would be highest in La Nina years. (It would be zero in glacial climate phases when the Holloway Current is not active.)

Larval exchanges between reef communities of Ashmore, Scott, and the Rowley Shoals (distances apart >400 km), and between the latter and the reefs of the West Pilbara and Ningaloo, are likely to be infrequent and require either pelagic larval durations in excess of 4 weeks or oceanographic conditions outside their model parameters (e.g., cyclonic storm conditions).[111] Even under favorable seasonal conditions of strong current flow, the time required for such long-distance dispersal is likely to be greater than the normal competent pelagic life of most species. In those circumstances, there is unlikely to be regular delivery of sufficient numbers of immigrant larvae to have a demographic impact, i.e., connectivity in ecological time is unlikely.

Studies of genetic structure in Scott Reef, Browse Reef, and Rowley Shoals populations of two coral species found significant differences between a broadcast spawner (*Acropora tenuis*) and a brooder (*Seriatopora hystrix*).[115] There was a clear lack of panmixis between populations of both species, much more evident in the brooder. Populations of both these corals self-seed, and in spite of potential larval duration of several weeks and potential for long-distance pelagic dispersal, they do not regularly disperse between the remote reef systems of the North West Shelf margin. This result aligns with the conclusions of Condie and Andrewartha.[114]

In a similar study of genetic structure in populations of a reef fish (*Chromis margaritifer*) at Scott Reef and the Rowley Shoals, no evidence was found of regular exchange of demographically significant numbers of larvae between these reefs.[116] This fish has a mean pelagic larval duration of 35 days that would be long enough to travel the distance from Scott to the Rowley Shoals but that potential is not utilized at the demographic level. There is evidence of sporadic long-distance dispersal that maintains genetic connections between these reefs, but populations of this fish on these geographically isolated reefs are demographically closed.

These studies suggest that the idea of regular, linear, "stepping stone" connectivity between the coral reefs of the North West Shelf margin is probably not sustainable at the demographic level. However, this does not preclude the possibility of episodic, infrequent connectivity, with pulses of larvae delivered from one reef to another by means of exceptional oceanographic events.

The existence of these isolated reefs, *ipso facto*, demonstrates that connectivity at geological timescales must have taken place in the past. These oceanic reef communities suffered severe disturbance, and probably some local species extinction, as a result of changes of sea level and metocean conditions during the Holocene. During the LGM when sea level was around 125 m below the present, these reefs were destroyed and, for several thousand years, coral communities were restricted to those species that could live in the exposed, high energy, and possibly nutrient-enriched conditions of the very steep perimeter wall (see Section 2.1.6.3 and Figure 2.12 of Chapter 2). Reestablishment of their species' assemblages must have occurred during the final phase of the post-LGM transgression, and they are little more than 6000 years old. Establishment of the present reef ecosystems depended not only on the sea level still stand but also on the resumption of the ITF-Holloway Currents and oligotrophic conditions.

The source population from which the contemporary slope atoll reef communities were reconstructed is problematical. An hypothesis has been proposed that deep populations may serve as refuges and a source of recruits for shallow reef communities.[119] While this clearly has relevance to contemporary reef communities under stress, it could only have been important in recovery following a low sea level event if the deep refuge habitats supported the same suite of species as the modern shallow reef habitats. That would seem to be unlikely. A stepping stone connectivity process that reestablished these populations from source populations on the Sahul Shelf, or even Indonesia, as rare events over time, though of little ecological (demographic) relevance, remains a credible hypothesis in terms of evolutionary and biogeographic processes and geological time.

The process of reestablishment of shelf-margin reef assemblages may have been an erratic process over the past 6000 years. Whether it has reached a point of relative stability at the present time is a moot point. Circumstantial evidence supporting present stability may be seen in the presence at all the reefs in the Oceanic Shoals Bioregion of a suite of predatory gastropod species that characterize reef-front habitats of Indo-West Pacific oceanic reefs (Table 4.4 of Chapter 4 and Table 5.3 of Chapter 5). These species are all planktotrophs, and the presence of the same suite of them at all the oceanic reefs along the shelf margin demonstrates that delivery of pulses of larvae must have occurred, at least over geological time sufficient to establish or reestablish stable populations. If these species communities are not sustained demographically by regular recruitment from elsewhere, they must at least have accumulated species by rare recruitment events over time, to the point that all the species of the typical reef-front assemblage in the regional recruitment pool are now present and functioning demographically as self-seeding breeding populations. (This raises the interesting matter of rules of ecological assembly and succession but that is another story.)

Cyclonic storms are probably the extraordinary oceanographic events that achieve rare but rapid long-distance transport of pulses of pelagic larvae along the shelf margin. The North West Shelf is a region prone to cyclonic storms and surface storm water may transport suspended particles long distances very quickly.

8.4.4.2 Benthic Shelf Communities

Unlike the linearly arranged series of oceanic reefs along the shelf margin, lying in the path of a major ocean current, inner and middle shelf benthic communities are normally arranged in mosaic patterns determined by substrate type and seabed topography. Water circulation capable of transporting pelagic larvae is multidirectional and subject to daily change. Tidal flux and surface wind waves are the primary connectivity drivers in these zones. Networks of interbreeding metapopulations are likely to be extremely complex, especially in macrotidal areas, and source-sink relationships would be very difficult to identify.

An excellent example of patchy benthic habitats on the inner shelf is seen in a cluster of small, nearshore patch reefs in the West Pilbara off Onslow (Figure 4.38 of Chapter 4). The distances apart of these reefs are rarely more than 8 km, and the velocities of tidal and wind-driven currents are such that regular, demographic connectivity among them is probably well within the normal larval dispersal capacities of most species. This complex of patch reefs is very likely to operate, under normal conditions, as a metapopulation in the way illustrated by Figure 8.6. But this is not the only consideration. A study of the trajectory of coral larvae around inshore and mid-shelf reef areas of the West Pilbara found that extreme events such as cyclonic storms may have significant importance in the connectivity between the fringing and patch reefs of that region.[120]

Connectivity among the fringing coral reefs of the Kimberley Bioregion is likely to be effected by similar processes of tidal flux and cyclonic storms. In this case, tidal current velocities are extreme and their flow is multidirectional, depending on ebb and flow conditions. Even late-stage fish larvae may have no control over where they go. An explanation of the high coral diversity of Kimberley Reefs and relatively low diversity of reef fish and other invertebrates may be found in this phenomenon. Under such tidal conditions, reef populations of fish and most invertebrates may have difficulty in sustaining regular recruitment but corals may manage because of their asexual reproduction and a capacity of their communities to survive for long periods without a need for regular larval recruitment.

Intertidal mangrove, beach, and rocky shore populations present yet another scenario. Here, species' populations are arranged more or less linearly (more complex mosaics in the archipelagos). Characteristically, the plants and animals of these habitats have long-distance dispersal abilities. For example, rocky shore littorinid gastropods of the genus *Echinolittorina* have planktotrophic larvae with a larval life of up to 4 weeks.[84,85] Connectivity between populations of these very common little snails on rocky shores of the North West Shelf is feasible on these figures, even past long stretches of the shoreline that have no rocky shore habitat, but the rate of it and its importance in sustaining larval exchange at the demographic level is unknown.

Connectivity of benthic invertebrates that lack a pelagic larval stage is more problematical. For example, the many endemic species of the gastropod family Volutidae that are found on the North West Shelf are burrowers in fine carbonate sands, a substrate type that is widespread but patchy on the inner and middle shelf. The dispersal of these animals is unlikely to be affected at all by wind-driven and tidal currents, and it is difficult to see how any level connectivity is maintained between their isolated populations, except perhaps by adults or juveniles rolled across the seabed by cyclonic storms, or over geological timescales by rearrangement of seabed sediments during phases of sea level change. No doubt this is the explanation of so many polytypic species and short-range endemics in this family.

8.4.4.3 Cross-Shelf Connectivity

An important question that is difficult to answer is the extent of connectivity between the outer shelf margin coral reefs of the Oceanic Shoals Bioregion and the fringing and patch reefs of coastal waters of the inner shelf in the Kimberley and Pilbara Bioregions. Referring to larval recruitment into coral reefs of the West Coast, Hatcher[113] wrote "Mixing of Leeuwin Current water onto the shelf has a similar effect as advected water masses, but it is attenuated by the quality and quantity of coastal water with which it mixes." Delivery of coral and other reef larvae onto the outer shelf of the West Coast by this mixing process is an explanation of the existence and maintenance of the coral reef communities of the Abrolhos Islands that are located close to the shelf break. A similar process may operate in the West Pilbara whereby oceanic corals and other reef species that are tolerant to turbid, nutrient-rich conditions, mix with the continental fringing and patch reef species of the shelf. At the western end of the North West Shelf, there is little distance between the Holloway-Leeuwin Current flowing along the shelf break and the middle shelf coral reef communities of the Pilbara (Offshore) Bioregion. The mixing zone of oceanic and continental water probably extends well onto this southern end of the shelf. This mixing process is a possible explanation of the presence of oceanic reef species on the Pilbara (Offshore) bioregional fringing reefs, mixing with continental elements of the shelf benthic fauna and, thereby, enriching the reef faunas in the West Pilbara. The concept is expressed graphically in Figure 8.9. The extent to which the mixing process and the pelagic larvae these waters carry may influence the inner shelf reef communities of the Pilbara (Nearshore) Bioregion remains uncertain.

Mixing of oceanic and coastal water and connectivity between the off-shore Oceanic Shoals and coastal Kimberley Bioregions might be expected to be similar to the circumstances in the west Pilbara but the distance from the shelf margin to the shore is very much greater (Figure 8.9).

Browse and Cartier Reefs lie on the outer shelf margin, and they have the suite of oceanic reef species typical of the Oceanic Shoals Bioregion indicating that they lie within the path of ITF water or within the mixing zone with coastal water. But this is not the case with the fringing and plat-form reefs of the nearby inner shelf of the Kimberley coast. There are marked faunistic differences between the reef assemblages of the Oceanic Shoals and Kimberley Bioregions that may be because of significant differ-ences between the clear water, oligotrophic conditions of the former, and the turbid, nutrient-rich conditions of the latter. However, poor connecti-vity is a likely contributory factor.

There is circumstantial evidence suggesting that cross-shelf connectiv-ity between the Oceanic Shoals and coastal Kimberley Bioregions may be very restricted. Apart from the corals, the invertebrates of the Kimberley fringing reefs are predominantly continental rocky shore species, many of them regional endemics. Few of the oceanic mollusc species that are char-acteristic of the outer shelf and shelf-edge reefs are present on the fringing reefs of the Kimberley, and those that do occur there are in very low num-bers. Some records of oceanic mollusc on Kimberley Reefs are based on single live specimens suggesting that delivery of oceanic larvae there is infrequent.

At the center of the North West Shelf, between the Dampier Archipel-ago and the Kimberley, the Canning and Eighty Mile Beach Bioregions, and the eastern section of the Pilbara Bioregion, the distance from the shelf edge to the shore is very great (>300 km) and connectivity between coastal inner shelf and oceanic outer shelf habitats is likely to be nonexistent or limited. This may be part of the explanation for the absence of coral reefs in these bioregions.

8.4.5 Connectivity Between the North West Shelf and Other Regions

This matter relates directly to the questions of biogeographic affinities and latitudinal gradients in diversity and was discussed briefly in Section 8.1. Connectivity of the biota of the North West Shelf lies in four directions, to the north with the world's greatest center of marine diversity in the Central Indo-West Pacific (East Indies Triangle), to the east across the Gulf of Carpentaria to the east coast of Australia, to the south around the Cape Range Peninsula to the West Coast, and to the west across a wide stretch of open ocean to the oceanic islands of the Indian Ocean. The limits of the North West Shelf, separating it from adjacent shelf biogeographic regions, are set by geomorphic features that function as partial or intermittent biogeographic barriers.

In the north, there is the Timor Trough that lies between the North West Self and the complex of islands of eastern Indonesia. This deep-sea trough is passable by planktonic larvae and is no barrier to planktotrophic or swimming species but is impassable or an impediment to species that are incapable of pelagic dispersal. Since it existed it has been a partial biogeographic barrier between the marine biota of the Banda Sea and Arafura and Timor Seas. However, there is no barrier along the continental margin between the Sahul Shelf and the southwestern coast of Irian Jaya (Figure 8.1).

In the south, there is the Cape Range Peninsula, or rather the very narrow Ningaloo Shelf on its western side, that lies between the North West Shelf and the shelves of the West Coast. This is also a partial biogeographic barrier, passable by species capable of pelagic dispersal but not so easily by others.

The Gulf of Carpentaria in the east has been subject to a dramatic eustatic history whereby, throughout most of the Quaternary, there has been a wide land connection between Australia and New Guinea completely blocking marine connectivity between the Arafura Sea and the Coral Sea of the southwestern Pacific and the east coast of Australia (Figure 8.1B and C). This has been an intermittent biogeographic barrier, breached briefly several times during periods of high sea level like the present.[121]

Connectivity between the western end of the North West Shelf and the central Indian Ocean is only a very minor circumstance, applying to a few species. There is no physical barrier in this direction other than distance. There are current flows that could facilitate pelagic dispersal but distance probably exceeds the dispersal capacity of most species to sustain connectivity in this direction.

The Timor Trough and Cape Range Peninsula are partial physical barriers that are outcomes of tectonic activity. The Carpentaria land bridge, when it existed, was an intermittent physical barrier that was a product of eustatic sea level change.

8.4.5.1 The Central Indo-West Pacific Connection

The predominance of Indo-West Pacific species in the reef and benthic shelf faunas of the North West Shelf leads to recognition of direct past and contemporary affinity and connectivity of this biogeographic region with the species-rich bioregions of the vast Indo-West Pacific realm.[11–13] It is a central thesis of this study that connectivity with that biogeographic realm has been, and is still, the primary factor that has determined the biogeographic character of the modern marine fauna of the North West Shelf.

As discussed in Section 8.1, the affinity of the North West Shelf marine fauna with that of the Indo-West Pacific realm derives from a common ancestry in the Middle Tertiary and close proximity. It was initially established in the Miocene when the Australian plate was brought into direct contact with the Eurasian plate by continental drift[1,2] and the northern coastline of Australasia was invaded by the diverse tropical marine fauna of the Sea of Tethys.[8] The majority of tropical invertebrates being planktotrophic, free passage among the islands of Indonesia and across to the northwestern coast of Australasia has been the norm since that time. However, the Middle Tertiary fauna of the Sea of Tethys included some groups that did not have pelagic larvae. In the Late Miocene, as the deep Timor Trough and its northern extensions developed, it became a partial biogeographic barrier, impassable to invertebrates like the species of *Zoila* that lack pelagic larval dispersal ability.[87,122] Since then, connectivity between the Central Indo-West Pacific region and the Australian continental shelf has been sustained by species with a larval dispersal capacity and able to cross between the Band Sea to the Arafura and Timor Seas but was severed for species that are philopatric.

At the present time, it appears that the means of connectivity of the North West Shelf marine fauna with the Central Indo-West Pacific region is the southerly flow of water known as the Indonesian ThroughFlow (Section 2.2.2 of Chapter 2) and larval dispersal is one-way traffic. The distances involved are such that connectivity is likely to be at the "geological timescale" with limited ongoing ecological outcomes. However, this could not have been the same during Quaternary periods of eustatic low sea level. During those periods, the northwestern coastline of Australasia was continuous all the way to Irian Jaya and the northern section of it was in very close proximity to the extremely biodiverse East Indies Triangle of the Central Indo-West Pacific (Figures 8.1B and C). This situation prevailed through most of the Quaternary. The conclusion is drawn that in the Late Tertiary and much of the Quaternary, connectivity with the eastern Indonesian region was much more integral than it is today.

Through most of the last phase of the Late Pleistocene, until the onset of the Last Glacial, sea level oscillated mostly between -12 and -50 m (Figure 2.12 of Chapter 2), the Torres Strait was closed, and the position of the western coastline changed back and forth across the width of the space now occupied by the Gulf of Carpentaria. This was not simply a "land bridge" but a vast tract of land nearly 1000 km wide, almost as big as New Guinea itself (Figure 8.1B). At the nadir of the glacial with sea level at -125 m, the coastline lay even further west, along the margin of the Timor Trough (Figure 8.1C).

There is a sill, known as the Arafura Sill, between the Wessel Islands of Arnhem Land and Yos Sudarso of southeastern Irian Jaya that is 53 m below present sea level.[121,123] During the 50,000 years prior to the LGM, with sea level fluctuating but averaging around −50 m, the coastline lay mostly across what is now the Arafura Sea. There was a wide gulf of shallow benthic shelf habitat between what are now the Aru Islands and the Sahul Shelf, and at best, the Gulf of Carpentaria was a lagoon opening to the Arafura Sea by a narrow channel. With the onset of the Last Glacial period, at around BP 30,000, sea level fell below the level of the Arafura Sill and the floor of the Gulf of Carpentaria and the eastern half of the Arafura Sea were exposed. The central basin of the gulf became an isolated inland lake.

The modern bathymetry along the shelf margin of the region indicates that the coastline during most of the Quaternary was complex with many islands and shores that were probably suitable for coral reef development. Throughout this period, the northwestern coastline of Australasia curled around the arc of islands (Timor, Tanimbar, Kai, Ceram) bordering the Banda Sea (Figures 8.1B and C).

The probability that the ITF was weak or nonexistent during the Last Glacial period was discussed in Section 2.1.5 of Chapter 2. That being so, southerly flow of ITF water along the Rowley Shelf margin probably did not occur and connectivity of the coral reefs there with the ancestral source of recruitment was weak or nonexistent. However, the continuous Australasian coastline being so close to the eastern boundary of the Banda Sea, it is hard to imagine that the coastal marine fauna of the northern part of the Sahul Shelf northern section was anything but a part of the species-rich East Indies Triangle intimately connected with it in demographic ways by proximity and local currents. This level of connectivity would have prevailed through much of the Quaternary, and the present species richness of the Sahul Shelf and the Kimberley coast may be understood in that context.

It needs to be emphasized again that while pelagic connectivity with the modern Central Indo-West Pacific region, dependent on flow of the ITF, is ongoing but intermittent, connectivity may have ceased altogether after development of the Timor Trough for philopatric invertebrates like the gastropod genus *Zoila* and the species of Volutidae. *Zoila* has several extinct Middle Tertiary/Pliocene fossil species in Indonesia and Irian Jaya and several living species on the continental shelf of Australia. The volutid genera *Cymbiola*, *Volutoconus*, and *Amoria* have a similar evolutionary history and are discussed further in Section 8.5 in regard to distribution patterns and Section 8.7 in regard to endemism.

There are large historic collections of Indonesian Tertiary and Quaternary molluscs in various museum collections, especially the Rijksmuseum van Natuurlijke in Leiden. It would be a worthy project to examine them to test an hypothesis that taxa with direct development that lost connectivity with their Tethyan ancestors with the advent of the Timor Trough subsequently radiated more vigorously on the Australian shelf than did taxa with planktotrophic larvae that maintained connectivity via pelagic dispersal. Another interesting mystery to be resolved is the age of species like *Volutoconus grossi* and *Amoria canaliculata* and the route that their ancestors took that carried them to (or from) the Australian east coast.

8.4.5.2 The Eastern Australian Connection

At the present time, Torres Strait is an open passage connecting the Coral Sea with the Gulf of Carpentaria, and the Northern Australian coastline is continuous from east to west (Figure 8.1A). However, between the top of Cape York and the southern coast of Papua New Guinea, there is a sill about 12 m below contemporary sea level.[121,123] Inspection of Figure 2.12 of Chapter 2 shows that through the last of the Quaternary climate cycles, this sill was breached by sea level rise three times. First for a little more than 10,000 years at the peak of the last interglacial (BP 115,000-125,000 years), again for about 5000 years (around BP 100,000 years), and again during the present Holocene high sea level stand (BP 0-6000 years). This amounted to a total of about 21,000 years when the Torres Strait was open and more than a hundred thousand years when it was closed. Similar situations occurred through the earlier climate cycles of the Pleistocene. In short, potential connectivity between the western Arafura Sea and the eastern Coral Sea has been restricted to the relatively brief periods of time when the Torres Sill has been breached and the Torres Strait has been open.

We need to be thinking of the Arafura and Timor Seas as an extension of the Central Indo-West Pacific region, and the Coral Sea and the coast of North Queensland as part of the southwestern Pacific with a separate biogeographic history, the two having had only brief periods of direct contact. There is a strong case to be made for greater biogeographic affinity of the North West Shelf with the Central Indo-West Pacific region than with eastern Australia. Nevertheless, those periods of direct connectivity have been a significant factor in the evolution of the Northern Australian marine fauna.

Water movement through the Torres Strait at the present time is a two-way flow, from the Gulf of Carpentaria eastward into the Coral Sea in December-March and in the reverse direction in April-December, the latter being the weaker flow.[124] Therefore, when it is open, there is a potential for current-assisted migration of planktotrophic, demersal, and nektonic marine species in either direction. However, it needs to be recognized that the shelf habitats on the Coral Sea side of the Torres Strait and those of the Gulf of Carpentaria are very different, the former predominantly oceanic coral reef habitats and the latter a wide expanse of shallow benthic shelf habitat. Larvae of coral reef species would have to travel 1000 km across the gulf to get from coral reef habitats on one side to those of the other. (Although coral communities have been described from Arnhem Land[125] and coral reefs have been discovered in the southern part of the Gulf[126] that may provide "stepping stones" for species tolerant to coastal conditions.)

From this, we may conclude that there may have been episodes of connectivity via an open Torres Strait, between marine species of the tropical east and west coasts of Australia, but they have been brief. Connectivity would be most effective for species with long-lived pelagic larvae (or adult swimming abilities) and those requiring sandy benthic habitats like those prevalent in the gulf. Among species that have crossed from one side to the other, we should expect to find evidence of genetic divergence of eastern and western populations because the periods of isolation have been much longer than the periods of connection.

These considerations are consistent with the recognition long ago by Charles Hedley, Gilbert Whitley, and others (see Chapter 1) of a distinctive biogeographic province, which they named the Damperian, west of the Torres Strait. Because of the Torres Sill, inundation of the Gulf of Carpentaria by the Holocene transgression must have been from the west and most of the shore and benthic shelf organisms that recolonized it must have come from source populations in the Arafura and Timor Seas. In other words, the cyclical reconstructions of the Gulf's marine fauna were the extensions of the northwestern biogeographic region.

It is possible that, in the final few thousand years of the transgression, once the Torres Sill was breached, some western species "escaped" into the Coral Sea. There they may have diverged and established incipient new species in that neighboring biogeographic region. The reverse is also feasible of course. Coral Sea species could just as easily have passed in the other direction and established new breeding populations in the gulf or on the northwest coast.

It also needs to be remembered that the Holocene eustatic high sea level event was not the first of its kind. There were several similar episodes in the Late Pleistocene and earlier, possibly even in the Late Tertiary. Species finding their way through the Torres Strait into the opposite bioregion during those earlier events may well have established new lineages in isolation (see discussion above). We should expect to find examples of such vicariant events of different ages that have led to evolutionary outcomes ranging forms distinctive genetic forms to the evolution of new species and diversification of new lineages.

In fact, there are many examples of western and eastern species and subspecies pairs that may be explained in this way. They occur in most of the major taxa of marine invertebrates and fishes.[127,128] Some molluscan examples are discussed in the next section. Some of these are readily recognizable morphologically. There may be others that are "cryptic" species, morphologically similar but genetically divergent and reproductively incompatible. This is a classical vicariant speciation process (but due to eustacy not tectonism). In biogeographic terminology, the Torres Sill operates as an intermittent biogeographic barrier.

However, fragmentation of ancestral species range by this means may not be the explanation of all of the examples of west and east species pairs. For many of them, the responsible vicariant events may have occurred much earlier, when the connectivity regime was very different. After the initial Miocene collision of the Australian and Eurasian plates, there was a wide epicontinental sea and carbonate platform between Northern Australia and the island arc terranes that later formed New Guinea.[3] For several million years, there was free connectivity between the two sides of Australia. That situation prevailed until the Pliocene when tectonic and sedimentary events rearranged the distribution of land masses and the Arafura Sea and Coral Sea became isolated from each other, except for those brief periods when the Torres Strait was open.

8.4.5.3 *The West Coast Connection*

There is a two-way connectivity between the North West Shelf and the Dirk Hartog Shelf of the West Coast, although it is principally from north to south, at least at present. At the present time, the main direction of it is a late summer-winter connection around the Cape Range Peninsula to the West Coast driven by the Leeuwin Current (Figure 2.14 of Chapter 2). But there is a minor, inshore reversal of connectivity in early- to mid-summer with inshore flow from south to north (Figure 2.15 of Chapter 2). Cape Range Peninsula and the constriction of the shelf along the Ningaloo coast play a fundamental role as a partial biogeographic barrier.

(a) *North-South Connectivity.* In the Early-Middle Miocene, the continental shelf formed a wide arch around the northwestern corner of the continent (Figure 2.1 of Chapter 2). The Cape Range of today did not exist, and while the shelf in that latitude was a little narrower than further north and south (<100 km), it was continuous, with a smooth transition from what is now the North West Shelf to the Dirk Hartog and Rottnest Shelves. This was a time of global warming, massive evolutionary radiation, and latitudinal expansion of the shallow marine fauna of the pantropical Sea of Tethys.[129] As noted above, it was also the time of a shallow sea connection between the Australian plate and the Asian plate and there was unimpeded passage for marine benthic shelf species with any development mode, not only from the Central Indo-West Pacific to the North West Shelf but all the way down the West Coast. In fact, the fossil record shows that this Miocene tropical marine fauna spread onto the southern shores of Australia and as far east as South Australia.

Subsequent periodic phases of global cooling and warming in the Pliocene and Quaternary saw the tropical northern fauna repeatedly expand south and contract again on the West Coast. However, a tectonic event in the Pliocene[130,131] that elevated Cape Range created the Cape Range Peninsula and a constriction of the continental shelf at that point, i.e., the very narrow shelf between North West Cape and Point Cloates that divided the northwestern shelf into northern and western sectors (now the North West and Dirk Hartog Shelves). The narrow Ningaloo shelf became an impediment to connectivity between the northern and western North West Shelf and the West Coast.

In interglacial periods of global warming and high sea level like the present, when the Leeuwin Current flows strongly, the peninsula is no impediment to species with long-lived planktotrophic larvae but may severely impede the passage of species with short-lived pelagic larvae and philopatry. In Quaternary glacial periods of low sea level, and when the Leeuwin Current weakened or failed, the Ningaloo shelf was even narrower and must have been a severe impediment to the passage of all marine benthic and demersal shelf species. It seems that, since the Pliocene, connectivity between the North West Shelf and the West Coast has been intermittent in geological time—most effective in periods of global cooling and only partially effective in warm times like the present. Here again is a vicariant speciation mechanism, this time a result of tectonism.

It should be noted that there is another section of the shelf-margin, northwest of the Montebello Islands, where the middle and outer continental shelf is extremely narrow and the continental slope is very steep (Figure 2.5 of Chapter 2). In glacial periods of low sea level, this may also have impeded dispersal of marine species lacking a pelagic larval stage. During the last glacial, for example, with sea level around 125 m below the present, it would have virtually isolated an "embayment" of shelf habitat 150 km long from another one east of the modern Montebellos.

(b) *South-North Connectivity.* The Leeuwin Current flows strongly from late summer through winter, but in spring and early summer, it weakens or ceases and at that time pelagic connectivity from north to south is less or not effective. The reverse appears to be the case. The summer, near-shore, and wind-driven Ningaloo Current (Section 2.2.2 of Chapter 2) carries small volumes of relatively cool West Coast water around North West Cape and onto the western end of the North West Shelf, providing an opportunity for dispersal of summer breeding, pelagic species from south to north. There is faunistic evidence that this does occur. For example, there is a living population of the temperate gastropod *Thais orbita* (Figure 8.10) on the seaward shore of Barrow Island and the Montebellos that may be explained this way. Invertebrate species such as the littorinid

FIGURE 8.10   The thaid *Thais orbita*, Barrow Island. This is a southern Australian temperate species that has isolated populations on Barrow and the Montebello Islands, perhaps as a result of the summer Ningaloo Current that flows around the top of North West Cape. (This species has late stage pelagic veliger larvae.)

gastropods *Nodilittorina australis* and *Nodilittorina nodosa* and the mytild bivalve *Brachidontes ustulatus* are examples of species that are endemic to the West Coast and the Pilbara. These molluscs are rocky shore intertidal species. North-south connectivity between their northern and western populations is unlikely to be maintained by the offshore Leeuwin Current. Assuming that they are summer breeders, summer connectivity is more likely with larval carriage from south to north via the Ningaloo Current.

(c) *Has There Ever Been a Seaway Connection?* A complication is that there may have been a Pliocene or Early Pleistocene seaway connection along the syncline between the Cape Range and Giralia Range anticlines, connecting the shallow waters of Exmouth Gulf to a West Coast gulf now represented by Lake Macleod. In that circumstance, Cape Range would then have been an island (like Barrow). There is no geological evidence of this although the lowlands of the area in question, where stratigraphic and fossil evidence of it might be expected, are now covered by Late Pleistocene sand dunes. Though conjectural, this possibility is worthy of exploration because if a seaway connection existed, it would have had major evolutionary consequences for the West Coast and North West Shelf coastal marine faunas, rather like the transient openings of the Torres Strait. This possibility was discussed by Wilson and Clarkson[122] in the context of vicariant speciation in the direct-developing cowry genus *Zoila* which has sister species on the North West Shelf and the West Coast.

8.4.5.4 A Central Indian Ocean Connection

Potential for minor pelagic connectivity exists between coral reefs of the western Pilbara and the central Indian Ocean Reefs. A planktotrophic, predatory, reef-front muricid *Drupina lobata* is present on reefs of the Montebellos, Barrow Island, and Ningaloo. This species is commonly found at Cocos Keeling, Christmas Island, and other reefs of the central Indian Ocean. It indicates past and possibly contemporary pelagic connectivity, presumably by means of offshoots from the South Java Current or Eastern Gyral Current. Input of Indian Ocean species onto the North West Shelf from this source, however, is rare and clearly overwhelmed by input of Pacific species through the ITF-Holloway Currents.

The possibility of connectivity of corals and reef fauna in the other direction, from the North West Shelf to Cocos Keeling, presumably by means of the South Equatorial Current, was noted by Veron.[35]

8.5 SPECIATION

Based on a study of distribution patterns of shallow water echinoids, Mayr[132] proposed that "the prevailing speciation process in sexually reproducing marine animals is that of geographic speciation" just as it is on land. Populations of species become geographically separated from each other by barriers to dispersal and, in this state of spatial isolation, diverge genetically in response to varied local environmental conditions. This concept has become known as the allopatric speciation model. It depends upon changing physical circumstances that sever connectivity and result in fragmentation of populations, a process known as vicariance.

Marine biologists have had difficulties with the allopatric speciation model because of a perception that most sexually reproducing marine animals have pelagic larvae and a potential, at least, for wide dispersal in the sea where robust barriers to dispersal are uncommon.[15,132–136] The marine environment has been regarded as "open" in terms of the large dispersal capacity of its inhabitants. With these perceptions, alternative ecological modes of speciation have been proposed and there is ongoing debate about the processes involved and a vast literature.[137–142]

The generality of the assumption that marine systems are open and characterized by wide pelagic dispersal has been questioned in an earlier section of this book (Section 8.4). While a majority of tropical marine fishes and invertebrates do have pelagic dispersal capacity, there is a current view that it is rarely utilized in full. The concept that marine populations are generally open as a result of wide pelagic dispersal appears to be overstated. There is growing evidence that coral reef communities tend to be self-recruiting in ecological timescales while in benthic shelf communities there are many species that do not practice wide dispersal and confinement to patchy benthic habitats is commonplace.

The idea that robust barriers to dispersal are uncommon in the sea has little credibility in the case of the benthic shelf environment. On the contrary, topographic barriers and oceanographic barriers, both prone to geological and climate change events, are prevalent and foster allopatric speciation in ways comparable to geographic speciation on land.

In this section, modes of speciation in the sea are reviewed in the context of observed distribution patterns of reef and benthic invertebrates. While it is possible that sympatric ecological speciation processes have occurred, evidence of vicariance, fragmentation of species populations, and divergence in allopatry are prevalent. The conclusion is drawn that biogeographic barriers, vicariance, and allopatric speciation have been significant factors in the evolution of the diverse marine fauna of the North West Shelf.

8.5.1 Modes of Speciation in the Shallow Marine Environment

Allopatric marine speciation, requiring complete isolation of incipient species, is sometimes seen as an enigma, especially in tropical coral reef ecosystems where biodiversity is so high and co-occurrence of many congeneric species may be commonplace. Alternative models of sympatric or ecological speciation hold that species may divide locally into diverging populations adapted to different habitats without a requirement for spatial separation or total isolation.[143–145] However, the ecological-genetic mechanisms that might drive such a process remain poorly understood[133,139] and are difficult to identify and test.[141]

Another approach to this matter is the notion of parapatric ecological speciation where adjacent sister populations do not co-occur in the same habitat but share a common boundary that may be "porous," allowing limited gene flow between them. It is thought that in this situation, natural selection within the different habitats may be strong enough to override limited gene flow across the porous boundary and drive the populations along separate evolutionary pathways leading to reproductive isolation.[139] Parapatric speciation may be prevalent and rapid in centers of high diversity, particularly in topographically complex shelf areas, like the Central Indo-West Pacific, where there are complex mosaics of habitat types. However, species occupying adjacent habitats like this are in fact spatially separated and the difference between this parapatric situation and classical allopatry is a matter of degree with wide spatial distributions and complete isolation at one end of the spectrum and restricted distributions and incomplete isolation at the other.

While extremely wide marine species' geographic distribution patterns are characteristic of the vast Indo-West Pacific realm, the supposed impediment to allopatric speciation imposed by pelagic dispersal and genetic panmixis of marine species may be overstated. Traditional studies of marine species' distribution within this realm long ago established the prevalence of geographically disjunct (allopatric) populations of many species within their realm-wide distributions, with varying degrees of morphological divergence. Also, at smaller geographic scales, there is increasing evidence from molecular genetic studies that many reef species, even those with pelagic larvae do not, in fact, disperse widely except on rare events (see Section 8.4). Local retention of larvae, that is, their settlement in their natal sites, and reduced gene flow over short geographical distances are commonplace.[108,146–149]

In most coral reef fishes studied to date, effective (demographic) dispersal is much smaller than potential dispersal extrapolated from current speed and duration of the larval stage.[150] It has been said that "reef fish populations are probably more toward the closed than the open end of the demographic spectrum."[151] These considerations suggest that genetic differentiation and reproductive isolation of reef populations, and consequential allopatric (complete isolation) and parapatric (incomplete isolation) speciation, may occur at small geographic scales much more frequently than has been appreciated. Sympatric speciation may be possible, but it is probably not common, while parapatric speciation is common and may be the prevalent mode of diversification in coral reef fishes.[152]

Another aspect is the relative importance of *vicariance* versus *dispersal* models of allopatric speciation. Vicariance refers to the fragmentation of species' geographic distributions by episodic environmental events, such as tectonic episodes and climate change that create barriers to dispersal and consequential reproductive isolation. The dispersal model refers to episodic dispersal events that allow the colonization of new habitats remote from a center of origin.

In the context of allopatric speciation among reef ecosystems of the prolific Indo-West Pacific oceanic islands, a "soft vicariance" model has been proposed where speciation involves a large, widespread ancestral species range connected by long-distance gene flow, being fragmented by oceanographic changes (rather than hard barriers created by tectonic activity) leading to peripheral isolation and endemism.[141,153]

Given the vast variability in dispersal capacity and duration of pelagic larval life, in distances between breeding populations, in the strength and seasonality of ocean currents and their changeability over geological time, and in the different ages and tectonic and climatic histories of benthic eco-systems, a wide spectrum of potential for allopatric and parapatric speciation in marine benthic animals should be expected over both broad regional and local scales.

Planes[150] divided factors affecting genetic structure of coral reef populations into three major categories:

1. functional factors related to the ecology and biology of species, including reproduction, behavior, and pelagic larval duration;
2. historical factors, such as colonization or extinction consequent to sea level variation;
3. physical factors, including the effect of oceanic currents and/or geomorphic structure of the reef limiting the potential to disperse.

To these should be added tectonic histories that create classical biogeographic barriers to dispersal, for these certainly do occur in continental shelf areas, as will be shown in the case of the North West Shelf. Historical factors predominate, and it is necessary to determine the historic origin of a species and its habitat prior to attempts to understand mechanisms driving genetic differentiation [speciation] within its contemporary geographic and ecological range.[150]

8.5.2 Speciation Processes on the North West Shelf

Examples of allopatric speciation are rife on the North West Shelf, and parapatric speciation is likely to occur in the complex mosaics of benthic shelf habitats. The factors involved are varied.

8.5.2.1 Tectonism

Mid-Tertiary tectonism that carried the northern half of the Australian continental plate into the equatorial tropic zone, and briefly into shallow sea contact with the Eurasian plate (mid-Miocene), is discussed in Section 8.1 and 8.4.5.1. Contact allowed the invasion of the Australian continental shelf by the vastly rich pantropical Tethyan fauna. The subsequent disconnection of the Australian and Eurasians plates, by means of the subduction of the former's shelf margin, forming the Timor Trough and its northern extensions, created a partial geographic barrier. It is partial in the sense that the deep trench now forms a barrier to species without pelagic dispersal capacity but no barrier, other than distance, to planktotrophic species where and when there are suitable current sets. Both the initial plate contact and subsequent disconnection were major vicariant events that had a profound impact on the marine biogeography of Northern Australia.

In the Pliocene tectonism and associated sedimentation also closed the space between Northern Australian and New Guinea except for the brief periods of high sea level in the Quaternary when the Torres Strait was open, severing the Mid-Late Tertiary connectivity between the west and east coasts.

Another tectonic event with a big impact on speciation of the North West Shelf fauna was the emergence of the Cape Range Peninsula in the Pliocene (Section 8.3.5.3). Located at the northwestern corner of the continent, near the limit of the present tropic zone, it forms a partial geographical barrier in a critical position in regard to change of species' distributions in response to climate change. It also is partial in the sense that there is a taxonomic differential in its effects, being a barrier to species lacking pelagic dispersal capacity and no or limited barrier to plankto-trophic species whose larvae may be dispersed by ocean currents. Evidence of its effect as a biogeographic barrier is seen in the presence of allopatric sister populations on either side of the peninsula. Molluscan examples are given in Section 8.7.

8.5.2.2 *Sea Level Change*

The importance of the Torres land bridge between Cape York and New Guinea as a biogeographic barrier operating intermittently through the cycles of sea level change in the Pleistocene, for both terrestrial and marine biota (in mirror image), was discussed in Section 8.4.5.2. Opening and closing of the Torres Strait has been a factor in the evolution of species pairs in the eastern Queensland and North West Shelf benthic (and demersal) faunas, comparable in its effects to that of the Panama Isthmus but younger. Molluscan examples are given below.

More difficult to document and interpret are the effects of sea level change that resulted in the series of transgressions and regressions across the North West Shelf during the Pleistocene (Section 2.1.2 of Chapter 2). Each transgression would create a mosaic of new marine benthic habitats, replacing the terrestrial ones before it, and there would be a period after inundation during which new benthic communities were established. It would not be a simple matter of immigrant species invading an established community, but the arrival and establishment of entire marine assemblages filling (and creating) the available ecological niches. The principles of ecological succession would apply, and even once sea level stability was established, there would probably be a period of demographic instability while the multispecies pool of immigrants sorted themselves into appropriate ecological functions and relationships. It is a moot point whether the present 6000 years of "still stand" has been long enough, and stable enough, for ecological stability to be established. But in any case, the contemporary period of relative stability is not the norm—the sea level graph (Figure 2.12 of

Chapter 2) indicates that through the Quaternary sea level has never stood still for very long and shelf habitats and communities must have been in a constant state of flux.

Substrate being the key determinant of community structure, and that being patchy across the continental shelf, there would be an element of chance in the eventual assemblages that were established and strong inter-species competition and selection pressures would certainly apply. Many species would fail, by chance, in the face of competition from later arrivals, or as a consequence of replacement in the succession of species as the habitat matured.

The source of immigrants that established these new marine communities would have been twofold—remnant populations in comparable habitats persisting in refugia within the region, and immigrants from further afield carried into the region by ocean currents.

There were five small embayments of benthic shelf habitat along the LGM coastline, isolated from each other by stretches of coast with a very narrow continental shelf.[155] Apart from their small size, these areas of benthic shelf habitats may not have been exactly like their contemporary equivalents of the middle and inner shelf. It is certain that there would have been rocky shore, estuary, and beach habitats like the modern ones but the soft substrate benthic shelf habitats are unlikely to have been the same (see Section 2.1.3.4 and Figure 2.10 of Chapter 2). Furthermore, it is likely that during the period of the LGM, benthic shelf species inhabiting the five persisting, semi-isolated areas of benthic shelf habitat may have genetically diverged to some extent. As each transgression commenced, they and the remnant populations of benthic species they supported would have expanded across the shelf and merged laterally with each other, establishing zones of secondary sympatry and hybridization.

There is some field evidence (personal observation) that successive reinvasions of inner shelf benthic and shore species did not always establish the same communities that had been in place during the cycles before. In the vicinity of the Ashburton Delta, at the western end of the North West Shelf, there is a complex fossil history of Quaternary coastal benthic communities, preserved in a series of raised benches beginning in the Early or Middle Pleistocene. The oldest and furthest inland, on Urala Station, contains fossils of the arcid *Anadara trapezia*, a temperate estuarine species now extinct in the region.[154] Closer to the modern coast, near the Urala homestead, is another high, fossiliferous limestone bench, probably Late Pleistocene, that is dominated by a different arcid species, *Anadara crebricostatum*—a common modern species that lives in shallow sandy coastal habitats. Then, close to present sea level throughout the coastal lands adjacent to the delta, there is a series of Holocene deposits with a very diverse molluscan fauna, including the mud-dwelling arcid *Anadara granosa* in great abundance. None of these Quaternary fossil assemblages have been studied to date, but they have potential to tell a tale of successive transgressions, establishing habitats for shallow coastal communities of varying species composition and structure with slightly different physical conditions each time, and local extinctions and replacements.

The reverse process, when sea level fell during regression periods, would have had even greater ecological impacts as the seabed of most of the shelf became terrestrial habitat and the area of benthic shelf habitats contracted. Local extinctions were total and regional extinctions may have been commonplace.

Therefore, environmental change of this kind, on this spatial scale and over these timescales, must have been highly inducive to genetic change, natural selection, and small-scale allopatric and parapatric speciation with vicariant and dispersal processes operating together, in a dynamic paradigm. Historical evidence of it is no doubt preserved in the modern shelf fauna as well as the fossil record, although the speciation events must have been so complex that teasing out their stories would be a formidable challenge.

8.5.2.3 Ocean Currents and Dispersal

Early in the Pleistocene transgressive cycles, the ITF, Holloway, and Leeuwin Currents did not flow, or their flow was restricted although there may have been local along-shelf currents. Current flows were reinitiated later in the transgressive cycles[155] with, presumably, renewed introduction of pelagic larvae from northern populations. This influx of propagules of Indonesian and western Pacific reef species switched on and off in pulses through the Pleistocene with a periodicity of 100,000 years or more.

During the cool cycles, the shelf-edge reefs would have been isolated from each other and from the Indonesian source by distance, without current-driven connectivity, and their communities relied upon local sources of larvae for their ecological persistence. Conditions may have been those of coastal waters with high nutrient levels as a result of upwelling along the shelf break and regional species extinction within the oceanic coral reef fauna may have occurred.

During the interglacial periods, with renewed flow of the ITF and shelf-edge currents, there would have been new waves of immigrants and genetic as well as ecological replenishment. The repeated cycles of isolation and reconnection of the shelf-edge reef communities, between themselves and with a "parental" Indonesian-West Pacific source, are a typical vicariant speciation process. Yet there is little evidence to date of any local endemism or genetic divergence in the contemporary reef fauna. Perhaps, with isolation intervals of 100,000 years or so, any local genetic divergence has been overcome by the weight of the northern recruitment pulses.

Benthic shelf and shore communities of the North West Shelf present a different circumstance. The cyclic influx of northern pelagic larvae may well have caused ecological upheaval in the newly established benthic communities of the outer shelf, as well as the reef communities, especially in the West Pilbara where the contemporary shelf is relatively narrow and mixing of oceanic and coastal water takes place along the shelf margin. However, the middle and inner shelf benthic and shore communities over much of the shelf are probably not greatly affected by pulsing of the major oceanic currents and influx of exogenous immigrants. Local wind-driven currents and tidal fluxes are likely to determine the dispersal rates and directions of pelagic dispersal of the benthic shelf species within the shelf regime. These complex shelf water movements are likely to confound the complexity of the matrix of substrate types and the interconnection networks of benthic metapopulations. Coupled with the cyclic creation and destruction of benthic shelf marine habitats, this is a formula for small-scale allopatric and parapatric speciation.

8.5.2.4 Isolation by Habitat Specialization

For most of the length of the North West Shelf, the outer and middle shelf is characterized by biogenic carbonate sediments. However, there are extensive stretches of terrigenous sediments on the inner shelf and shore. These are in the central Pilbara (Figure 2.10 of Chapter 2), Kimberley coast, and Bonaparte Gulf. To the extent that substrate is a key factor that determines the composition of benthic communities, such inner shelf areas of diverse substrate stand in relation to each other as remote island-like territories and the connectivity between them, in the demographic sense, are likely to be restricted.

Rocky shore habitats of the coastline are also split into widely separated regions (Chapter 3). The Kimberley Bioregion is characterized by rocky shores, with a diverse rocky shore fauna, as far south as the Buccaneer Archipelago. The Pilbara Bioregion also has extensive development of rocky shore habitat. But the Canning Bioregion has only a few rocky headlands, and the Eighty Mile Beach is entirely lacking in rocky shore habitat. The distance between Kimberley and Pilbara rocky shores is around 600 km, and genetic connectivity between rocky shore faunas of these bioregions may be limited.

Similarly, subtidal rock substrates on the shelf are distinctly patchy in their distributions. On the inner shelf, there are wide areas of exposed rock platform in the Pilbara (nearshore) Bioregion and a mosaic of rocky ridges and pinnacles in the Kimberley Bioregion. Bordering the middle and outer shelf for much of the length of the North West Shelf, there is a rocky ridge thought to represent the LGM shoreline (Section 2.1.5 of Chapter 2). These habitats support diverse communities of epibenthic invertebrates. Connectivity between the inner shelf subtidal rock substrate habitats is driven largely by complex tidal fluxes, while the outer shelf ridge, at the present time, may be affected by the along-shore Holloway-Leeuwin Currents and internal tidal currents.

In all these examples, disjunct habitats support allopatric populations of different species and morphologically distinctive forms that may be thought of as "incipient" species.

8.6 TYPICAL DISTRIBUTION PATTERNS OF NORTH WEST SHELF BENTHIC INVERTEBRATES

In this section, typical geographic species distributions are described, illustrating some recurrent patterns. The modern reef and benthic shelf faunas of the North West Shelf comprise species that are predominantly widespread in the Indo-West Pacific realm or are closely related to species that are so. Nevertheless, within that vast biogeographic realm recurrent patterns of species distribution are evident and the distribution patterns of Northern Australian marine species need to be considered and interpreted in that wider biogeographic context. Their study provides insights into environmental history of the region and the possible ecological and evolutionary responses of its biota to ongoing climate change.

Contemporary distribution patterns we see are a complex tangle resulting from multiple causes and histories and varied biological responses to environmental change. At any one point in geological time, patterns of distribution of marine species are transient outcomes of the interactions between

- habitat requirements of the species and the distribution in space of habitats that are suitable;
- adult and/or larval dispersal mechanisms and the environmental means of delivery;
- the history of tectonic, sedimentary, and climatic events;
- the evolutionary responses of species to environmental change;
- chance.

A striking feature of Indo-West Pacific species' distributions is the fragmentation of many widespread species into isolated (allopatric) populations. Morphologically distinct populations may be recognized in species complexes, variously given species or subspecies taxonomic rank. These fragmented distributions are outcomes of diverse vicariant and dispersal events and reflect the history of regional environmental change. Especially revealing are the many examples of allopatric sister species/subspecies pairs whose ancestral range has been split by physical events that have broken connectivity and led to genetic divergence of the daughter populations in isolation.

There is not sufficient information on the distribution of North West Shelf species to enable a catalog and comprehensive interpretation of their diverse distribution patterns. Mainly molluscan examples are chosen here to illustrate some recurrent patterns that probably have general relevance. Typical distribution patterns of widespread Indo-West Pacific species are illustrated by examples from the gastropod family Strombidae, which includes mainly coral reef species that lay gelatinous egg capsules and have planktotrophic veliger larvae. A very different perception is provided by Northern Australian examples from the gastropod family Volutidae which are benthic shelf predators that lay gelatinous capsules, nurtured by the mother until hatching, and there is no pelagic larval stage.

Examples of genera and species complexes, drawn from the families Strombidae and Volutidae, that have split into isolated sister species/subspecies among the bioregions of the North West Shelf and on the eastern and western sides of Australia are discussed. These examples illustrate the fragmentation of many species complexes and show that the evolutionary process in this region has been dynamic and responsive to environmental change, as it has throughout the Indo-West Pacific realm.

8.6.1 Distribution Patterns of Indo-West Pacific Strombs

Distribution patterns of polytypic species complexes in the Indo-West Pacific realm are well illustrated by the tropical gastropod family Strombidae which had an Early Tertiary origin in pantropical Sea of Tethys.[11] There are four living genera (*Strombus, Lambis, Rimella, Terebellum*). The family flourished in the Pliocene and Early Pleistocene but has diminished since then, and there are now 7 living species in the Caribbean region, 4 in the Panamic region, 1 in tropical West Africa, and 38 named species in the Indo-West Pacific, 28 of which are found on the North West Shelf (Table 8.7).

Strombs are herbivores and most live in intertidal or subtidal sandy habitats. The exceptions are species of *Lambis* that live on coral rock and rubble flats. They lay gelatinous egg capsules and have long-lived planktotrophic veliger larvae. With the exception of the endemics, all of the North West Shelf strombs are widely distributed in the Indo-West Pacific including eastern Queensland. The most conspicuous feature of the regional stromb fauna is that the majority of the species (19-60%) live on the oceanic shelf margin coral reefs and do not occur in the coastal bioregions. This family may be regarded as comprising predominantly oceanic coral reef species belonging to the group of "oceanic strombs" referred to by Abbott.[11]

Only nine strombs are found in the coastal bioregions of the North West Shelf. All of them are found on intertidal sand flats or sandy rock plat-forms of the coastal bioregions but not in the Oceanic Shoals Bioregion. One of them (*Strombus campbelli*) is also found in eastern Queensland. The others, *Strombus iredalei* and *Strombus orrae*, are regional endemics that belong to Indo-West Pacific species complexes.

Polytypic superspecies complexes are a feature of Indo-West Pacific strombs, many divided into morphologically distinct allopatric popula-tions in various stages of genetic divergence. Very often, there are Indian Ocean and western Pacific Ocean species or subspecies pairs, sometimes overlapping with zones of sympatry in the Indo-Malay Archipelago but sometimes completely separated. Such distribution patterns seen in the Strombidae are typical of Indo-West Pacific molluscs and most other marine macroinvertebrates.

8.6.1.1 Monotypic Species That Range Widely in the Indo-West Pacific Realm

An example of a species that ranges virtually throughout the realm, without division into morphological distinct populations, is the intertidal *Strombus mutabilis* which is common on oceanic coral reefs and rocky shores of Asia, East Africa, and both sides of Australia (Figure 8.11). On the North West Shelf, it is common on the oceanic coral reefs of the shelf margin, on the fringing reefs and rock platforms of the outer Pilbara islands and at Ningaloo and localities on the West Coast as far south as Cape Naturaliste. But it is not found in the Kimberley or Pilbara (near-shore) Bioregions. It is one of many tropical invertebrates referred to as a "rider of the Leeuwin Current" that penetrates far into higher latitudes of the temperate zone.

Another stromb with an undivided, though narrower tropical Indo-West Pacific range, is *Lambis lambis* that lives on coral reefs and rocky shores in both oceanic and continental environments. It is found on all the shelf-edge reefs and on the fringing reefs of the Kimberley coast, but not in the Pilbara or further south.

8.6.1.2 Indo-West Pacific Taxa Split into Allopatric Sister Populations

Many strombs exhibit conspicuous regional polytypy, with distinctive allopatric sister species and subspecies. Several examples illustrate such distributions.

Lambis chiragra is a large, very well known oceanic reef-front species, which has two allopatric subspecies (Figure 8.12A). The nominate form *L. chiragra chiragra* occurs in the western Pacific and eastern Indian Ocean, including the shelf-edge reefs of the North West Shelf Ningaloo Reef (but not on the coastal reefs). The allopatric subspecies *L. c. arthritica* occurs in

FIGURE 8.11   The distribution of *Strombus mutabilis*, an example of a species ranging throughout most of the Indo-West Pacific realm, extending into temperate waters of the African, Asian, and Australian coasts. It is an oceanic species and, in Western Australia, is found on the offshore coral reefs of the North West Shelf margin and open coast shores of the West Coast. *Drawn by Chevron.*

the central and western Indian Ocean. *Lambis truncata*, found on oceanic reefs and atolls throughout much the Indo-West Pacific realm, also has two subspecies, *L. truncata truncata* in the Indian Ocean, including Cocos Keeling, and *L. t. sebae* in the western Pacific, the East Indies Triangle, and at the North West Shelf atolls.

Strombus luhuanus is a common stromb on coral reefs throughout the western Pacific, including the Great Barrier Reef and the Central Indo-West Pacific, and has recently been found on the North West Shelf shelf-edge reefs (Figure 8.12B). Its closely related sister species, *Strombus decorus*, is found in the northern and western Indian Ocean.

Strombus gibberulus is a polytypic oceanic reef species with a Pacific form, *S. g. gibbosus*, that also occurs on the North West Shelf atolls, and an Indian Ocean form, *S. g. gibberulus* (Figure 8.12C).

There are very many other examples of polytypic gastropods with distributions of this kind. Importantly, the representatives of oceanic polytypic species complexes found on the coral reefs of the Oceanic Shoals Bioregion are invariably the EIT-Pacific species or subspecies, not the Indian Ocean ones.[38] This confirms that the primary connectivity of the North West Shelf

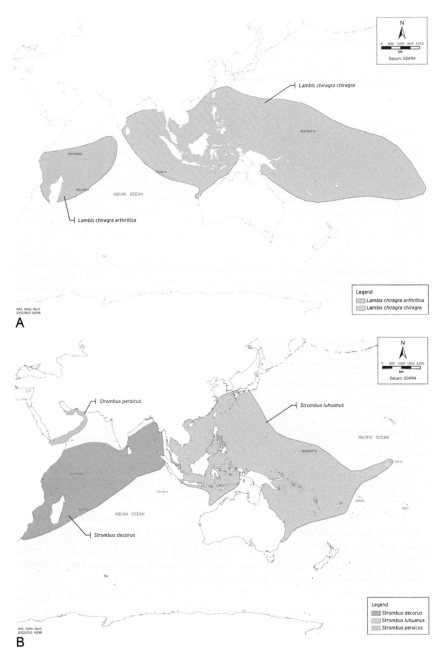

FIGURE 8.12 Three stromb examples of Indo-West Pacific species or superspecies that are polytypic, with distinct Indian Ocean and Pacific Ocean morphological forms. (A) *Lambis chiragra chiragra* and the Indian Ocean subspecies *L. c. arthritica*; (B) *Strombus luhuanus* and its western Indian Ocean sister species *S. decorus*;

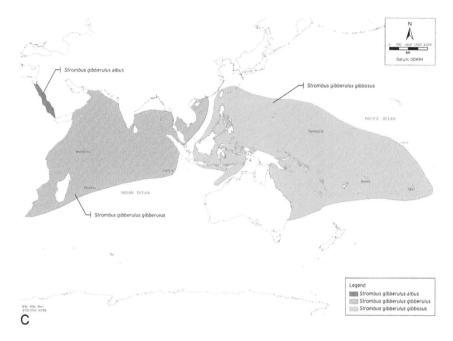

C

FIGURE 8.12—cont'd   (C) *Strombus gibberrulus* which has a northern and western Indian Ocean distribution and subspecies in the Red Sea (*S. g. albus*) and western Pacific (*S. g. gibbosus*). In all these examples, the form found on the North West Shelf shelf margin reefs is the Pacific one. *Drawn by Chevron.*

oceanic reef fauna lies with the Central Indo-West Pacific region and the western Pacific, not with the central and western Indian Ocean.

8.6.1.3 Strombs with Central Indo-West Pacific and Northern Australian Distributions

There are many examples of species or species complexes that extend their range from the East Indies Triangle and the Western Pacific into higher latitudes of either or both the western and eastern Australian coasts.

Strombus campbelli is a northern Australian endemic, closely related to the Central and Western Pacific stromb *S. vittatus* (Figure 8.13). It is a continental species that lives in intertidal and shallow subtidal habitats of the coastal bioregions but not on the offshore coral reefs. These two strombs are sometimes treated taxonomically as allopatric subspecies.

The Central and Western Pacific strombs *S. latissimus* (an oceanic reef species) and *S. epidromus* (a benthic shelf species) extend onto the shelf margins of the North West Shelf, but are not found in Queensland. Conversely, *S. labiatus* is an oceanic stromb that extends down the Queensland coast but is not found on the North West Shelf. The very

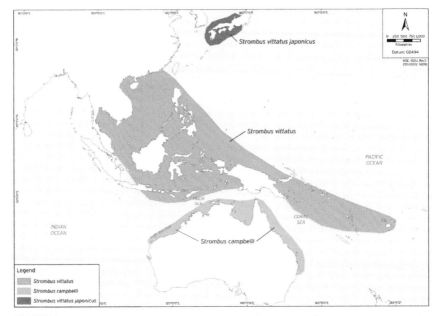

FIGURE 8.13   Some stromb examples of central Indo-West Pacific species or superspecies (EIT) that extend down one or both sides of the Australian coast. (A) The Australian *Strombus campbelli* and the nominate subspecies in the EIT; (B) the EIT species *Strombus epidromis* that is found on the Oceanic Shoals atolls of the North West Shelf and North Queensland. *Drawn by Chevron.*

common EIT shelf reef species *S. canarium* is fished commercially in South East Asia and is found in North Queensland but not on the north-western coast.

8.6.1.4 *Strombs with Damperian Distributions*

Many years ago, marine biogeographers recognized the Torres Strait as a significant biogeographic boundary, marking the ends of range of many eastern and Western Australian marine species (Section 8.4.5.2). There are many molluscs and other invertebrates that have distributions ranging from the Gulf of Carpentaria to the North West Shelf or further south on the West Coast. Two of the North West Shelf coastal strombs have geographic distributions like this.

S. orrae is an intertidal, continental stromb that ranges from Exmouth Gulf to the Northern Territory. It is a member of a species complex that includes *S. urceus* in the Central and Western Pacific and *S. klineorum* in the northern Indian Ocean (Figure 8.14A). *S. iredalei* is a sand-dwelling endemic of the region, ranging from Shark Bay to the Northern Territory

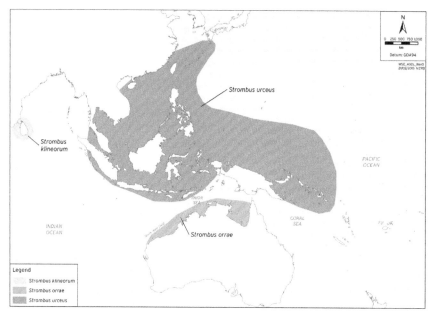

FIGURE 8.14   Stromb examples of species with a "Damperian" distribution and close EIT relatives. (A) *Strombus orrae* found from Shark Bay to the Gulf of Carpentaria, and the its sister species *S. urceus* in the EIT; (B) *Strombus iredalei* found from Shark Bay to the Gulf of Carpentaria and its three sister species in the EIT. *Drawn by Chevron.*

(Figure 8.14B) with sister subspecies in New Caledonia, Ryuku Islands, and Hawaii.

Like *S. campbelli*, these two strombs belong to polytypic species complexes of the Central Indo-West Pacific region, but unlike that species, they have not extended their range onto the eastern Queensland coast (or they may have became extinct there). They are examples of molluscs for which closure of the Torres Strait has been an impassable barrier.

8.6.2 Distribution Patterns of Northern Australian Volutid Gastropods

The continental shelves surrounding Australia are centers of evolution and endemism for the Volutidae, with high generic and species diversity compared to other regions of the world. The Australian genera have either of two primary centers of origin. In the temperate waters of southeastern Australia, there are a number of genera that had their Early Tertiary origins in the Palaeoaustral province of the high latitudes in the southern hemisphere.[156] The origins of the genera of volutes found on the North

West Shelf are ambivalent, but the ancestors of at least some of them probably lived on the Sunda Shelf, spreading to the Arafura Shelf in the Late Oligocene or Miocene and from there down both sides of Australia. Perhaps they originated on the Australasian shelf and spread the other way? There is an important story here to be told about the evolution of the Australian continental and its marine biota.

In this section, a brief outline is given of the largely allopatric distributions of Northern Australian species of *Cymbiola*, *Volutoconus*, and *Amoria*. Like all volutes, these snails do not have a pelagic larval stage. Their distributions show striking similarities and illustrate patterns of distribution, resulting from vicariant speciation, that are very different to those of molluscs with planktotrophic larvae.

8.6.2.1 *The Genus* Cymbiola (Figures 8.15 and 8.16)

The large genus *Cymbiola* has a Central Indo-West Pacific and Australian distribution.[91,156] It has a long fossil record, first known in the Late Oligocene of southeastern Australia and with several Miocene and Pliocene species described from Indonesia. While present in the warm

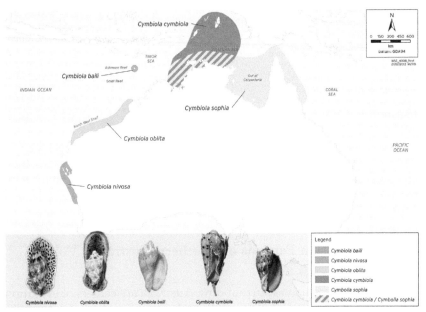

FIGURE 8.15   Distribution of sister species of the volutid genus *Cymbiola*. *C. cymbiola* is from the Arafura Sea, ranging from Joseph Bonaparte Gulf and the Northern Territory of Northern Australia to the south coast of West Irian and the Wallacea province of eastern Indonesia. *C. sophia* has established a population on the northern coast of eastern Queensland. The three allopatric Western Australian species indicate fragmentation of the ancestral range, probably in the Late Tertiary. *Drawn by Chevron, photos by Barry Wilson.*

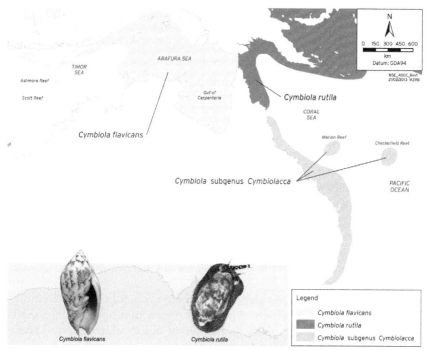

FIGURE 8.16   Distribution of *Cymbiola flavicans* and *C. rutile*, species that occupy the Arafura Sea and Coral Sea, respectively, but neither of which crosses through the Torres Strait. The eastern Australian contemporary range of the subgenus *Cymbiolacca* is also shown in the Figure. There is a Pliocene record of the type species *Cymbiola (Cymbiolacca) pulchra* from Java.[158]

waters of southern Australia in the Middle Tertiary, it became extinct there when sea temperatures cooled and today it persists on the tropical and warm temperate coasts of the northern half of the continent.

The taxonomy of this genus is controversial and some authors recognize several subgenera. One of these is named *Cymbiolacca*, a subgenus with several living species on the eastern coast of Australia[157] (Figure 8.16) and a fossil species in the Middle Pliocene of West Java.

In Northern and Western Australia, there is a group of five medium-sized species belonging to the nominate subgenus that, on shell morphology and body color pattern, appear to be closely related. The group includes the type of the genus *Cymbiola cymbiola*. This volute is found in the Arafura Sea, the north coast of the Kimberley and Northern Territory, the southwestern coast of Irian Jaya, and the Moluccu Province of eastern Indonesia.

C. cymbiola is partly sympatric with a closely related sister species, *Cymbiola sophia* that is found in similar habitats across Northern Australia from Joseph Bonaparte Gulf to about Townsville on the east Queensland coast.

There are three similar species on the North West Shelf and West Coast. From Cape Leveque to Exmouth Gulf, *Cymbiola oblita* lives in intertidal and shallow subtidal sandy habitats. It is polytypic with varied morphology throughout its range. It has a closely related sister species, *C. baili*, on the banks surrounding Ashmore Reef. These two volutes have very similar shell morphology and probably diverged recently, perhaps as a result of Holocene sea level change that isolated the Ashmore population on the shelf margin banks about 600 km from its source population.

The fifth member of this group is *C. nivosa* that has a moderately restricted West Coast range from Fremantle to Shark Bay. Its shell morphology and body color patterns suggest an early divergence, perhaps as a result of separation of the North West Shelf and West Coast populations by emergence of the Cape Range Peninsula in the Pliocene.

The distribution patterns exhibited by these five species of *Cymbiola* suggest a common Middle Tertiary ancestry with divergence on the North West Shelf, and one of the daughter species, *C. sophia*, extending its range through the Torres Strait onto the east coast of Queensland in the Quaternary or Holocene.

In addition to these five sister species, there is another pair of medium-sized *Cymbiola* in the Northern Australian region that have heavier shells with stronger columellar plaits and different body color patterns. These are *Cymbiola flavicans* that ranges from Moluccu to the Northern Territory and the Gulf of Carpentaria as far east as the Torres Strait and *Cymbiola rutila* that ranges from the Torres Strait to North Queensland and the western Pacific islands of the Solomon Sea. Here again we see a northern species whose range extends into eastern Indonesian and an eastern species with a southwestern Pacific range. Neither of them has passed the Torres Strait. In this case, the sharp boundary may have an ecological basis—*C. rutila* is a coral reef volute, while *C. flavicans* lives in muddy sand habitats.

8.6.2.2 *The Genus* **Volutoconus** (Figure 8.17)

The endemic Northern Australian genus *Volutoconus* has five named living species in benthic shelf habitats of the North West Shelf. Several of them are polytypic and may need further division. It is evident that the genus has speciated and spread across Northern Australia and southward on both sides of the continent.

The type of the genus is *Volutoconus coniformis*, a rare and morphologically distinct species with a restricted range on the North West Shelf and no close relative within the genus.

Volutoconus bednalli is widespread in the Arafura Sea and Northern Australia from the north Kimberley to the Torres Strait. It is also present in the Moluccu and southwestern Irian Jaya provinces of eastern Indonesia.

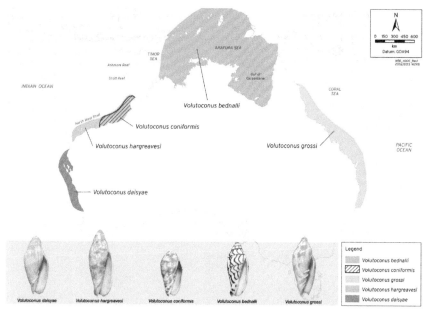

FIGURE 8.17   Distribution of the six living species of the volutid genus *Volutoconus*. *V. hargreavesi* has a fossil record in the Miocene of West Irian but is now restricted to the southern North West Shelf. It has allopatric sister species on the west coast (*V. daisyae*) and east coast (*V. grossi*) the latter itself polytypic. *Drawn by Chevron, photos by Barry Wilson.*

The North West Shelf species *Volutoconus hargreavesi* has a closely related allopatric sister species *Volutoconus daisyae* on the West Coast and another *Volutoconus grossi* on the Queensland east coast. There is also a partially sympatric deepwater form on the North West Shelf with a prominent protoconch, which has been named as a subspecies *Volutoconus h. calcarelliformis*. There are fossil records of similar species, a Middle Miocene form from Barrow Island and a Upper Miocene/Pliocene form from Irian Jaya. Both are closely related to the living North West Shelf species *V. hargreavesi*. The living species appear to be descendants of a once widespread benthic shelf species now fragmented into allopatric populations.

8.6.2.3 *The* **Amoria grayi** *Complex* (Figure 8.18)

The genus *Amoria* has speciated extensively in Northern Australia and has spread onto both the western and eastern sides of the continent and even into the southern Ocean on the south coast. In contemporary times, *Amoria* is circum-Australian and speciose with representatives in both the tropical north and temperate south. There is a Miocene/Pliocene species

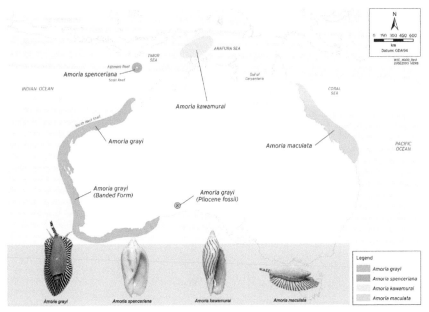

FIGURE 8.18   Distribution of sister species of the volutid genus *Amoria*. *A. grayi* has a large, polytypic modern range on the southern North West Shelf and the west and south coasts of Western Australia and fossil records in the Upper Pliocene of the Roe Plains beds at the Head of the Bight, and the Upper Miocene of West Irian. *A. kawamurai* has a restricted range in the Arafura Sea. *A. maculata* is a Queensland species. *Drawn by Chevron, photos by Barry Wilson.*

in Irian Jaya, identified as *A. canaliculata*.[6] This is a living species in southern Queensland, indicating a Pliocene contraction of range, perhaps as a result of the development of the land bridge between Australia and New Guinea.

The origins of this large genus are unclear. A Tethyan ancestry is likely with vigorous radiation in the Australian benthic shelf habitats in the Middle and Late Tertiary. There are several Miocene species in the fossil beds of southeastern Australia and one Middle Miocene record of the genus from Barrow Island on the North West Shelf.

Figure 8.18 illustrates the allopatric distributions of species in the *A. grayi* complex. *A. grayi* is a common, polytypic benthic shelf Western Australian volute that ranges from Cape Leveque to the South Coast with living populations as far east as the Recherche Archipelago. In the Late Pliocene, it extended even further east where it is a common fossil in the shell beds of the Roe Plain. There is an allopatric sister species, *Amoria kawamurai*, apparently with a very restricted range in the Arafura Sea, and another, *Amoria maculata*, that inhabits the east Queensland coast.

A tropical Miocene ancestry on the Australian-New Guinea carbonate platform seems likely.

8.6.2.4 *Summary of Volutid distribution Patterns*

The distributions of these three direct-developing genera illustrate several patterns:

- Miocene origins, most likely in the central part of the pantropical Sea of Tethys (antecedent of the modern Indo-West Pacific region) at the time of contact of the Australasian and Eurasion plates.
- *Cymbiola* with Middle Miocene species in Australia and the Sunda province of central Indonesia; *Volutoconus* and *Amoria* with Upper Miocene/Pliocene representatives in Irian Jaya on the Australasian side of Wallace's line.
- All three genera radiated on the Australasian continental shelf during the warm Miocene period and spread around both sides of the Australian continent.
- *Cymbiola* and *Amoria* extended to the southeastern Australian province in the Miocene. *Cymbiola* became extinct in southern Australia during the cooling phase that followed the Miocene and contracted its range to the tropical and subtropical waters of the northern part of Australia. Some species of *Amoria* adapted to the cooler conditions and persisted in the temperate waters of southern Australia.
- All three genera suffered fragmentation and contraction of range that resulted in allopatric sister species populations.
- All three genera produced species that established populations on both the west coast and the east coast of Australia.

Prevalence of west and east coast species pairs in Australian temperate marine fauna is well known, produced by phases of contraction and expansion of range in response to climate change. The same process in reverse would produce west and east coast populations of tropical species as well, easily accounted for on the west coast but problematical on the east coast because of the massive land connection between Australia and New Guinea through most of the Quaternary (Figure 8.1B and C). A Quaternary route through an open Torres Strait would be feasible during the brief phases of opening of the Torres Strait. Perhaps the *V. hargreavesi/grossi* and *A. grayi/maculata* species pairs split during one of those Quaternary eustatic episodes. However, it seems more likely that establishment of western and eastern sister populations occurred in the Miocene and subsequent fragmentation resulted from the Pliocene demise of the epicontinental sea between Australian and New Guinea. This is one of the most intriguing phenomena in Australian marine biogeography and is worthy of further study.

8.6.3 North West Shelf Species and Species Groups That Range into the Temperate Zone

Many fishes and invertebrates common to the North West Shelf and the subtropical West Coast extend onto the temperate South West and South Coasts of Western Australia. Molluscan examples include species that exhibit interpopulation variability in morphology but apparently without regional polytypy. Conspicuous intertidal examples are the bivalves *B. ustulatus* (Kimberley to the South Coast) and *Fragum erugatum* (Pilbara to the South Coast), two rocky shore littorinid gastropods, *N. australis* and *N. nodosa* (Pilbara to the South Coast) and the muricid *Cronia avellana* (Kimberley to the South Coast).

8.6.4 North West Shelf and West Coast Species Pairs

Intermittent southern expansion and speciation of tropical species on the West Coast occurred during phases of global warming in the Miocene, Pliocene, and Quaternary (Section 8.1.1) and are evident in the prevalence of species and subspecies pairs on the North West Shelf and West Coast, with the partial biogeographic barrier imposed by the Cape Range Peninsula playing a key role. Two gastropod examples follow:

(i) Turban shells of the subgenus *Marmarostoma*
Several large species of this subgenus of *Turbo* live on coral reefs in the Indo-West Pacific realm, and there is one on the subtropical and temperate shores of southwestern Australia that is evidently derived from them or a common ancestor. *Turbo (Marmarostoma) pulcher* (Figure 8.19A) is a very common snail on rocky shores on the South Coast and lower West Coast and also occurs abundantly in coral reef habitats of the Abrolhos. One of its Indo-West Pacific congeners, *T. (M.) argyrostomus* (Figure 8.19B), is a key herbivore in reef-front habitats on the oceanic coral reefs of the North West Shelf and Ningaloo. While shell morphology indicates that these two species are closely related, they are sufficiently distinct that a recent common ancestry seems unlikely. A Tertiary origin for *T. (M.) pulcher*, from a North West Shelf ancestor, as the result of a dispersal event into southern attitudes during a warming phase is indicated.

(ii) Sister species of *Zoila* (Gastropoda, Cypraeidae)
The nine nominal living species of *Zoila* fall into two geographic groups, five on the North West Shelf and four on the West and South Coasts, separated by the Cape Range Peninsula. While totally allopatric, one species pair appears to share a relatively recent common ancestry. On the basis of studies on mitochondria, *Z. eludens* (North West Shelf) and *Z. friendii jeaniana* (Dirk Hartog

FIGURE 8.19   Two sister species of *Turbo (Marmarostoma)*. (A) The endemic West Coast species *T. (M.) pulcher* that ranges from Shark Bay to the South Coast; (B) the Indo-West Pacific coral reef species *T. (M.) argyrostomus* that is found as far south as Ningaloo Reef. *Photos: Barry Wilson.*

Shelf) are believed to be sister species that separated probably in the Early Pleistocene[153] and are now isolated from each other by the very narrow shelf on the western side of the Cape Range Peninsula.

8.7 ENDEMISM

Endemic refers to a species (or higher taxon) that occurs naturally in a given area. In this sense, all species are endemic to somewhere but in common usage, especially in conservation science, the term mostly refers to species with regionally or locally restricted contemporary geographical ranges. Although many planktotrophic, tropical marine species have wide geographic ranges, most species that inhabit the planet are more or less narrowly endemic.[158–161]

Restricted geographical range commonly comes about through

(a) speciation, i.e., by way of dispersal (colonization of remote areas);
(b) contraction or fragmentation of ancestral range and local extinction.

Endemism in the North West Shelf benthic shelf and shore fauna is moderate in molluscan families characterized by planktotrophic larvae (Tables 8.7–8.9), while it is very high in families that lack pelagic larvae (Table 8.10). This probably applies to all the major invertebrate taxa in the region.

It is a central hypothesis of this study that the benthic and shore fauna of the North West Shelf exhibits a high level of endemism brought about by frequent vicariant and parapatric speciation processes, while the oceanic reef fauna of the shelf margin, serviced by the ITF delivery of planktotrophic larvae, exhibits little or no endemism.

Overall endemism (including the shelf margin reef faunas) is around 10–12% for most invertebrate taxa. However, when the reef fauna of the shelf margin is excluded, endemicity of the benthic shelf and shore fauna is of the order of 20%. If this is confirmed by further study, the North West Shelf could be regarded as a center of endemism within the Indo-West Pacific realm.

8.7.1 The Importance of Endemism in Conservation

8.7.1.1 Centers of Endemism

Regional conservation programs generally give priority to the identification of areas of high endemicity[162,163,168] as well as areas where there are concentrations of species richness, sometimes referred to as biodiversity hot spots.[164,165] Centers of endemism are commonly interpreted as past centers of cladogenesis, that is, places where high levels of speciation and evolutionary radiation have occurred.[162]

It is sometimes argued that centers of endemism are also centers of biodiversity (because they have been places of cladogenesis) and that they may be considered together in conservation programs.[165,166] However, that conclusion has been disputed for both terrestrial and marine biota. There is no concordance of areas of high endemicity and biodiversity "hot spots" in reef environments in the Indo-West Pacific realm.[61] It may be concluded that centers of endemism and centers of high biodiversity need to be treated separately in conservation programs.

An area of endemicity commonly (but not always) includes the place of evolutionary origin of many taxa. Contemporary environmental and historical factors are both important in the identification of centers of endemism, and it is necessary to define both the spatial and geological history contexts.

8.7.1.2 Short-Range Endemics

At species level, conservation management priority is also given to *short-range endemics* because whole species are vulnerable to extinction in the face of local environmental change that they are unable to avoid or tolerate. Small geographic range most commonly occurs when disturbance reduces or fragments the habitat of species that have limited dispersal capacity. The importance of life history was emphasized in the discussion of connectivity (Section 8.4.2), and it is also a key factor in the speciation process and the maintenance of geographic range. Because of a notion that marine species are predominantly widespread, the importance and prevalence of short-range endemism in the sea is generally overlooked. Yet, many benthic shelf invertebrates have restricted dispersal capacity and some, especially those that have no pelagic larval stage, are prone to fragmentation of range and local extinction.

In the Western Australian terrestrial environment, short-range endemism refers to species with naturally small ranges of less than 10,000 km.[2,167] A key word is "naturally" for this definition would exclude a species whose wide natural range has contracted in recent times as a result of anthropogenic disturbance. Perhaps restriction of the term in this way would be unwise. Contraction of range is an issue that needs management whatever its cause.

In this account, short-range endemism is applied to species that are confined to a single bioregion.

8.7.2 Endemism of Marine Species on the North West Shelf

8.7.2.1 Varied Endemicity Among Invertebrate Taxa

There is great variation in the levels of endemicity in different marine taxa on the North West Shelf. Corals, for example, exhibit almost none. At the other extreme, in benthic shelf representatives of the gastropod family Volutidae species endemicity in the Northern Australian shelf fauna is almost 100%.

To illustrate this variability between major taxa, Tables 8.7–8.10 list the North West Shelf species of the molluscan families Mytilidae (marine mussels), Cardiidae (heart cockles), Strombidae (stromb conchs), and Volutidae (volutes) and identify the endemics.

If the data for the three planktotrophic families (Mytilidae, Cardiidae, Strombidae) are pooled and the strictly oceanic species excluded, the regional endemicity for bivalve and gastropod benthic shelf molluscs would be roughly 19%. Inclusion of the direct developing family Volutidae in which all shelf species are either regional or Northern Australian endemics would reduce this figure only slightly. Although this is not an accurate estimate, regional endemicity is probably of this order for the benthic shelf and shore fauna of the North West Shelf. Endemicity in the reef communities of the shelf margin coral reefs is virtually zero.

(a) *Mytilidae* (Table 8.7). The mytilds are byssate nestlers, borers in carbonate rocks (including live corals), or infaunal burrowers. They are mostly free spawners although some mytilids brood their larvae. The family may be regarded as predominantly continental. With the exception of three endemic species, all the North West Shelf mytilids are widely distributed in the Indo-West Pacific and eastern Queensland.

About three quarters of the mytilids (20 species) are found in the coastal bioregions but not on the coral reefs of the Oceanic Shoals Bioregion. Conversely, all but one of the species found on the oceanic reefs also occur in the coastal zone.

There are three regional endemic species (10% of the regional total), all of them confined to the coastal bioregions. Two of them also occur on the West Coast. *Modiolus* sp. indet. appears to be an undescribed species belonging to a group of species that live on rocky shores in the central Indo-West Pacific region. It is abundant on rock platforms in both the Pilbara and Kimberley and probably extends into the Northern Territory. *Lioberus* sp. indet. is also undescribed. It builds byssal nests and lives half buried in weedy intertidal sand flats from Cape Leveque to Shark Bay on the West Coast. *B. ustulatus* is a mat-forming mussel that lives in the mid-littoral zone of rock platforms from the Kimberley to the Western Australian south coast.

(b) *Cardiidae* (Table 8.8). The cardiids are infaunal, short-siphoned burrowers. They are free spawners with relatively short-lived pelagic veliger larvae. The species of this family are predominantly infaunal species of the coastal bioregions. Except for six endemics, all the North West Shelf cardiids are found in the Indo-West Pacific realm and most of them also in eastern Queensland. Eleven of the species occur in the reef faunas of the Oceanic Shoals Bioregion, four of them occurring only there.

There are six endemic species (19% of the regional total): two of them in all the North West Shelf coastal bioregions, three endemic to the Pilbara, and one to the Pilbara and West Coast. None of the endemics occur in the Oceanic Shoals Bioregion, and none are short-range endemics. One of them, *Acrosterigma reeveanum*, has a "Dampierian" distribution from Fremantle on the West Coast to the Gulf of Carpentaria. Another, *Acrosterigma fultoni*, ranges from Exmouth Gulf to the Kimberley. *Acrosterigma wilsoni*, *Acrosterigma rosemariensis*, and *Acrosterigma dupuchense* are limited to the sandy coasts of the Canning, Eighty Mile Beach, and Pilbara regions. The last-named species was common in Shark Bay during the Last Interglacial but contracted its range in the LGM and has not recovered its original distribution. *F. erugatum* has an amazing distribution from South Australia to the Pilbara.

(c) *Strombidae* (Table 8.9). Distribution patterns and the planktotrophic life history of Indo-West Pacific strombs have been discussed in Section 8.7.1. On the North West Shelf the majority of this family live in oceanic coral reef habitats (19 of the 28 species) and have wide spread Indo-West Pacific distributions. There are nine species in benthic shelf and coastal intertidal habitats. Six of these continental species are also widespread

Indo-West Pacific, one (*Strombus campbelli* - Figure 8.13) is a Northern Australian endemic and two (*S. iredalei, S. orrae* - Figure 8.14;) are endemic to the North West Shelf and Gulf of Carpentaria region. All of these continental species, including the endemics, belong to polytypic Indo-West Pacific species complexes. Thus, regional endemicity, at 7%, is low and confined to the coastal habitats.

(d) *Volutidae* (Table 8.10). Distribution patterns of selected volutid genera are discussed in Section 8.6.2. These predatory gastropods are direct developers. The females lay gelatinous capsules that they nurture until the veligers hatch as shelled snails (Figure 8.7).

All of the volutes found on the North West Shelf are Australian endemics, and the level of species endemicity is 100%. (Three widespread deep-sea species of the genus *Teramachia* are not part of the continental shelf benthic fauna and are not included in Table 8.10.) However, there is a group of species from the Arafura Sea that have ranges that extend to the southern coast of Irian Jaya and the easternmost islands of the Moluccu province of Indonesia. The extent to which these species cross the deep troughs that separate the Australasian Shelf from Wallacea is still uncertain.

The level of endemicity varies. *Melo amphora* has a wide Northern Australian distribution from Shark Bay to Moreton Bay and north to the shores of Irian Jaya. At the other extreme is the short-range endemic *Amoria macandrewi* that is found only on the shores of Barrow and Montebello Islands.

Most of the Northern Australian volutes are geographically polytypic or belong to "superspecies" complexes and have become endemics through fragmentation of their ancestral range. Several of them are short-range endemics. Their sister-species relationships with congeners beyond the borders of the North West Shelf are discussed in the previous section.

There are no volutes on the Rowley Shelf atolls, but there are two species (*Amoria hansenae* and *V. calcarelliformis*) described from the subsided continental slope that are closely related to benthic shelf congeners.

8.7.2.2 *Varied Endemicity Among Ecosystems*

There is also significant variation between levels of endemicity in different ecosystems within the region. While volutid gastropods of the continental shelf are regionally endemic, bathyal volutids of the continental slope have widespread Indo-West Pacific or cosmopolitan distributions.

A conspicuous feature of the North West Shelf is the very low level of endemicity in the coral reef assemblages of the Oceanic Shoals Bioregion. Overall, species endemicity of the shelf-edge coral reef invertebrates is virtually zero, while that of the shelf benthic invertebrates (other than corals) is of the order of 10-20%. The species comprising the oceanic reef communities are almost all widespread in the Indo-West Pacific realm. Several fish species have been described from the shelf-edge atolls of the Rowley Shelf but, so far, are not known from other localities.

8.7.3 Endemic Species of the Western Australian West and South Coasts

A host of temperate shelf taxa that are endemic and now isolated in allopatry on the West and South Coasts are thought to have been derived from tropical North West Shelf ancestors. West Coast endemism is beyond the scope of the study of the North West Shelf but is noted here for its significance in the regional context.

8.7.4 Species Restricted to Bioregions of the North West Shelf

There are major environmental differences between the coastal bioregions of the North West Shelf. A significant change takes place at Cape Leveque, sharply demarcating the boundary between the Canning Bioregion to the south with its carbonate sediments and the Kimberley Bioregion to the north where the inner shelf is dominated by terrestrial sediments (Section 2.1.4.1 of Chapter 2). Corresponding with these ecological changes, there is an abrupt change at Cape Leveque in the composition of benthic shelf and shore biotic assemblages. This has been demonstrated for fish assemblages,[169–171] and there are many examples of species and lesser evolutionary units that are endemic to one side or the other of Cape Leveque.[12,171]

A species pair that illustrates this biogeographic boundary is seen in the volutid genus *Amoria*, with *A. praetexta* ranging from Exmouth Gulf to Cape Leveque, replaced by *A. turneri* in the Kimberley Bioregion and further north (Figure 8.20). The former species is a North West Shelf endemic, found only in the Canning, Eighty Mile Beach, and Pilbara Bioregions. The range of *A. turneri* extends beyond the North West Shelf across the Gulf to the northeast coast of Queensland and north to the south coast of Irian Jaya and the islands along the margin of the Banda Sea in eastern Indonesia.

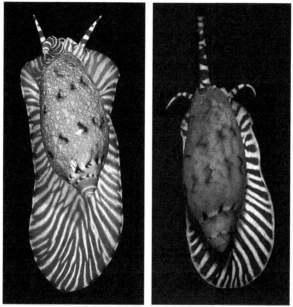

FIGURE 8.20 Two sister species of the volutid genus *Amoria*: A) *A. praetexta*, Dampier Archipelago, Pilbara Bioregion. B) *A. turneri*, Arafura Sea. The boundary between these two species is sharply demarcated at Cape Leveque. *Photos: Barry Wilson.*

Another example illustrating a similar distribution pattern is seen in the small, sand-dwelling biscuit urchin genus *Arachnoides* with *A. tenuis* in the south, replaced by *A. placenta* north of Cape Leveque.

8.7.5 Short-Range Endemism Within the Bioregions of the North West Shelf

Although marine species are often geographically widespread as a result of pelagic dispersal (see Section 8.3), short-range endemics are commonplace in benthic shelf and shore faunas. They are most common in groups that lack pelagic dispersal. One example is the volute species *A. macandrewi* which is known only from intertidal sand flats and shallow sublittoral sand banks that surround Barrow and the Montebello Islands (Figure 8.21). Others are *Amoria ellioti* restricted to sandy habitats in the 200 km stretch of coast between Port Hedland and Port Sampson and *A. dampieria* restricted to the inner shelf of West Pilbara coast between Barrow Island and the Dampier Archipelago.

Three other examples of volutid gastropods that are short-range endemics are *A. spenceriana*, *C. baili*, and *M. ashmorensis* that are confined

FIGURE 8.21 The volutid *Amoria macan-drewi* is a short-range endemic species found only on the shores and sandy shallows surrounding Barrow Island and the Montebello Islands in the west Pilbara. *Photo: Barry Wilson.*

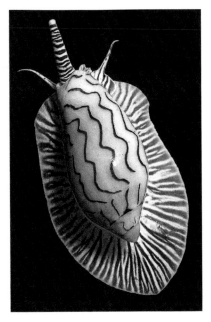

to Ashmore Reef and the shallow banks surrounding it on the margin of the Sahul Shelf.[40] These short-range endemic volutes appear to be relict, allopatric, daughter populations, derived from ancestors that were once more widely distributed on the North West Shelf and suffered range contractions as a result of sea level change.

There are some mollusc species that are endemic to the Kimberley Bioregion alone, for example, the upper-littoral littorinid *Tectarius rusticus* which is very common from Cape Leveque to Cape Londonderry. This is an allopatric member of a complex of species otherwise inhabiting the Central Indo-West Pacific region and is one of many examples of a North West Shelf endemic that is derived from a genus that flourishes in the East Indies Triangle.

In the cypraeid genus *Zoila*, there is a sister pair of "evolutionary units," inhabiting very restricted, contiguous, inner shelf areas between North West Cape and Peak Island at the western end of the North West Shelf. These cowries are sponge eaters that inhabit the filter-feeding communities on hard substrates. They have been given species or subspecies rank with the names *eludens* and *delicatura* (Figure 8.22A and B).[122] Interpretation of their restricted, contiguous ranges is problematical. They may represent post-LGM invasions of inner shelf habitats from separate shelf-edge populations of a common ancestor.

TABLE 8.7   Species of the Bivalve Family Mytilidae from the North West Shelf

Species	Habitat	Species status	Indo-West Pacific	Oceanic Shoals	Pilbara and Canning	Kimberley
Septifer bilocularis	Reef flat		✓	✓	✓	✓
Botula fusca	Reef flat—borer		✓	✓	✓	✓
Lithophaga teres	Reef flat—borer		✓	✓	✓	✓
Amygdalum watsoni	Deep lagoon—infaunal		✓	✓	✓	✓
Lioberus flavidus	Deep lagoon—infaunal		✓	✓	✓	✓
"Lithophaga" malacanna	Reef flat—borer		✓	✓	✓	✓
Modiolus auriculatus	Reef flat		✓	✓	✓	
"Lithophaga" hanleyana	Reef flat—borer		✓	✓		
"Lithophaga" nasuta	Reef flat—borer		✓		✓	
"Lithophaga" lima	Reef flat—borer in live coral		✓		✓	✓
"Lithophaga" simplex	Reef flat—borer in live coral		✓		✓	✓
Stavelia horridus	Reef flat		✓		✓	✓
Modiolus philippinarum	Reef flat		✓		✓	✓
Modiolus micropterus	Muddy infaunal		✓		✓	✓
"Lithophaga" obesa	Reef flat—borer		✓		✓	✓
Musculista glaberrima	Subtidal soft substrate —infaunal		✓		✓	✓

Continued

TABLE 8.7   Species of the Bivalve Family Mytilidae from the North West Shelf—cont'd

Species	Habitat	Species status	Indo-West Pacific	Oceanic Shoals	Pilbara and Canning	Kimberley
Brachidontes curvatus	Reef flat and mangal		✓		✓	✓
Modiolous proclivis	Sand flat—epifaunal		✓		✓	
Lioberus lignea	Sand flat—epifaunal		✓		✓	
"Lithophaga" divaricalx	Reef flat—borer		✓		✓	
"Lithophaga" laevigatus	Reef flat—borer in live coral		✓		✓	
"Lithophaga" lessepsianus	Reef flat—borer in live coral		✓		✓	
Modiolusia elongatus	Subtidal soft substrate—infaunal		✓		✓	
Modiolusia nitidus	Subtidal soft substrate—infaunal		✓		✓	
Arenifodiens vagina	Sand flat—infaunal		✓		✓	
Modiolus sp. indet.	Rocky shore—intertidal	NWS endemic			✓	✓
Brachidontes ustulatus	Rocky shore—intertidal	NWS and West Coast endemic			✓	✓
Lioberus sp. indet.	Sand flat—infaunal	Pilbara and West Coast endemic			✓	
			25	8	27	17

The species of this family live predominantly in benthic shelf and coastal habitats. Only seven species are found on the oceanic reefs and all but one of these also live in the coastal bioregions.

TABLE 8.8   Species of the Bivalve Family Cardiidae from the North West Shelf

Family Cardiidae— species	Species status	Indo- West Pacific	Oceanic Shoals	Pilbara and Canning	Kimberley
Corculum cardissum	Sandy reef flat	✓	✓		
Fragum fragum	Sandy reef flat	✓	✓		✓
Microfragum festivum	Sandy reef flat	✓	✓		
Acrosterigma orbita	Sandy reef flat	✓	✓		
Acrosterigma mendanaense	Sandy reef flat	✓	✓		
Fragum unedo	Intertidal sand flats	✓	✓	✓	✓
Acrosterigma alternatum	Sandy reef flat	✓	✓	✓	✓
Acrosterigma angulata	Sandy reef flat	✓	✓	✓	✓
Acrosterigma elongata	Sandy reef flat	✓	✓	✓	✓
Nemocardium lyratum	Subtidal sand	✓	✓	✓	✓
Fulvia australe	Subtidal sand	✓	✓	✓	
Acrosterigma dampierense	Subtidal sand	✓		✓	✓
Acrosterigma ? transcendens	Subtidal sand	✓		✓	
Ctenocardia fornicata	Subtidal sand	✓		✓	
Ctenocardia perornata	Subtidal sand	✓		✓	✓
Fragum (Lunulicardia) retusum	Intertidal sand flats	✓		✓	✓
Fragum (Lunulicardia) hemicardium	Intertidal sand flats	✓		✓	✓

Continued

TABLE 8.8 Species of the Bivalve Family Cardiidae from the North West Shelf—cont'd

Family Cardiidae— species	Species status		Indo- West Pacific	Oceanic Shoals	Pilbara and Canning	Kimberley
Fulvia aperta	Subtidal sand		✓		✓	✓
Laevicardium attenuatum	Subtidal sand		✓		✓	✓
Laevicardium biradiatum	Subtidal sand		✓		✓	✓
Nemocardium bechei	Subtidal sand		✓		✓	✓
Nemocardium exasperatum	Subtidal sand		✓		✓	
Nemocardium torresi	Subtidal sand		✓		✓	✓
Plagiocardium setosum	Subtidal sand		✓		✓	✓
Vepricardium multispinosum	Subtidal sand		✓		✓	✓
"Cardium" victor	Subtidal sand		✓		✓	✓
Acrosterigma reeveanum	Sandy reef flat	NWS endemic			✓	✓
Acrosterigma fultoni	Intertidal sand flats	NWS endemic			✓	✓
Acrosterigma wilsoni	Sandy reef flat	Pilbara endemic			✓	
Acrosterigma rosemariensis	Intertidal sand flats	Pilbara endemic			✓	
Acrosterigma dupuchense	Sandy reef flat	Pilbara endemic			✓	
Fragum erugatum	Intertidal sand flats	Pilbara and West Coast endemic			✓	
			26	11	27	20

The species of this family live predominantly in benthic shelf and coastal habitats with only five living on the oceanic coral reefs. All the endemics live in the coastal bioregions.

TABLE 8.9 Species of the Gastropod Family Strombidae from the North West Shelf

Family Strombidae—species	Habitat	Species status	Indo-West Pacific	Oceanic Shoals	Pilbara and Canning	Kimberley	Gulf of Carpentaria
Lambis chiragra	Oceanic reef—reef front		✓	✓			
Lambis crocata	Oceanic reef—reef flat		✓	✓			
Lambis truncata	Oceanic reef—lagoon		✓	✓			
Lambis scorpius	Oceanic reef—reef flat		✓	✓			
Strombus dentatus	Oceanic reef—lagoon		✓	✓			
Strombus fragilis	Oceanic reef—lagoon		✓	✓			
Strombus gibberulus	Oceanic reef—sandy reef flat		✓	✓			
Strombus latissumus	Oceanic reef—lagoon		✓	✓			
Strombus lentiginosus	Oceanic reef—sandy reef flat		✓	✓			
Strombus luhuanus	Oceanic reef—sandy reef flat		✓	✓			

Continued

TABLE 8.9 Species of the Gastropod Family Strombidae from the North West Shelf—cont'd

Family Strombidae—species	Habitat	Species status	Indo-West Pacific	Oceanic Shoals	Pilbara and Canning	Kimberley	Gulf of Carpentaria
Strombus microrourceus	Oceanic reef—sandy reef flat		✓	✓			
Strombus aurisdianae	Oceanic reef—sandy reef flat		✓	✓			
Strombus bulla	Oceanic reef—sandy reef flat		✓	✓			
Strombus erythrinus	Oceanic reef—sandy reef flat		✓	✓			
Strombus haemastoma	Oceanic reef—sandy reef flat		✓	✓			
Strombus pipus	Oceanic reef—sandy reef flat		✓	✓			
Strombus plicatus pulchellus	Oceanic reef—subtidal sand		✓	✓			
Strombus sinuatus	Oceanic reef—sandy reef flat		✓	✓			
Strombus labiatus	Oceanic reef—sandy reef flat		✓	✓			

Species	Habitat	Endemism					
Terebellum terebellum	Subtidal sand		✓			✓	
Rimella cancellata	Subtidal sand		✓	✓	✓		
Lambis lambis	Reef flat		✓	✓	✓		
Strombus mutabilis	Intertidal reef flat		✓	✓	✓		
Strombus dilatatus	Subtidal sand		✓		✓	✓	
Strombus vittatus	Subtidal sand		✓		✓		
Strombus campbelli	Subtidal and intertidal sand flat	Northern Australia endemic			✓	✓	✓
Strombus orrae	Sandy reef flat	Regional endemic			✓	✓	✓
Strombus iredalei	Subtidal and intertidal sand flat	Regional endemic			✓	✓	✓
			25	23	8	6	3

The species of this family are predominantly oceanic and live on the coral reefs of the shelf margin.

TABLE 8.10 Species of the Gastropod Family Volutidae from the North West Shelf (S.R. Means Short Range Endemic)

Volutidae—species	Habitat	Species status	Oceanic Shoals	West Coast	Pilbara and Canning	Kimberley	Arafura Sea	Gulf Carpentaria
Melo amphora	Intertidal and subtidal sand	Regional endemic Northern Australia	✓	✓	✓	✓	✓	✓
Cymbiola sophia	Subtidal sand	Regional endemic Northern Australia				✓	✓	✓
Cymbiola flavicans	Intertidal and subtidal sand	Regional endemic Northern Australia				✓	✓	✓
Volutoconus bednalli	Subtidal sand	Regional endemic Northern Australia				✓	✓	✓
Amoria turneri	Intertidal and subtidal sand	Regional endemic Northern Australia				✓	✓	✓
Cymbiola cymbiola	Subtidal sand	Regional endemic Northern Australia				✓	✓	
Amoria ryosukei	Subtidal sand	? S.R. Arafura Sea					✓	
Amoria kawamurai	Subtidal sand	? S.R. Arafura Sea					✓	
Amoria damoni	Intertidal and subtidal sand	Regional endemic Northern Australia		✓	✓	✓	✓	✓
Amoria grayi	Intertidal and subtidal sand	Regional endemic Canning-Pilbara, West and South coast		✓	✓			
Volutoconus hargraevesi	Intertidal and subtidal sand	Regional endemic Canning-Pilbara			✓			

Species	Habitat	Endemism		
Amoria praetexta	Intertidal and subtidal sand	Regional endemic Canning-Pilbara		✓
Amoria jamrachi	Intertidal and subtidal sand	Regional endemic Canning-Pilbara		✓
Cymbiola oblita	Intertidal and subtidal sand	Regional endemic Canning-Pilbara		✓
Amoria dampieria	Intertidal and subtidal sand	S.R. endemic Pilbara		✓
Amoria ellioti	Intertidal and subtidal sand	S.R. endemic Pilbara		✓
Volutoconus coniformis	Intertidal and subtidal sand	S.R. endemic Pilbara		✓
Volutoconus calcarelliformis	Deep sea muddy sand	S.R. endemic (deep sea)	✓	
Amoria hansenae	Deep sea muddy sand	S.R. endemic (deep sea)	✓	
Cymbiola baili	Intertidal and subtidal sand	S.R. endemic Ashmore	✓	
Melo ashmorensis	Intertidal and subtidal sand	S.R. endemic Ashmore	✓	
Amoria spenceriana	Intertidal and subtidal sand	S.R. endemic Ashmore	✓	
			5 3 9	5 7 9 6

All of these species are endemic to Northern Australia (although several also live in the Arafura Sea and eastern Indonesia) and many are short range endemics.

FIGURE 8.22   Two closely related "species" of the sponge-eating cowry genus *Zoila*, with contiguous and very restricted, short-range distributions: (A) *Z. delicatura*, 36 m off Peak Island; (B) *Z. eludens*, 23 m off North Muiron Island. *Photos: Peter Clarkson.*

References

Biogeographic Affinities

1. Hall R. Cenozoic tectonics of SE Asia and Australasia. In: Howes JVC, Noble RA, editors. *Proceedings of the international conference on petroleum systems of SE Asia and Australia.* Jakarta: Indonesian Petroleum Association; 1997. p. 47–62.
2. Hall R. The plate tectonics of Cenozoic SE Asia and the distribution of land and sea. In: Hall R, Holloway D, editors. *Biogeography and geological evolution of SE Asia.* Leiden: Backhuys Publishers; 1998. p. 99–131.
3. Sandiford M. The tilting continent: a new constraint on the dynamic topographic field from Australia. *Earth Planet Sci Lett* 2007;**261**:152–63.
4. Martin K. Die Fossilien von Java—Samml. Geology Reichsmus. Leiden, N.S., 1: 1-332, pis 1-45; 333-386, pis 46-50.123; 1891-1922.
5. Visser WA, Hermes JJ. Geological results of the exploration for oil in Netherlands New Guinea, carried out by the 'Nederlandsche Nieuw Guinea Petroleum Maatschappij' 1935-1960—Verhand. Geology Mijnbouw. Gen. Nederland & Kolon. *Geology* 1962;**20**: 11–265, 18 encl.
6. Beets C. Neogene Mollusca from the Vogelkop (Bird's Head Peninsula) West Irian, New Guinea. *Scr Geol* 1986;**82**:107–8.
7. Kauffman EG. Cretaceous Bivalvia. In: Hallam A, editor. *Atlas of palaeobiogeography.* Amsterdam: Elsevier; 1973. p. 353–83.
8. Zinsmeister WJ. Late Cretaceous-Early Tertiary molluscan biogeography of the southern circum-Pacific. *J Paleontol* 1982;**56**:84–102.
9. Fleming CA. The nomenclature of biogeographic elements in the New Zealand biota. *Trans R Soc N Z* 1963;**1**:13–22.
10. Darragh TA. Molluscan biogeography and biostratigraphy of the Tertiary of southeastern Australia. *Alcheringa* 1985;**9**:83–116.
11. Abbott RT. The genus *Strombus* in the Indo-Pacific. *Indo-Pacific Mollusca* 1960;**1**(2):33–146.
12. Wilson BR, Allen GA. Major components and distribution of marine fauna. In: Dyne GR, Walton DW, editors. *The fauna of Australia.* Canberra: Australian Government Publishing Service; 1987. p. 43–68, General Articles [chapter 3].
13. Taylor JD. Reef associated molluscan assemblages in the western Indian Ocean. *Sympos Zool Soc Lond* 1971;**28**:501–34.
14. Reid DG. *The littorinid molluscs of mangrove forests in the Indo-Pacific region. The genus Littoraria.* British Museum (Natural History) Publication No. 978. London; 1986.
15. Vermeij GJ. The dispersal barrier in the tropical Pacific: implications for molluscan speciation and extinction. *Evolution* 1987;**41**:1046–58.
16. Williams ST, Reid DG. Speciation and diversity on tropical rocky shores: a global phylogeny of snails of the genus *Echinolittorina. Evolution* 2004;**58**(10):2227–51.
17. Endean R. The biogeography of Queensland's shallow-water echinoderm fauna (excluding Crinoidea) with a rearrangement of the faunistic provinces of tropical Australia. *Aust J Mar Freshw Res* 1957;**8**:233–73.
18. Marsh LM, Marshall JI. Some aspects of the zoogeography of northwestern Australian echinoderms (other than holothurians). *Bull Mar Sci* 1983;**33**(3):671–87.
19. Rowe FWE. Preliminary analysis of distribution patterns of Australia's non-endemic, tropical echinoderms. *Proceedings of the 5th echinoderm conference, Galway, Ireland, 1984*; 1985.

Biodiversity

20. Veron JEN. *Corals in space and time. The biological evolution of the Scleractinia.* Sydney: UNSW Press; 1995, 321 pp.

21. Veron JEN, Marsh LM. Hermatypic corals of Western Australia. *Records of the Western Australian Museum.* Supplement No. 29; 1988. 136 pp.
22. Veron JEN. Reef-building corals. Part II. In: Berry PF, editor. *Faunal surveys of the Rowley Shoals, Scott Reef and Seringupulam Reef, North-western Australia.* Records of the Western Australian Museum. Perth: Western Australian Museum, Supplement No. 25; 1986. p. 27–35.
23. McKinney D. A survey of scleractinian corals at Mermaid, Scott and Seringapatam Reefs, Western Australia. In: Bryce C, editor. *Marine Biodiversity Survey of Mermaid (Rowley Shoals), Scott and Seringapatam Reefs.* Records of the Western Australian Museum. Perth: Western Australian Museum, Supplement No. 77; 2009. p. 105–43.
24. Veron JEN. Hermatypic corals of Ashmore Reef and Cartier Island. Part 2. In: Berry PF, editor. *Marine faunal surveys of Ashmore Reef and Cartier Island, North-Western Australia.* Records of the Western Australian Museum. Perth: Western Australian Museum, Supplement 44; 1993. p. 13–20, 1–91.
25. Griffith JK. The corals collected during September/October 1997 at Ashmore Reef, Timor Sea. Unpublished W.A. *Museum Report to Parks Australia;* 1997. 57 pp.
26. Richards Z, Beger M, Hobbs J-P, Bowling T, Chong-Seng K, Pratchett M. Ashmore Reef national nature reserve and Cartier Island Martine reserve. *Marine Survey 2009.* Report to Department of the Environment, Water, Heritage and the Arts; 2009. 81 pp.
27. Marsh LM. Scleractinian and other hard corals. In: Morgan GJ, editor. *Survey of the aquatic fauna of the Kimberley Islands and reefs, Western Australia.* Unpublished report Western Australian Museum. Perth: Western Australian Museum; 1992. p. 15–22.
28. Veron JEN. A biogeographic database of hermatypic corals. *Australian Institute of Marine Science Monograph Series.* No. 10; 1993. 433 pp.
29. INPEX. Biological and Ecological studies of the Bonaparte Archipelago and Browse Basin. Inpex, Perth: Western Australia; 2013. In press.
30. Marsh LM. Scleractinian corals of the Montebello Islands. In: Berry PF, Wells FE, editors. *Survey of the Marine Fauna of the Montebello Islands and Christmas Island, Indian Ocean.* Records of the Western Australian Museum. Perth: Western Australian Museum, Supplement No. 59; 2000. p. 15–19.
31. Griffith JK. Scleractinian corals collected during 1998 from the Dampier Archipelago, Western Australia. In: Jones DS, editor. *Marine Biodiversity of the Dampier Archipelago, Western Australia 1998-2002.* Records of the Western Australian Museum. Perth: Western Australian Museum, Supplement No. 66; 2004. p. 101–20.
32. Richards Z. Hard and soft corals. Section 6. In: *Gorgon Gas Development and Janz feed gas pipeline.* Coastal and marine baseline State and environmental impact report. Chevron Document G1-NT-REPX0001838; 2010.
33. Richards ZT, Rosser NL. Abundance, distribution and new records of scleractinian corals at Barrow Island and Southern Montebello Islands, Pilbara (Offshore) Bioregion. *J R Soc West Aust* 2012;**95**:155–165.
34. Cairns SD. Azooxanthellate Scleractinia (Cnidaria:Anthozoa) of Western Australia. *Rec West Aust Museum* 1998;**18**:361–417.
35. Veron JEN. Re-examination of the reef corals of Cocos (Keeling) Atoll. *Rec West Aust Museum* 1990;**14**(4):553–81.
36. Clark HL. Echinoderms from Australia. *Mem Museum Comp Zool Harvard* 1938;**55**:1–595, 28 pls.
37. Clark HL. The echinoderm fauna of Australia. *Publ Carnegie Inst* 1946;**566**:1–567.
38. Wilson BR. Notes on a brief visit to Seringapatam Atoll, Sahul Shelf, Western Australia. *Atoll Res Bull* 1985;**292**:83–99.
39. Wells FE. Molluscs of Ashmore Reef and Cartier Island. Part 4. In: Berry PF, editor. *Historical background, description of the physical environments of Ashmore Reef and Cartier Island and notes on exploited species;* 1993. p. 25–45; In: Berry PF, editor. *Marine Faunal Surveys of Ashmore*

Reef and Cartier Island, North-Western Australia. Records of the Western Australian Museum. Perth: Western Australian Museum, Supplement 44; 1993. p. 1–11, 1-91.

40. Willan RC. The molluscan fauna from the emergent reefs of the northernmost Sahul Shelf, Timor Sea—Ashmore, Cartier and Hibernia Reefs; biodiversity and biogeography. In: Russell BC, Larson CJ, Glasby RC, Willan RC, Martin J, editors. *Understanding the cultural and natural heritage values and management challenges of the Ashmore Region.* Records of the Museums and Art Galleries of the Northern Territory. Darwin: Museums and Art Galleries of the Northern Territory, Supplement 1; 2005. p. 51–81.

41. Wells FE, Slack-Smith SM. Molluscs. Part IV. In: Berry PF, editor. *Faunal surveys of the Rowley Shoals, Scott Reef and Seringapatam Reef.* Records of the Western Australian Museum 25. Perth: Western Australian Museum; 1986. p. 1–106.

42. Bryce CW, Whisson C. The macromolluscs of Mermaid (Rowley Shoals), Scott and Seringapatam Reefs, Western Australia. In: Bryce C, editor. *Marine biodiversity survey of Mermaid (Rowley Shoals), Scott and Seringapatam Reefs.* Records of the Western Australian Museum. Perth: Western Australian Museum, Supplement No. 77; 2006. p. 177–208.

43. Wells FE. Molluscs. Part IV. In: Morgan GJ, editor. *Survey of the aquatic fauna of the Kimberley Islands and Reefs, Western Australia.* Report of the Western Australian Museum Kimberley Island and Reefs Expedition, August 1991. Perth: Western Australian Museum, unpublished report No. UR8; 1992. p. 30–42.

44. Wells FE, Bryce CW. Molluscs. Part 8. In: Wells FE, Hanley JR, Walker DI, editors. *Marine biological survey of the Southern Kimberley, Western Australia, 1994.* Perth: Western Australian Museum, unpublished report No. UR286; 1995. p. 101–17.

45. Wells FE, Bryce CW. Molluscs. Part 7. In: Walker DI, Wells FE, Hanley JR, editors. *Marine biological survey of the Eastern Kimberley, Western Australia.* Perth: Western Australian Museum, unpublished report No. UR353; 1996. p. 54–67.

46. Slack-Smith SM, Bryce CW. Molluscs. In: Hutchins JB, Slack-Smith SM, Bryce CW, Morrison SM, Hewitt MA, editors. *Marine biological survey of the Muiron Islands and the eastern shore of Exmouth Gulf, Western Australia.* Unpublished report, Perth: Western Australian Museum, Western Australia; 1996. p. 64–106, 135.

47. Wilson BR. Molluscs. Tables 15-34. In: Bowman Bishaw Gorham, editor. *Survey of the intertidal shores of the eastern side of Barrow Island.* Report No. R16335, unpublished report West Australian Petroleum Pty Ltd. Perth: Western Australia; 1997. p. 36–7.

48. Wells FE, Slack-Smith SM, Bryce CW. Molluscs of the Montebello Islands. In: Berry PF, Wells FE, editors. *Survey of the Marine Fauna of the Montebello Islands and Christmas Island, Indian Ocean.* Records of the Western Australian Museum. Perth: Western Australian Museum, Supplement No. 59; 2000. p. 21–46.

49. Slack-Smith SM, Bryce C. A survey of the benthic molluscs of the Dampier Archipelago, Western Australia. In: Jones D, editor. *Marine biodiversity of the Dampier Archipelago Western Australia 1998-2002.* Records of the Western Australian Museum. Perth: Western Australian Museum, Supplement No. 66; 2004. p. 2121–245.

50. Taylor JD, Glover EA. Diversity and distribution of subtidal benthic molluscs from the Dampier Archipelago, Western Australia; results of the 1999 dredge survey (DA2/99). *Records of the Western Australian Museum.* Perth: Western Australian Museum, Supplement No. 66; 2004. p. 247–91.

Latitudinal Biodiversity Gradient

51. Hillebrand H. On the generality of the latitudinal diversity gradient. *Am Nat* 2004;**163**(2):192–211.

52. Hillebrand H. Strength, slope and variability of marine latitudinal gradients. *Mar Ecol Prog Ser* 2004;**273**:251–67.

53. Crame JA. Evolution of taxonomic diversity gradients in the marine realm: evidence from the composition of recent bivalve faunas. *Paleobiology* 2000;**26**:188–214.
54. Bellwood DR, Hughes TP. Regional-scale assembly rules and biodiversity of coral reefs. *Science* 2001;**292**:1532–4.
55. Willig MR, Kaufman DM, Stevens RD. Latitudinal gradients of biodiversity: patterns, process, scales, and synthesis. *Annu Rev Ecol Evol Syst* 2003;**34**:273–309.
56. Colwell RK, Lees DC. The mid-domain effect: geometric constraints on the geography of species richness. *Trends Ecol Evol* 2000;**15**:7–76.
57. Colwell RK, Rahbeck C, Gotelli N. The mid-domain effect and species richness patterns: what have we learned so far? *Am Nat* 2004;**163**:E1–E23.
58. Jablonski D, Hunt G. Larval ecology, geographic range, and species survivorship in Cretaceous Mollusks: organismic versus species-level explanations. *Am Nat* 2006;**168**(4):556–64.
59. Jablonski D, Roy K. Geographical range and speciation in fossil and living molluscs. *Proc R Soc B* 2003;**270**:401–6.
60. Martin PR, Bonier F, Tewksbury JJ. Revisiting Jablonski (1993): cladogenesis and range expansion explain latitudinal variation in taxonomic richness. *J Evol Biol* 2007;**20**:930–6.
61. Hughes TP, Bellwood DR, Connolly SR. Biodiversity hotspots, centres of endemicity and the conservation of coral reefs. *Ecol Lett* 2002;**5**:775–84.
62. Price ARG. Simultaneous "hotspots" and "coldspots" of marine biodiversity and implications for global conservation. *Mar Ecol Prog Ser* 2002;**241**:23–7.
63. Worm B, Lotze HK, Myers RA. Predator diversity hotspots in the blue ocean. *Proc Natl Acad Sci USA* 2003;**100**:9884–8.
64. Gray JS. Antarctic marine benthic diversity in a world-wide latitudinal context. *Polar Biol* 2001;**24**:633–41.
65. Ellingsen KE, Gray JS. Spatial patterns of benthic diversity: is there a latitudinal gradient along the Norwegian continental shelf? *J Anim Ecol* 2002;**71**:373–89.

Connectivity

66. Thorson G. Reproductive and larval ecology of marine bottom invertebrates. *Biol Rev* 1950;**125**:1–45.
67. Scheltema RS. Long-distance transport by planktonic larvae of shoal-water benthic invertebrates among central Pacific islands. *Bull Mar Sci* 1986;**39**:241–56.
68. Scheltema RS. On dispersal and planktonic larvae of benthic invertebrates: an eclectic overview and summary of problems. *Bull Mar Sci* 1986;**39**:290–322.
69. Roberts CM. Connectivity and management of Caribbean coral reefs. *Science* 1997;**278**:1454–7.
70. Crowder LR, Lyman S, Figueira WF, Priddy J. Source-sink population dynamics and the problem of siting marine reserves. *Bull Mar Sci* 2000;**66**:799–820.
71. Karunda-Arara B, Rose GA. Long-distance movement of coral reef fishes. *Coral Reefs* 2004;**23**:410–2.
72. Cowen RK, Sponaugle S. Relations between early life history traits and recruitment among coral reef fishes. In: Chambers RC, Trippel EA, editors. *Early life history and recruitment in fish populations.* London: Chapman and Hill; 1997. p. 423–49.
73. Leis JM, McCormack MI. The biology, behaviour and ecology of the pelagic stage of coral reef fishes. In: Sale PF, editor. *Coral Reef fishes, dynamics and diversity in a complex ecosystem.* San Diego: Academic Press; 2002. p. 171–99.
74. Bayne BL. Growth and the delay in metamorphosis of the larvae of *Mytilus edulis* (L.). *Ophelia* 1965;**2**:1–47.
75. Kritzer JP, Sale PF, editors. *Marine metapopulations.* San Diego: Academic Press; 2006.
76. Kritzer JP, Sale PF. The future of metapopulation science in marine ecology. In: Kritzer JP, Sale PF, editors. *Marine metapopulations.* San Diego: Academic Press; 2006. p. 517–29.

77. Kritzer JP, Sale PF. The metapopulation ecology of coral reef fishes. In: Kritzer JP, Sale PF, editors. *Marine metapopulations.* San Diego: Academic Press; 2006. p. 31–67.
78. Sale PF, Kritzer JP. Connectivity: what it is, how it is measured, and why it is important for management of reef fishes. In: Grober-Dunsmore R, Keller BD, editors. *Caribbean connectivity: implications for marine protected area management. Proceedings of a special symposium, 9-11 November 2006, 59th annual general meeting of the Gulf and Caribbean Fisheries Institute, Belize City, Belize;* 2008. p. 12–26.
79. Ovenden JR, Lloyd J, Newman SJ, Keenan CP, Slater LS. Spatial genetic subdivision between northern Australian and southeast Asian populations of *Pristipomoides multidens*: a tropical marine reef fish species. *Fish Res* 2002;**59**:57–69.
80. Underwood JN, Smith LD, Van Oppen MJH, Gilmour JP. Multiple scales of genetic connectivity in a brooding coral on isolated reefs following catastrophic bleaching. *Mol Ecol* 2007;**16**:771–84.
81. Veron JEN. Reef-building corals, Part II. In: Berry PF, editor. *Faunal surveys of the Rowley Shoals, Scott Reef and Seringapatam Reef, North-western Australia;* 1986. p. 27–35. Records of the Western Australian Museum. Supplement No. 25.
82. Harrison PL, Wallace CC. Reproduction, dispersal and recruitment of scleractinian corals. In: Dubinsky Z, editor. *Ecosystems of the world. 25: Coral Reefs.* Amsterdam: Elsevier Science; 1990. p. 133–207.
83. Miller K, Mundy C. Rapid settlement in broadcast spawning corals: implications for larval dispersal. *Coral Reefs* 2003;**22**:99–106.
84. Babcock RC, Heyward AJ. Larval development of certain gamete-spawning scleractinian corals. *Coral Reefs* 1986;**5**:111–6.
85. Wilson JR, Harrison PL. Settlement-competency periods of larvae of three species of scleractinian corals. *Coral Reefs* 1998;**131**:339–45.
86. Ostergaard JM. Spawning and development of some Hawaiian marine gastropods. *Pac Sci* 1950;**4**:75–115.
87. Wilson BR. Direct development in southern Australian cowries (Gastropoda; Cypraeidae). *Aust J Mar Freshw Res* 1985;**36**:267–80.
88. Reid DG. The comparative morphology, phylogeny and evolution of the gastropod family Littorinidae. *Philos Trans R Soc Lond B* 1989;**324**:1–110.
89. Reid DG. Morphological review and phylogenetic analysis of *Nodilittorina* (Gastropoda: Littorinidae). *J Molluscan Stud* 2002;**68**:259–81.
90. Scheltema RS. Evidence for trans-Atlantic transport of gastropod larvae belonging to the genus *Cymatium. Deep Sea Res* 1966;**13**:83–95.
91. Wilson BR. *Australian marine shells.* Perth: Odyssey Publishing; 1994, volume 2: 370 pp, 53 col. pls.
92. Thorson G. Reproductive and larval development of Danish marine bottom invertebrates, with special reference to the planktoniç larvae in the sound (Oresund). *Medd Komm Havundersog Kjob Plankton* 1946;**4**:343–67.
93. Marsh LM. Spawning of coral reef asterozoans coincident with mass spawning of tropical reef corals. In: Burke, et al. editors. *Echinoderm biology.* Rotterdam: Balkema; 1988. p. 187–92.
94. Emlet RB. Developmental mode and species geographic range in regular sea urchins (Echinodermata: Echinoidea). *Evolution* 1995;**49**(3):476–89.
95. Allen GR. *Altrichthys*, new genus of damselfish (Pomacentridae) from Philippines seas with descriptions of a new species. *Rev Fr Aquariol* 1999;**26**:23–8.
96. Brothers EB, Thresher RE. Pelagic duration, dispersal, and the distribution of Indo-pacific coral-reef fishes. In: Reaka ML, editor. *The ecology of Coral Reefs. NOAA symposia series for undersea research,* 3 **(1)**:Washington: NOAA; 1985. p. 53–69.
97. Sale PF. The ecology of fishes on coral reefs: what has the last decade taught us? *Fish Sci* 2002;**68**:113–8.

98. Thresher RE. Distribution, abundance, and reproductive success in the coral reef fish *Acanthochromis polycanthus*. *Ecology* 1985;**66**:1139–50.
99. Thresher RE, Colin PL, Bell LJ. Planktonic duration, distribution and population structure of western and central Pacific damselfishes (Pomacentridae). *Copeia* 1989;**2**: 420–34.
100. Ogden JC, Quinn TP. Migration in coral reef fishes: ecological significance and orientation mechanisms. In: McCleave JD, Arnold GP, Dodson JJ, Neil WH, editors. *Mechanisms and migrations in fishes*. New York: Plenum Press; 1984. p. 293–308.
101. Scheltema RS, Williams IP. Long distance dispersal of planktonic larvae and the biogeography of some Polynesian and Western Pacific Mollusks. *Bull Mar Sci* 1983;**33**:545–65.
102. Bohonack JA. Dispersal, gene flow, and populations structure. *Q Rev Biol* 1999;**74**:21–45.
103. Shanks AL, Grantham BA, Carr MH. Propagule dispersal distance and the size and spacing of marine reserves. *Ecol Appl* 2003;**13**(Supp.):S159–68.
104. Mortensen T. Contributions to the study of the development and larval forms of echinoderms. 111. *K Danske Vidensk Sesk Skr 9 Raekke* 1937;**7**(1):1–65.
105. Levin LA. Recent progress in understanding larval dispersal: new directions and digressions. *Integr Comp Biol* 2006;**46**:282–98.
106. Underwood JN, Travers MJ, Gilmour JP. Restricted connectivity in a coral atoll fish: genetic structure reveals restricted population connectivity in a coral reef fish; 2008.
107. Cowan RK, Lwiza KMM, Spongaugle S, Limouzy-Paris CB, Olson DB. Connectivity of marine populations: open or closed? *Science* 2000;**287**:857–9.
108. Jones GP, Milicich MJ, Emslie MJ, Lunow C. Self-recruitment in a coral reef fish population. *Nature* 1999;**402**:802–4.
109. Swearer SE, Caselle JE, Lea DW, Warner RR. Larval retention and recruitment in an island population of a coral-reef fish. *Nature* 1999;**402**:799–802.
110. Hughes TP, Baird AH, Dinsdale EA, Moltschaniwskyi NA, Pratchet MS, Tanner JE. Supply-side ecology works both ways: the link between benthic adults and larval recruits. *Ecology* 2001;**81**:775–84.
111. Ayre DJ, Hughes TP. Climate change, genotypic diversity and gene flow in reef-building corals. *Ecol Lett* 2004;**7**:273–8.
112. Underwood JN, Smith LD, Van Oppen MJH, Gilmour JP. Multiple scales of genetic connectivity in a brooding coral on isolated reefs following catastrophic bleaching. *Mol Ecol* 2007;**16**:771–84.
113. Hatcher BG. Coral reefs in the Leeuwin Current—an ecological perspective. *J R Soc West Aust* 1991;**74**:115–27. Smith L. Long-term effects of a severe bleaching event at an isolated reef system. In: Pattiaratchi C, editor. *Indian Ocean marine environment conference, Perth*. ; 2005.
114. Condie SA, Andrewartha JR. Circulation and connectivity on the Australian North West Shelf. *Continental Shelf Res* 2008;**28**:1724–39. Underwood JN, Smith LD, Van Oppen MJH, Gilmour JP. Ecologically relevant dispersal of a brooding and a broadcast spawning coral at isolated reefs: implications for managing community resilience. *Ecol Appl* 2009;**19**:18–29.
115. Underwood JN. Genetic diversity and divergence among coastal and offshore reefs in a hard coral depend on geographic discontinuity and ocean currents. *Evol Appl* 2009;**2**(2):222–33.
116. Underwood JN, Travers MJ, Gilmour JP. Subtle genetic structure reveals restricted connectivity among populations of a coral reef fish inhabiting remote atolls. *Ecol Evol* 2012;**2**(3):666–79.
117. Montaggioni LF. History of Indo-Pacific coral reefsystems since the last glaciation: development patterns and controlling factors. *Earth Sci Rev* 2005;**71**:1–75.
118. James NP, Bone Y, Kyser TK, Dix GR, Collins LB. The importance of changing oceanography in controlling late Quaternary carbonate sedimentation on a high-energy, tropical, oceanic ramp: north-western Australia. *Sedimentology* 2004;**51**(6):1179–205.

119. Van Oppen MJH, Bongaerts P, Underwood JN, Peplow LM, Cooper TF. The role of deep reefs in shallow reef recovery: an assessment of vertical connectivity in a brooding coral from west and east Australia. *Evol Appl* 2011;**20**(8):1647–60.

120. Radford B, Babcock R, Van Neil KP, Done T. Is there potential for cyclonic activity to enhance coral larval connectivity between reefs? A modelling approach. *Coral Reefs.* In press.

121. Chivas AR, Garcia A, van der Kars S, Couapel MJJ, Holt S, Reeves JM, *et al.* Sea-level and environmental changes since the last interglacial in the Gulf of Carpentaria, Australia: an overview; 2001.

122. Wilson BR, Clarkson P. *Australia's Spectacular Cowries. A review and field study of two endemic genera, Zoila and Umbilia.* San Diego: Odyssey Publishing; 2004, 396 pp.

123. Jones M, Torgersen T. Late Quaternary evolution of Lake Carpentaria on the Australia-New Guinea continental shelf. *Aust J Earth Sci* 1988;**35**(3):313–24.

124. Jennings JN. Some attributes of Torres Strait. In: Walker D, editor. *Bridge and barrier: the natural and cultural history of Torres Strait.* Canberra: Publ. BG/3 Australian National University; 1972.

125. Veron JEN. *Coral survey at selected sites in Arnhem Land.* Report to National Oceans Office; 2004. 214 pp.

126. Harris PT, Heap AD, Marshall JF, McCulloch M. A new coral reef province in the Gulf of Carpentaria, Australia: colonisation, growth and submergence during the early Holocene. *Mar Geol* 2008;**251**:85–97.

127. Last PR, Stevens JD. *Sharks and Rays of Australia.* 2nd ed. Melbourne: CSIRO Publishing; 2009.

128. van Herwerden L, Aspden WJ, Newman SJ, Pegg GG, Briskey L, Sinclair W. A comparison of the population genetics of *Lethrinus miniatus* and *Lutjanus sebae* from the east and west coasts of Australia: evidence for panmixia and isolation. *Fish Res* 2009;**100**(2):148–55.

129. Kay EA. Evolutionary radiations in the Cypraeidae. In: Taylor J, editor. *Origin and evolutionary radiations in the Mollusca.* Oxford, UK: Oxford University Press; 1996. p. 211–20.

130. Malcolm RJ, Pott MC, Delfos E. A new tectono-stratigraphic synthesis of the North West Cape area. *APEA J* 1991;**31**:154–76.

131. Allen AD. Outline of the geology and hydrogeology of Cape Range, Carnarvon Basin, Western Australia. In: Humphreys WF, editor. *The biogeography of Cape Range, Western Australia.* Records of the Western Australian Museum. Supplement No. 45; 1993. p. 25–39.

Speciation

132. Mayr E. Geographic speciation in tropical echinoids. *Evolution* 1954;**8**(1):1–18.

133. Palumbi SR. Reproductive isolation, genetic divergence, and speciation in the sea. *Annu Rev Ecol Syst* 1994;**25**:547–72.

134. Lessios HA, Kessing BD, Robertson DR. Massive gene flow across the world's most potent marine biogeographic barrier. *Proc R Soc Lond B* 1998;**265**:583–8.

135. Mora C, Sale PF. Are populations of coral reef fishes open or closed? *Trends Ecol Evol* 2002;**17**:422–8.

136. Paulay G, Meyer C. Diversification in the Tropical Pacific: comparisons between marine and terrestrial systems and the importance of founder speciation. *Integr Comp Biol* 2006;**42**(5):922–34.

137. Coyne JA, Orr HA. *Speciation.* Sunderland: Sinauer Associates; 2004.

138. Ridley M. *Species concepts and intraspecific variation.* Oxford: Blackwell; 2004.

139. Rocha LA, Robertson DR, Roman J, Bowen BW. Ecological speciation in tropical reef fishes. *Proc R Soc Lond B Biol Sci* 2005;**272**:573–9.

140. Butlin RK, Galindo J, Grahame JW. Sympatric, parapatric or allopatric: the most important way to classify speciation? *Philos Trans R Soc* 2008;**363**:2997–3007.

141. Hickerson MJ, Meyer C. Testing comparative phylogeographic models of marine vicariance and dispersal using a hierarchical Bayesian approach. *Evol Biol* 2008;**8**:11471–2148.
142. Prada C, Schizas V, Yoshioka PM. Phenotypic plasticity or speciation? A case from a clonal marine organism. *Evol Biol* 2008;**8**:47.
143. Ruber L, Van Tassell JL, Zardoya R. Rapid speciation and ecological divergence in the American seven-spined gobies (Gobiidae, Gobiosomatini) inferred from a molecular phylogeny. *Evolution* 2003;**57**(7):1584–98.
144. Duran S, Rutzler K. Ecological speciation in a Caribbean marine sponge. *Mol Phylogenet Evol* 2006;**40**:292–7.
145. Carlin DB, Budd AF. Incipient speciation across a depth gradient in a scleractinian coral. *Evolution* 2007;**56**(11):2227–42.
146. Cowan RK. Larval dispersal and retention and consequences for population connectivity. In: Sale PF, editor. *Coral Reef fishes, dynamics and diversity in a complex ecosystem*. San Diego: Academic Press; 2002. p. 149–70.
147. Swearer SE, Shima JS, Hellberg ME, Thurrold SR, Jones GP, Robertson DR, et al. Evidence of self-recruitment in demersal marine populations. *Bull Mar Sci* 2002;**70**:251–71.
148. Bierne N, Bonhomme F, David P. Habitat preference and the marine speciation paradox. *Proc R Soc Lond B* 2003;**270**(1522):1399–406.
149. Taylor MS, Hellberg ME. Genetic evidence for local recruitment of pelagic larvae in a Caribbean reef fish. *Science* 2003;**299**:107–9.
150. Planes S. Biogeography and larval dispersal inferred from population genetic analysis. In: Sale PF, editor. *Coral Reef Fishes. Dynamics and diversity in a complex ecosystem*. San Diego: Academic Press; 2002. p. 201–2230.
151. Leis JM. The pelagic stage of reef fishes: the larval biology of coral reef fishes. In: Sale PF, editor. *The ecology of fishes on Coral Reefs*. San Diego: Academic Press; 1991. p. 183–230.
152. Rocha LA, Bowen BW. Speciation in coral-reef fishes. *J Fish Biol* 2008;**72**(5):1101–21.
153. Paulay G, Meyer C. Diversification in the Tropical Pacific: comparisons between marine and terrestrial systems and the importance of founder speciation. *Integr Comp Biol* 2006;**42**(5):922–34.
154. Murray-Wallace CV, Beu AG, Kendrick GW, Brown LJ, Belperio AP, Sherwood JE. Palaeoclimatic implications of the occurrence of the arcoid bivalve *Anadara trapezia* (Deshayes) in the Quaternary of Australasia. *Quatern Sci Rev* 2000;**19**:559–90.
155. James NP, Bone Y, Kyser TK, Dix GR, Collins LB. The importance of changing oceanography in controlling late Quaternary carbonate sedimentation on a high-energy, tropical, oceanic ramp: north-western Australia. *Sedimentology* 2004;**51**(6):1179–205.

Distribution Patterns

156. Darragh TA. A revision of the Tertiary Volutidae (Mollusca: Gastropoda) of Southeastern Australia. *Mem Museum Vict* 1989;**49**(2):195–307.
157. Wilson BR. *Australian Marine Shells, Prosobranch Gastropods. Part 2*. Perth: Odyssey Publishing; 1994.
158. Dharma B. *Recent and Fossil Indonesian Shells*. Hackenheim, Germany: Conch Books; 2005, 424 pp.

Endemism

159. Cracroft J. Species diversity, biogeography and the evolution of biotas. *Am Zool* 1994;**34**:33–47.
160. Rosenzweig ML. *Species diversity in space and time*. Cambridge University Press; 1995.
161. Gaston KJ. Global patterns in biodiversity. *Nature* 2000;**405**:220–7.
162. Platnick NI. On areas of endemism. *Aust Syst Bot* 1991;**4**:xi–xii.

163. Stattersfield AJ, Crosby MJ, Long AJ, Wege DC. *Endemic bird areas of the World: priorities for conservation.* Cambridge, UK: Bird Life International; 1998.

164. Meyers N, Mittermeier RA, Mittermeier CG, da Fonseca GA, Kent J. Biodiversity hotspots for conservation priorities. *Nature* 2000;**403**:853–8.

165. Roberts M, McClean CJ, Veron JE, Hawkins JP, Allen GR, McAllister DE. Marine biodiversity hotspots and conservation priorities for tropical reefs. *Science* 2002;**295**: 1280–4.

166. Harvey MS. Short-range endemism amongst the Australian fauna: some examples from non-marine environments. *Invertebr Syst* 2002;**16**(4):555–70.

167. Jetz W, Rahbek C, Colwell RK. The coincidence of rarity and richness and the potential signature of history in centres of endemism. *Ecol Lett* 2004;**7**:1180–91.

168. Ceballos G, Rodriguez P, Medellin RA. Assessing conservation priorities in megadiverse Mexico: mammalian diversity, endemicity and endangerment. *Ecol Appl* 1996;**8** (1):8–17.

169. Hutchins JB. Biodiversity of shallow reef fish assemblages in Western Australia using a rapid censusing technique. *Rec West Aust Museum* 2001;**20**:247–70.

170. Travers MJ, Newman SJ, Potter IC. Influence of latitude, water depth, day *v.* night and wet *v.* dry periods on the species composition of reef fish communities in tropical Western Australia. *J Fish Biol* 2006;**69**:987–1017.

171. Travers MJ, Potter IC, Clarke KR, Newman SJ, Hutchins JB. The inshore fish faunas over soft substrates and reefs on the tropical west coast of Australia differ and change with latitude and bioregion. *J Biogeogr* 2010;**37**:148–69.

172. Marsh LM. Western Australian Asteroidea since H.L. Clark. *Thalassia Jugoslavica* 1976;**12** (1):213–25.

An Overview of the Historical Biogeography of the North West Shelf

Australia is unique among the nations of the world in being an island continent bordered by three oceans. This circumstance dominates the geopolitics of the country and is the primary factor in the biogeography of the marine plants and animals that inhabit its coastlines. It came about in the Early-Middle Tertiary when the Australian plate broke away from Gondwanaland, the Southern Ocean was formed, and the new Australian continent was moved by continental drift half into the equatorial zone. Its northwestern margin was thrust against Sundaland, a wide continental shelf of the Eurasian plate, and the northern margin against island arcs of the Pacific plate. This was one of the world's great vicariant events with profound biogeographic consequences. It thrust together two marine shelf faunas with different evolutionary histories and established the basic character of the modern Australian coastal marine biota with tropical elements in the north and a very different temperate flora and fauna on the southern shores.

Subsequent Australian marine biogeographic history has involved repeated expansions and contractions of the northern tropical elements along the west and east coasts. These fluctuations in latitudinal distribution were driven by climate and sea-level change and associated ocean circulation patterns, impeded in places by biogeographic barriers to dispersal. The outcome on the North West Shelf has been a biogeographic region with very high biodiversity and, in benthic shelf and shore habitats, a high level of regional endemism.

The contemporary biogeography of the Australian marine flora and fauna needs to be considered in this historical context of physical change in the environment, for many aspects of modern species' distribution patterns make no sense without it. It is also necessary to consider the modes and means of dispersal and connectivity because marine animals that have planktotrophic larval development respond to change and barriers to dispersal differently to those that lack pelagic larvae.

With the information now available, it is possible to construct a plausible, provisional overview of the evolution of the benthic shelf, shore, and reef marine fauna of the North West Shelf and interpret the observed distribution patterns in terms of evolutionary and biogeographic history. The following summary is presented as a series of hypotheses that fit observations on the physical environments and biota of the region and may be a stimulus for further studies.

9.1 BREAKUP OF THE SEA OF TETHYS AND ITS BIOGEOGRAPHIC CONSEQUENCES

In the Early Tertiary, the equatorial zone of the world was occupied by the pantropical Sea of Tethys, an open ocean system around the equatorial belt from the eastern Pacific coasts of the Americas, the Atlantic, and Indian Oceans to the western Pacific. On continental shelves and shores throughout this region, there was an immensely species-rich tropical marine fauna including the newly successful and diverse reef-building scleractinian corals. Dispersal of most coastal marine animals around this broad pantropical sea was uninterrupted except by mid-ocean distance barriers, especially the mid-Pacific.

A record of the Mid Tertiary tropical marine fauna of the Sea of Tethys is preserved in the extraordinarily rich fossil beds of the Central American, Mediterranean, and South East Asian regions. This was a time of global warming when the tropical zone of the world occupied a wider latitudinal span than it does today. It was also a time of extensive evolutionary radiation of marine species in shallow seas, resulting in perhaps the most diverse tropical marine fauna the world has seen. But like all else in this changeable world, it did not last. Two major changes in the global environment, occurring simultaneously, reduced the geographic scale of the Sea of Tethys and its biodiversity.

With fluctuations of great amplitude, the climate of the world "deteriorated," that is, it became progressively cooler through the Tertiary. The world's tropical belt contracted as a consequence. Through the same period, fragmentation of the south polar super-continent and subsequent continental drift repositioned the southern continental plates and this process broke up the Sea of Tethys into isolated parts. In the Miocene, the Mediterranean Sea became a remnant, narrowly open to the Atlantic in the west (Straits of Gibraltar) but completely shut off from the Indian Ocean.[1] Later, in the Pliocene, the Panamic isthmus formed the land bridge between the north and south American continents isolating the eastern Pacific from the Atlantic. The outcome of these tectonic events was the breakup of Tethys into the present four major tropical marine biogeographic realms—Eastern Pacific, Western Atlantic, Eastern Atlantic, and Indo-West Pacific—isolated from each other by land barriers.[2]

The subsequent fate of these four tropical marine realms was not uniform. Advent of the Panamic and West Asian land bridges and Late Tertiary changes of climate and ocean circulation patterns resulted in drastic changes in the marine faunas of the tropical realms. The eastern Atlantic tropical zone contracted to a narrow section of the West African coast between Cape Verde and the Congo, and there was mass regional extinction of many of the Tethyan elements in the coastal marine fauna (including coral reefs) of the Mediterranean and West Africa. The Eastern Pacific also lost much of its tropical biodiversity and coral reefs are now poorly developed there. The restricted West Atlantic tropical realm, essentially the region we now call the Caribbean with its diverse coral reefs, became a regional center of evolution although it never again achieved the levels of species-richness of Tethys. Nor did it equal the biodiversity that prevailed in the Indo-West Pacific.

Through these tectonic, climatic, and hydrological events, the tropical marine Tethyan biogeographic realm was reduced to a mere shadow of its former self, now best seen in the Indo-West Pacific realm. Biogeographer Sven Ekman[2] put it this way: "...the great climatic transformation of the middle and late Tertiary seems at first inexplicable. To one who looks on the world with the philosopher's eye it makes a profound impression of the sheer squandering of life ...."

The Indo-West Pacific realm remained by far the largest remnant of the Sea of Tethys. It retained its biodiversity and has grown to be a major center of ongoing evolution. Although all of the Tethyan families, most of the genera, and many of the species still survive in the region, and are common to both the Indo-West Pacific and the tropical Western Atlantic (Caribbean), many genera radiated vigorously which has led to the high species diversity of today, especially in the central East Indies Triangle (EIT). Nevertheless, the Indo-West Pacific realm was also impacted by the repositioning of the southern continental plates. The tropical Indian and western Pacific Oceans became partially separated by the thrust of Australasia (Australia and New Guinea) against South East Asia (Sundaland) with connections between the two oceans thereafter limited to narrow seaways between the islands of Wallacea (Figure 9.1C).

The point is that the modern Indo-West Pacific fauna is a remnant of the once much more widely spread, immensely diverse, and species-rich pantropical fauna of the Sea of Tethys. But at its center, the Indo-West Pacific biogeographic realm expanded its biodiversity and its area. The complex tectonic and topographic changes that took place in the Wallacean contact zone (see Section 8.1.1.1 in Chapter 8 and Figure 9.1C)

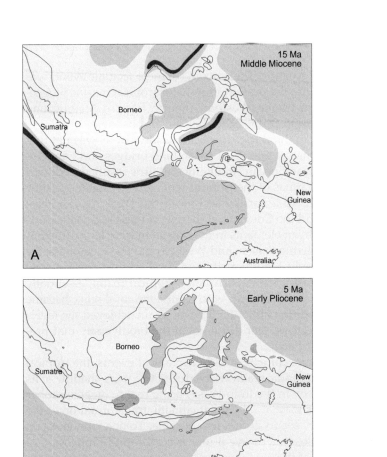

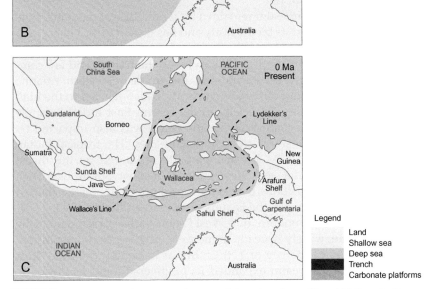

FIGURE 9.1 Contact of the Australasian and Eurasian plates. (A and B) Two shallow sea connections between the Australasian continental shelf and the Sunda Shelf: (A) Middle Miocene and (B) Early Pliocene. (C) The present situation—Wallacea, the transition zone between the Australasian and Sunda Shelves, bordered by Wallace's Line and the Weber-Lydekker Line. *After Ref. 10, amended from Ref. 3.*

provided many opportunities for vicariance and speciation[4] while the arrival of the Australian plate thrust against the center of the realm allowed the Indo-West Pacific shelf fauna to invade and occupy the northern continental shelves of this island continent. The sum outcome of all this activity was the positioning of the Australian northwestern margin proximal to the central part of the Indo-West Pacific Realm, the modern world's greatest center of marine biodiversity. This is the context in which we must consider the biogeographic history of the North West Shelf.

9.2 THE CONTACT PERIOD AND THE EXCHANGE OF SPECIES BETWEEN TWO CONTINENTS

In the Early Tertiary, the northern margin of Australia was a complex passive margin. At the end of the Oligocene, it collided with Sundaland[3–5] and the island arcs of the Western Pacific that were merging to form New Guinea.[6] In the Middle Tertiary, a shallow epicontinental sea, hundreds of kilometers wide, developed across northern Australia, broadly open at the ends to the Indian to the Pacific Oceans.[6] There was rapid marine sedimentation and development of a vast carbonate platform. At its western, Indian Ocean end, it joined the carbonate platform that extended southward along the northwestern margin of the continent with its chain of coral reefs (Figure 2.1 in Chapter 2). At its eastern end, it joined the carbonate shelf of the eastern Queensland coast along which the Great Barrier Reef developed. As a result, a U-shaped area of carbonate shelf was established across northern Australia, providing "new" tropical benthic and coral reef habitats that were occupied by immigrants from the species-rich marine fauna of Sundaland and the Western Pacific. There was direct connectivity between the Indian and Pacific Oceans and pulses of migration from this new center of evolution into higher latitudes on both sides of the island continent. It is likely that many benthic species ranged right across northern Australia from the west to east coasts. This was the situation through about 25 million years of the Late Tertiary.

That golden age of northern Australian marine diversity came to an end in the Pliocene when the central New Guinea orogenic mountain chain was uplifted. The northern part of the Papuan marine platform emerged to become the lowland plains of southern New Guinea. Terrestrial sediments poured off the mountains and covered the carbonate platform along the northern margin of Australia, building a vast area of emergent lowland that separated Australia and New Guinea and separated the Arafura Sea from the Coral Sea. Since then, this very wide land bridge has been briefly breached several times (the Torres Strait gap) during Quaternary periods of eustatic high sea level, but otherwise there has been little direct connectivity between the shallow marine faunas of western and eastern Australia.

It is unlikely that the process of colonization of the northern Australian continental shelves by Indo-West Pacific species was uniform across the taxa. As the Australian plate drifted northward during the Early Tertiary, the deep ocean gap between it and the Eurasian and Pacific plates slowly narrowed until in the Late Oligocene contact was made.[3–6] We may assume that, until then, the continental shelf of the northwestern Australian margin carried with it a diverse benthic fauna inherited from its higher latitude origins. What followed after contact was one of the most dramatic biogeographic events the world has seen, where two major marine continental shelf faunas with quite different evolutionary histories (Tethyan and Palaeoaustral) were thrust together.

Prior to actual contact of the two continental shelves of Australia and Sundaland, there may have been pelagic connectivity between them as Australia drifted "within range," depending on ocean circulation at that time. After contact and the establishment of shallow sea connections, connectivity would have increased by virtue of proximity and exchange of species with and without pelagic dispersal capacity. That situation prevailed until at least the Early Pliocene,[3–5] with possibly two migration routes for nonpelagic species through Wallacea between the Sunda Shelf and the northwestern corner of the Australasian shelf (Figure 9.1A and B).

There was probably never a land connection and, except for some "island hopping" across the Wallacean transition zone, terrestrial flora and fauna of the two continents remained separated.[3] But the shallow sea connection allowed free exchange of benthic shelf marine species including, for a time, species without pelagic larvae.

Although there are many rich marine fossil deposits of this period in the islands of Sundaland and western New Guinea, it may never be possible to reconstruct the detail of events of that time. Tectonic activity in this central part of the Indo-West Pacific realm was ongoing and intense, and there were very rapid changes in topography and distribution of land and sea.[3] Fragments of the several minor plates involved attached themselves to other land masses. Islands emerged, merged together, and disappeared again. Deep sea basins and troughs developed, expanded, and contracted. There must have been many opportunities for vicariant speciation as shelf areas were built, broken up and rearranged, or destroyed. In the warm and highly productive shallow sea conditions that prevailed, rapid speciation and cladogenesis ensued, producing the EIT center of marine biodiversity that we see today.

An important question is whether or not exchange of shallow sea taxa was in both directions. Was the outcome of this huge biogeographic event a mixing of two faunas or did one replace the other?

A biogeographic principle in such circumstances is that migration generally flows from the larger to the smaller area, from the most diverse to the least diverse faunas, and from low to higher latitudes. We know nothing about the biodiversity of northern Australian coastal waters as it approached Sundaland in the Oligocene (there is no northern Australian fossil record of that period), and the tropical continental shelves in the Miocene, with the additional vast area of the northern Australian epicontinental sea, could hardly be thought of as a small area. On biogeographic first principles, we might suppose that migration was mostly from the Central Indo-West Pacific region to northern Australia but the reverse was at least possible in some cases.

A key issue is the means by which dispersal across the Wallacean transition zone was accomplished. In the majority of shallow tropical marine species, the means of dispersal is provided by ocean circulation that controls directions and distances that may be traveled by pelagic larvae. The means of dispersal of benthic species that lack pelagic larvae is more ambivalent—for them shallow sea connections are required for they cannot float across deep-sea troughs.

9.2.1 Species with Pelagic Larvae

Proximity and local ocean circulation probably led initially to free, multidirectional exchange of reef and benthic shelf species with pelagic larvae between Wallacea and the northern part of the Australasian shelf (Arafura Shelf), and this situation probably continued intermittently through the Late Tertiary and Quaternary (Figure 8.1 in Chapter 8). The high biodiversity and close affinity of the modern reef and benthic shelf faunas of the North West Shelf and the Central Indo-West Pacific may be understood as an inheritance of that free-connectivity period.

There are indications that in regard to coral reef ecosystems, species exchange was unidirectional, from the Central Indo-West Pacific to northern margin of Australia. Reef-building scleractinian corals appeared on the North West Shelf in the Miocene (Chapter 4), undoubtedly along with the diverse assemblages of reef invertebrates and fishes that travel with them. The species were most likely the same as modern ones or their immediate ancestors. These reef animals are almost all planktotrophic with wide modern distributions in the Indo-West Pacific realm, and it is reasonable to assume that their introduction from Sundaland and the Western Pacific to both the west and east coasts of Australia was accomplished by pelagic dispersal as soon as suitable ocean circulation was established and the Australian plate was "within range."

The ocean currents that facilitated this process on the western side of the continent were the antecedents of the modern Indonesian Throughflow (ITF), that is, flow of Pacific water through the Indonesian archipelago into the Indian Ocean[7] and southward along the Arafura, Sahul, and Rowley Shelf margins, intermittently and subject to pulsing with climate change. This is the circumstance that prevails at present.

9.2.2 Species That Lack Pelagic Larvae

The circumstances applying to benthic species that lack pelagic larval dispersal capacity must have been different. These species disperse independently of ocean circulation but require shallow sea migration routes (continuous shorelines or benthic habitat at <200 m). Passable routes through the Wallacean transition zone were established in the Miocene and prevailed, in various arrangements, until the Pliocene (Figure 9.2A and B).

Consideration of direct-developing gastropods is informative in this context. Distribution patterns of the cowry genus *Zoila* and genera of the family Volutidae have been discussed earlier and warrant a little more attention.

Zoila is a benthic sponge-eating genus (Cypraeidae) that flourishes in both the tropical waters (three nominal species) and temperate waters (four nominal species) of Western and South Australia. There are fossil species in the Miocene of India, the Miocene/Pliocene of Irian Jaya (Figure 9.2C), and the Pliocene and Pleistocene of Java (Figure 9.2D and E). There are many Australian Tertiary fossil species, including one in the Middle Miocene at Barrow Island on the North West Shelf and two in the Late Pliocene of the South Coast (Figure 9.2F).

The Australian fossil species were reviewed by Darragh[8] who noted the existence of morphologically distinct western and eastern groups of species and assigned southern Australian cowries as old as Eocene to the genus. In this author's opinion, the morphological features (shell characters) of neither the eastern group nor the Eocene species demonstrate a close affinity with the three western fossils and the seven living species (which include the type species *Z. friendii*), and classification of them all in the same genus needs to be reconsidered.

Darragh[8] also noted that the Australian Eocene cowries were potential ancestors of *Zoila*. If that were the case, migration of the genus must have been from the Australian shelf across Wallacea to the Sunda Shelf, possible in the Early Miocene at the earliest when a shallow sea connection was made. I have previously expressed the alternative view, given the timing, that a Tethyan origin in the Indian Ocean region is more likely with migration in the opposite direction.[9] Genetic sequencing of living species has shown that the living *Zoila* is most closely related to the tropical Indian Ocean genus *Barycypraea*, and these two genera probably have a common ancestry.[10] *Barycypraea* has two remnant living species in the western Indian Ocean and a strong fossil record in the region from Oligocene to Recent,[11] with the two genera coexisting in Java in the Pliocene.

However, this argument applies only to the Miocene-Recent fossils that Darragh refers to as the "western group," meaning the living species and the Roe Plain and Barrow Island fossils. They and the fossils from Irian Jaya and Java share quite similar shell characters. They are unlike the older

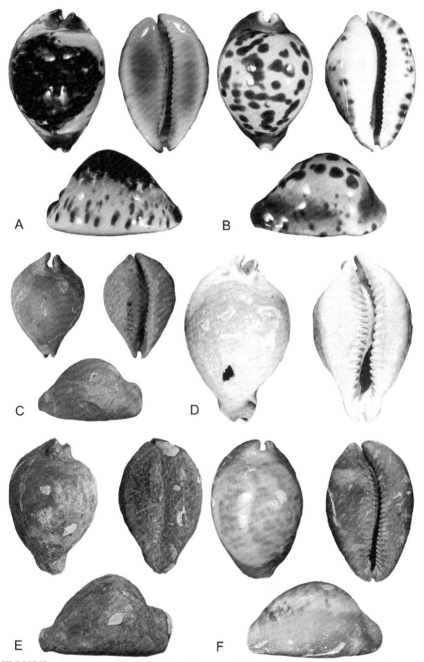

FIGURE 9.2   Recent and fossil shells of the genus *Zoila* (Cypraeidae): (A) *Z. eludens*. Living; Exmouth Gulf, North West Shelf, Western Australia. (B) *Z. venusta*. Living; Albany, South Coast, Western Australia. (C) *Z. caputavisensis*. Miocene/Pliocene, Vogelkop, West Irian. (D) *Z. gendinganensis*. Upper Pliocene, East Java. (E) *Z. kendengensis*. Pliocene/Pleistocene; East Java. (F) *Z. campestris*. Upper Pliocene, Roe Plain, South Coast, Western Australia. *Photos: (A), (B), and (F) by Barry Wilson; (C), (D), and (E) by Frank Wesselingh, Rijksmuseum of Geologie and Mineralogie, Leiden.*

southeast Australian Oligocene-Miocene species which appear to represent a separate, possibly autochthonous lineage or lineages. Pending resolution of this taxonomic issue, *Zoila* is here considered to have Tethyan ancestry, having migrated from the Sunda Shelf in the Middle Miocene.

An interesting contrast to *Zoila* with its tropical Tethyan origins is the unrelated genus *Umbilia* (Cypraeovulinae) which is also a direct-developing cowry but has a southern, high latitude Palaeoaustral origin. *Umbilia* species live on soft substrata and feed on benthic bryozoans. The genus has a strong Tertiary fossil record in southeastern Australia and several living species on the southern and eastern coasts. While from a northwestern source *Zoila* spread down the west coast to the south coast, *Umbilia* spread up the east coast from a southeastern source (Figure 9.3). These two genera exemplify the tropical Tethyan and temperate Palaeoaustral elements of the Australian marine fauna.

The fossil record of the volutid genus *Cymbiola* tells another story. It has many fossil and living species on either side of Wallacea. The place of origin is problematical but at some point, as was the case with *Zoila*, there must have been an ancestor on one side or other of the transition zone and there must have been at least one crossing.

A record of *Cymbiola* (*Cymbiolacca*) *pulchra* from the Middle Pliocene of Java[12] is of special interest because it and several congeners are conspicuous in the living fauna of eastern Queensland (Figure 8.16 in Chapter 8).[13] If identification of the Java fossil were correct, it would indicate a species (and subgenus) with a much wider Late Tertiary distribution than it has at present, either a Sunda or Australian origin, Late Tertiary migration through Wallacea in one direction or the other, and subsequent extinction except on the eastern coast of Queensland where it radiated into a number of polytypic species.

Melo is another volutid genus that presents challenges. There are two living species in the Central Indo-West Pacific region and two in northern Australia (plus another, *M. miltonis*, on the temperate south west and southern coasts of Australia). There are Pliocene records in Java of the two living northern Australian species, *M. amphora* and *M. umbilicatus*.[12] Again, crossing by these direct-developers between the Australian shelf and the Sunda Shelf is indicated but in which direction is problematical. And again, these two species became extinct on the Sunda Shelf but flourished on the shelves of northern Australia.

The volutid genera *Volutoconus* and *Amoria* have present distribution patterns that indicate vicariant speciation in northern Australia (Figures 8.17 and 8.18). The former genus has a fossil record in the Middle Miocene at Barrow Island on the North West Shelf, and both have fossil species in the Miocene/Pliocene of Irian Jaya. Neither is known from the Sunda Shelf, either living or fossil, and they may have originated on the northern Australian shelf from an unknown Tethyan or Palaeoaustral ancestor and failed to make the crossing to the other side.

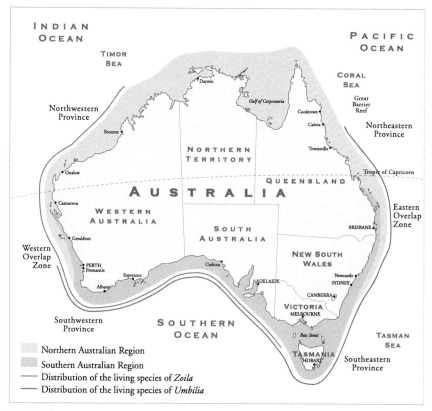

FIGURE 9.3   The known living distributions of the Australian cypraeid genera *Zoila* and *Umbilia* that have Tethyan and Palaeoaustral origins, respectively. *Source: Map 1, Ref. 9.*

Whatever the direction of crossing may have been, living and fossil distributions of direct-developers like the gastropod genera *Zoila*, *Cymbiola*, and *Melo* clearly indicate that migration across the transition zone did occur in the Miocene or Early Pliocene. But further tectonic activity in the Late Pliocene produced deep troughs separating the shelf areas around the islands of Moluku and Nusa Tenggara from the Arafura Shelf, and passage of species that lack a pelagic larval stage between the two regions is blocked at the present time.

9.3 TWO SIDES OF THE STORY

Biogeographically, the tropical western and eastern coasts of modern northern Australia have much in common. Both are clearly part of the present Indo-West Pacific with a common origin in that realm. In Section 9.2, the wide Miocene epicontinental sea that lays across the northern Australian margin, with shallow sea connections to the Sunda Shelf, was identified as the place of that common origin.

Today, many widely distributed tropical species range down both the western and eastern sides of Australia. Southerly flowing currents on both sides of the continent facilitate dispersal of species with pelagic larvae and maintain north-south connectivity but west-east connectivity is barred by the Australian-New Guinea land barrier (except during brief periods when the Torres Strait is open—see below). The Arafura Sea and Coral Sea have been isolated from each other since the Pliocene demise of the northern Australian sea.

Some common planktotrophic species range down one side of Australia but not the other (Section 8.5 in Chapter 8). Circumstances like this are likely to have arisen for historical reasons relating to successful or unsuccessful dispersal or regional extinction on one side or the other. Evidence of fragmentation of once continuous northern Australian distributions is seen in the presence of North West Shelf and eastern Queensland sister species pairs. These are a conspicuous biogeographic feature of the modern northern Australian region. Some examples are illustrated in Section 8.5 in Chapter 8. These distribution patterns prompt questions about places of origin, initial migration routes, and vicariant events that have led to range fragmentation and isolation of allopatric east and west populations.

Connectivity between the Indian and Pacific Oceans was unimpeded through the Miocene and Early Pliocene but blocked thereafter. Phases of connectivity in the Quaternary between the Arafura Sea and the Coral Sea through the Torres Strait during brief episodes of high sea level like the present were discussed in Section 8.4.5.2 in Chapter 8. There are examples of apparently recent migration by benthic shelf gastropods through an open Torres Strait. For example, *Cymbiola sophia* is a benthic species in the Arafura Sea and Gulf of Carpentaria that seems to have "escaped" through the Torres Strait onto the North Queensland coast (Figure 8.15 in Chapter 8). Another volute, *Amoria turneri,* has a similar distribution. These may have been recent events during the Holocene opening or an earlier Pleistocene one. Closure of the strait again by sea-level fall will fragment these species into isolated western and eastern populations. Quaternary sequences of opening and closure of the Torres Strait could be an explanation for some of the west-east sister species pairs.

However, a much more significant vicariant event was the demise of the Miocene epicontinental sea by sedimentation in the Pliocene, the process that built the broad land barrier between northern Australian and New Guinea and isolated the Arafura and Coral Seas. This event affected not only a few easy travelers. It split apart an entire, immensely species-rich benthic and reef fauna. Isolation since the Pliocene seems more in keeping with the degrees of morphological divergence and widely separated species pairs like the volutes like *Volutoconus hargreavesi/V. grossi* (Figure 8.17 in Chapter 8) and *Amoria grayi/A. maculata* (Figure 8.18 in Chapter 8).

Some provisional conclusions are

- The biota of the North West Shelf and Arafura Sea maintains pelagic connectivity with the Central Indo-West Pacific region. Separately, the biota of the Coral Sea and North Queensland maintains pelagic connectivity with the Western Pacific.
- The wide land connection between northern Australia and New Guinea has been a major biogeographic barrier responsible for separation of the marine faunas of northwestern and northeastern Australia since the demise of the northern Australian epicontinental sea in the Pliocene.
- At the present time and during previous Quaternary phases of high eustatic sea level, the Torres Strait has provided limited opportunities for direct connectivity and migration between the Australian northwest and northeast coasts but these periods have been brief and infrequent.
- Examples of northwestern and northeastern species pairs are probably results of vicariant speciation events in the Pliocene when the Australia-New Guinea land bridge was built, isolating the Arafura and Coral Seas from each other, although Quaternary origins relating to the closure of the Torres Strait is possible for some species.
- The marine faunas on the two sides of Australia have had a common Mid Tertiary origin, but at the present time, they occupy isolated biogeographic regions that are far apart, in different oceans, and they have little connectivity with each other.

This matter is discussed further in the following section dealing with the distinction of western and eastern biogeographical provinces.

9.4 THE MARINE FAUNA OF THE MODERN NORTH WEST SHELF

9.4.1 Biodiversity

The North West Shelf and its northern extension, the Arafura Shelf, share affinities and ongoing connectivity with the Indo-West Pacific biogeographic realm, especially the extremely biodiverse EIT at its center. This close affinity arose in the Late Oligocene or Early Miocene when the Australasian and Eurasian plates drew close and collided.

Recent survey work in the region has indicated that its biodiversity is very high. For example, over 400 species (69) genera of reef corals, 480 species (185 genera) of echinoderms, and about 2000 species of macromolluscs are presently recorded from the region, and these figures will increase as survey work progresses. Without doubt, the Kimberley Bioregion is a hotspot of coral diversity. The Pilbara coral fauna is almost as diverse, and it is also a major center of biodiversity in other invertebrate taxa. Details of the present state of knowledge may be found in Section 8.2 in Chapter 8.

As well as being a region of high marine diversity, the North West Shelf is a center of endemism in many invertebrate groups. While some genera radiated in the region, some became extinct in their ancestral range on the Sunda Shelf and the Wallacean transition zone. As a consequence of these evolutionary processes (cladogenesis on the North West Shelf and extinction on the Eurasian shelf), there are now many regionally endemic species and endemic genera. However, endemism is primarily in the benthic shelf and shore faunas and is lacking in the faunas of the oceanic reefs along the shelf margin.

The coral assemblages of reefs on the Kimberley coast are more species-rich in corals (more than 300 species) and more heterogeneous (69 genera) than the oceanic reefs offshore, but the species-richness of other invertebrates is low. The high species-richness of corals in the Kimberley is interpreted as a product of proximity to the ancestral source (the EIT) and the long history of the continental reef fauna on the adjacent Sahul Shelf transcending phases of climate and sea-level change. In the early Holocene during prolonged phases of lower sea level and closer proximity (Figure 8.1B and C in Chapter 8), connectivity at the demographic level probably occurred to a greater extent than it does today.

9.4.2 Two Ecological Elements in the Benthic Fauna of the North West Shelf

The proposition considered here is that offshore oceanic reef communities along the shelf margin not only live in different ecological conditions to those of communities of the coastal bioregions but also operate with different connectivity systems. Although there is considerable overlap, two different biogeographic elements may be recognized.

(i) A "permanent" benthic shelf and shore fauna, including assemblages of coastal fringing and patch reefs in the Kimberley and West Pilbara. This fauna may be regarded as "continental" in character. It has historic affinity with the EIT but now evolves largely independently except for episodic immigration of continental species that have pelagic larvae. It includes many endemic species and genera, especially prevalent in taxa that lack pelagic larvae.

(ii) An "itinerant" element of the oceanic reef fauna of the shelf margin (Oceanic Shoals Bioregion) that comes and goes through geological time as the ITF-Holloway Current system switches on and off. This fauna is replenished with immigrants of planktotrophic reef species in each phase of transgression following phases of global cooling. All of its species are found widely distributed on oceanic reefs of the Indo-West Pacific realm, and it contains no endemic species and genera. (The Ashmore complex is an exception where several short-range endemic mollusc species occur, evidently relict populations of North West Shelf benthic species, stranded in isolation by sea-level rise.)

There is around 70% commonality in coral species between the combined Kimberley and Pilbara Bioregions and those of the Oceanic Shoals Bioregion but much less in the case of other invertebrates. About 90 coral species known from Kimberley and Pilbara coastal reefs do not occur on the oceanic reefs. These species represent the continental coral fauna of the North West Shelf and this element is probably of long standing in the region. This hypothesis could be tested by further faunistic analysis of the coral reef assemblages on the North West Shelf.

At the level of functional group, the distinctiveness of the oceanic and coastal biota is striking. For example, an assemblage of widespread Indo-West Pacific reef-front predatory gastropods is found in this habitat on oceanic reefs throughout the Oceanic Shoals Bioregion (Section 4.5.3 (b); Figures 4.46–4.48 in Chapter 4). This suite of predators is virtually missing on fringing and patch reefs of the Kimberley Bioregion, represented only by odd individuals assumed to be stragglers delivered by ineffective connectivity. Reef-front invertebrate assemblages are very poorly developed in this Kimberley Bioregion. The few functional equivalents present are species that live on intertidal rocky shores as well as coral reefs of the coastal regions and include a high percentage of regional endemics.

Immigrant oceanic reef species are more frequent in shelf reef communities of the Pilbara (Offshore) Bioregion where fringing and patch reefs occur close to the shelf edge. This may occur by means of mixing of ITF and coastal water along the shelf margin (Montebello Islands). In this area, like the Kimberley, reef invertebrate assemblages are dominated by continental elements but they are enriched with a number of oceanic reef species.

The low species-richness of noncoral taxa on coral reefs of the Kimberley Bioregion needs explanation. Perhaps it relates to connectivity problems associated with the limited ability of planktotrophic reef species to settle in the extreme tidal range and the muddy conditions that prevail there, although neither seems to be a problem for many corals.

9.4.3 Environmental Change

Since the Mid Tertiary invasion by Indo-West Pacific species, three main kinds of environmental change have had impacts on the evolution of the benthic fauna of the northwestern continental shelf:

9.4.3.1 Sea-Level Change

Associated with periods of latitudinal expansion and contraction of the tropic zone, successive phases of sea-level rise and fall have resulted in transgression and regression of the sea across the continental margin. This process has resulted in repeated alternation of large areas between terrestrial coastal plain and marine benthic shelf habitat with profound ecological and evolutionary consequences, especially on what is now the inner and middle continental shelf. The most dramatic scale of this is evident in the Gulf of Carpentaria that was low woodland and grassy plain only a few thousand years ago (Figure 8.1B and C in Chapter 8) but which is now a vast area of shallow marine habitat (Figure 8.1A in Chapter 8). Similar change of this kind occurred on the Sahul Shelf and on the Rowley Shelf in the space overlaying the Canning Basin. Repeated local extinction and recolonization of such large areas of benthic habitat must have severely disturbed the regional composition of the benthic fauna.

Sea-level change has also resulted in repeated opening and closing of the Torres Strait, respectively connecting and isolating the northern Australian tropical shelf faunas of the eastern (Pacific) and western (Indian Ocean) coasts. The significance of this in regard to connectivity of the eastern and western marine faunas has been discussed earlier.

Perhaps the greatest effect of eustatic sea-level change on the biogeography of the North West Shelf has been the exposure of the shelf margin, especially in the north along the margin of the Sahul Shelf and Arafura Sea. In that area, low sea levels in the Quaternary exposed a shoreline likely to have been suitable for coral reef development in close proximity to the reefs and other benthic habitats of the EIT (Figure 8.1B and C in Chapter 8). Connectivity between the shores on either side of the Timor Trough and its northern extensions would have been intimate and demographic. With biodiversity of the order present in that part of eastern Indonesia, the Quaternary shores of the northern Arafura Sea and Sahul Shelf would have been a rich source of recruits for recolonization of populations further south that may have suffered more severely from lowered sea level.

9.4.3.2 Changes in Ocean Circulation

Through the Quaternary, circulation patterns in the eastern Indian Ocean alternated, also in response to oscillation of global climate. In particular, the ITF of warm water from the Western Pacific onto the North West Shelf, the primary force that established and maintains connectivity with the Indo-West Pacific realm, varies in intensity and switches on and off between warming and cooling phases of global climate.

9.4.3.3 Tectonism

Tectonism has affected the evolution of the marine fauna of the North West Shelf. In the south, Pliocene elevation of the Cape Range Peninsula created a partial biogeographic barrier separating the North West Shelf from the Dirk Hartog Shelf and, in the north, ongoing subsidence produced conditions favorable for development of coral reefs.

A tectonic process that has had fundamental impacts on the evolution of the North West Shelf fauna was the collision of the Indo-Australasian plate with that of Eurasia, the subduction of the margin of the former beneath the latter, and the development of the deep troughs that now separate the Sunda and Australasian shelves.

Pliocene tectonism in the New Guinea orogenic belt created the New Guinea mountain chain and subsequent erosion created the Australia-New Guinea land bridge that caused the demise of the Miocene northern Australian sea and isolated the Arafura and Coral Seas with profound biogeographic consequences.

9.4.4 Ongoing Pelagic Connectivity

At the geological timescale, connectivity between the EIT and the North West Shelf continues by means of pelagic larval dispersal, though episodically. It is most effective along the shelf margin but there is also mixing of ITF and coastal water on the outer shelf (Figure 8.9 in Chapter 8). This connectivity system relates only to species, like most coral reef invertebrates, that have planktotrophic larvae.

Pelagic connectivity functions during climatic periods, like the present, when the ITF operates with a seasonal flow of warm, low salinity, oligotrophic ITF water through the "gateways" between the islands of eastern Indonesia and down the length of the North West Shelf margin (the Holloway Current).

Pelagic connectivity has high relevance to the establishment and reestablishment of populations over geological time, but recruitment of planktotrophic reef species probably occurs as rare events and it is unlikely that it is effective at the short-term demographic scale. Two different scenarios may be recognized.

(a) Shelf-edge reefs (the oceanic elements)

During climatic periods of global cooling and low sea level, the ITF and Holloway Currents did not flow or were greatly weakened. During those phases of the Quaternary, pelagic connectivity with reefs further north was interrupted and shelf margin may have suffered repeated severe disturbance. If this was the case, we could regard the shelf-edge coral reefs and their biotic assemblages we see today as itinerant over roughly 100,000 year cycles, requiring periodic reconstruction from a northern source. Coral reefs may have persisted through the climate change and sea-level cycles along the margins of the Sahul and Arafura Shelves and provided refuge and the source of recruitment for the southern reefs.

(b) Benthic shelf species

When it flows, the ITF-Holloway Current system must also carry planktotrophic larvae of benthic and demersal species originating from northern areas of shelf habitat. Supply of these larvae to the North West Shelf may also switch on and off with phases of climate change. However, assemblages of these animals in southern benthic and shore habitats may not face the same cyclic phases of disturbance (at least to the same degree) as the shelf margin reefs. They may respond to change of benthic habitat brought on by sea-level change by spatial adjustments on the shelf. If this hypothesis were sound, we could regard the benthic shelf and shore faunas as more-or-less stable (though evolving) over geological time.

Inner shelf patch and fringing reefs of the Kimberley and Pilbara faced a different problem. They were episodically destroyed by low sea level, and rocky substrate refuges at lower levels on the outer shelf may have been lacking so that local spatial adjustments were not an option. For these reefs and their biotic assemblages, northern source populations for recruitment and recovery may have been required.

9.4.4.1 Speciation and Biogeographic Barriers

Vicariance associated with tectonic activity and "soft vicariant" events associated with climate and sea-level change, repeated at intervals through the Miocene, Pliocene, and the Quaternary, seem to have resulted in significant speciation on the North West Shelf manifest today in allopatric populations and sister species and subspecies on the shelf and adjacent bioregions. This applies to benthic shelf and shore species but not to the faunas of the offshore coral reefs.

The uplift of Cape Range, forming the Cape Range Peninsula in the Pliocene, created a partial biogeographic barrier impeding later dispersal of many marine benthic animals during phases of global warming and latitudinal expansion of the tropic zone. The barrier is most effective for species that lack pelagic dispersal and ineffective for planktotrophic species that "ride the Leeuwin Current." The result of this has been isolation of West Coast populations of nonplanktotrophic species from their North West Shelf ancestors and subsequent genetic divergence. Many cases of North West Shelf/West Coast species and subspecies pairs are evident.

The abrupt change of coastal landforms and inner shelf sediments at Cape Leveque marks a significant biogeographic ecological boundary, if not a partial barrier. There is a parallel abrupt change in composition of the benthic shelf and shore fauna. This environmental change also has been of long standing (and relates to fundamental differences in geology and climate). The relative importance of ecological and historical factors in the evolution of the differences in fauna on each side of it is a matter for further investigation.

9.5 BIOGEOGRAPHIC CLASSIFICATION OF THE NORTHERN AUSTRALIAN REGION

In Section 1.2 in Chapter 1, a three-level classification of biogeographic regions is proposed:

Realm (=Region in some biogeographic studies)
Province
Bioregion.

The position of the Northern Australian biogeographic region as a part of the Indo-West Pacific Realm is not in question. The Middle Tertiary tectonic events that produced this situation are discussed above. What needs to be considered is whether the marine biota of northern Australia is sufficiently uniform for it to be regarded as a province of the Indo-West Pacific Realm, or whether the differences between the western and eastern Australian tropical marine faunas and their histories meet the criteria (Section 1.2 in Chapter 1) for designation of separate northwestern and northeastern biogeographic provinces.

This requires reconsideration of the merits of the old notion of a northwestern Dampierian Province (Section 1.3.2 in Chapter 1), a biogeographic region that was originally defined as extending from the Torres Strait to Geraldton on the West Coast.

Modern taxonomic classifications have built in knowledge and interpretations of past evolutionary history and classifications of this nature are proven to be the most informative and useful. For the same reasons, biogeographical classifications should not be based only on assembled description of contemporary distribution patterns, no matter how refined the statistical analysis may be. To have maximum utility, they must have an historical component that includes consideration of past environmental changes, especially those that affect connectivity and the evolutionary processes of speciation and extinction. The four criteria for assessing biogeographical provinces (Section 1.2 in Chapter 1) take these considerations into account and the merits of the Dampierian Province are reviewed here in their context.

9.5.1 Geomorphic Characteristics and Geological and Climatic History

Hedley originally conceived the Dampierian Province as extending from Geraldton on the West Coast to the Torres Strait (Figure 1.4 in Chapter 1). Torres Strait, the eastern boundary of the province, is clearly delimited by geomorphic characteristics that are explainable in terms of geological and climatic history. The western boundary is set within the transition zone, midway between the northern tropical and southern temperate coastal faunas, at a latitude that includes the coral reefs of the Abrolhos Islands off Geraldton. It does not "clearly delimit" the western end of the province.

Until now, with a focus on the Australian coastline, Australian biogeographers have not considered a northern boundary to the Dampierian Province. But Figure 8.1B and C in Chapter 8 indicates that the coastline as well as the continental margin was continuous between Australia and New Guinea through most of the Quaternary. That being so, any northern Australian biogeographic region should embrace the Arafura Sea, including the Indonesian part of it. The northern boundary of the region should be the southern coast of New Guinea and the northwestern boundary, the shelf margin of the Timor Trough and its northern extensions.

Thus, the Dampierian Province (as here defined to include the Arafura Sea) is clearly delimited by geomorphology, oceanography, and habitats and biota at its eastern, northern, and northwestern boundaries but not so at the present southern boundary. Nevertheless, the first criterion is satisfactorily met.

9.5.2 Geographic Barriers: Impediments to Dispersal and Drivers of Vicariant Speciation

9.5.2.1 Eastern Barrier

The eastern boundary at Torres Strait is at present a potential route for marine organisms migrating between the Arafura Sea and Coral Sea and there is evidence of some passage through it. But through most of the Holocene and probably most of the Quaternary, there has been a very wide land connection between Australia and New Guinea that has been a total barrier to marine animals and there has been no direct route between the Arafura Sea and Coral Sea. The many examples of allopatric western and eastern species pairs may be attributed to the presence of that land barrier as an impediment to dispersal through most of the past several million years (since the Late Pliocene). Initially, the land barrier was created by a phase of sedimentation that followed the tectonic rise of the New Guinea mountain chain in the Pliocene. In the Quaternary, it has been episodically modified by eustatic sea-level change—widened during periods of low sea level and breached during periods of high sea level.

9.5.2.2 Northern Barrier

The New Guinea land mass is, of course, a total geographic barrier blocking contact with the Western Pacific. The northwestern side of the province is open and in close proximity to the Central Indo-West Pacific region but blocked to species that lack pelagic larvae by the deep-sea troughs. Those troughs represent a partial biogeographic barrier brought about by tectonism.

9.5.2.3 *Southwestern Barrier*

There are no geographic barriers at the southwestern end of the Dampierian region. Rather, there is an ongoing (in geological time) process of repeated advance and retreat of the tropical marine fauna on the Western Australian coastline in response to climate change. However, in the Pliocene, the emergence of the Cape Range Peninsula produced a partial biogeographic barrier that may impede the passage of species that lack pelagic larvae but not passage of planktotrophic species. The southwestern boundary of the region might be better located at North West Cape.

Thus, for most of the Quaternary, the Dampierian Province has been bounded in the east, north, and northwest by barriers to dispersal and the second criterion is met. The southern boundary is open but if the province was redefined as ending at the Cape Range Peninsula there would be a partial barrier to dispersal.

9.5.3 Species Endemism

Endemism on the North West Shelf is discussed in Section 8.7 in Chapter 8 based on species data presented in Tables 8.7–8.10.

The purpose here is to compare the northwestern fauna (west of Torres Strait) with the fauna of North Queensland. Accordingly, species that are endemic to the North West Shelf, Arafura Shelf, and the Gulf of Carpentaria are considered together as a percentage of the whole fauna.

In this study, it has not been possible to determine the total number of endemic marine species of the province. (That would be a worthy project for a team of taxonomic experts in the diverse groups.) Rather, an indicative estimate is offered (Table 9.1), based on four groups of molluscs with which this author is familiar (Mytilidae, Cardiidae, Strombidae, Volutidae). The selected families are representative of diverse reproductive and dispersal strategies.

Overall species endemicity of the three families that have planktotrophic larvae averages 12%. As noted previously, there are no endemic species in the invertebrate fauna of the oceanic coral reefs which operate on a different connectivity system to the inshore benthic fauna. When the species that are strictly shelf margin, oceanic coral reef inhabitants are removed, the average endemicity of benthic shelf and shore planktotrophic species is 18%. Overall regional endemicity in the direct-developing Volutidae is 86%. Three species (14%) of the western volutes are also found in Queensland, two of them apparently recent immigrants from the Arafura Sea via the Torres Strait.

Provisionally, on the basis of the four chosen families of molluscs, the requirement of criterion 3 for 10% or more regional endemicity is met. A broader sample of invertebrate taxa is needed to confirm this result.

TABLE 9.1   Endemism in Four Families of Molluscs in the North West Shelf, Arafura Sea, and Gulf of Carpentaria (See Tables 8.7–8.10 in Chapter 8 for Details)

Family	Larval Development	No. Species	Endemics	% Regional Endemicity
Mytilidae (Bivalvia)	Planktotrophic	28	3	11
Cardiidae (Bivalvia)	Planktotrophic	32	6	19
Strombidae (Gastropoda)	Planktotrophic	28	2	7
Volutidae (Gastropoda)	Direct (capsular)	22	19	86

9.5.4 Faunistic Dissimilarity to Neighboring Areas

This criterion requires comprehensive lists of the regional fauna and the faunas of adjacent regions. Compiling such lists is presently a work in progress involving several of the Australian museums. Nevertheless, inspection of Tables 8.7–8.10 in Chapter 8 provides a provisional conclusion. Much of the difference between the western and eastern faunas of these four molluscan families is accounted for by the endemics. However, there are species in the western fauna that are not endemic to that region (being found also in the Central Indo-West Pacific region) but do not occur in Queensland. Conversely, there are some species in Queensland that are also found in the Western Pacific but not on the western coast of Australia. This leads to a degree of dissimilarity that is greater than that indicated by the endemics alone but it would be pointless to attempt to estimate a figure until a better database is available. Further study is needed.

Consideration of the distribution data of these four molluscan families against the criteria for designating a biogeographic province leads to the provisional conclusion that the Dampierian region meets most of the criteria for designation as a Province but its boundaries need amendment. The southern boundary needs to be moved to the Cape Range Peninsula. In the north, the province should include the whole of the Arafura Sea and Gulf of Carpentaria so that the northern and northwestern boundaries are the southern coast of New Guinea and the Australasian shelf margin bordering the Timor Trough and its northern extensions.

This recognizes the western and eastern sides of the Australian continental shelves as distinct biogeographic provinces of the Indo-West Pacific Realm. The former faces northwest into the Indian Ocean and arches closely around the island arc of Moluku and Nusa Tenggara with direct connectivity with the Central Indo-West Pacific region. The latter faces east into the Western Pacific Ocean with which it has direct connectivity. While there was direct connectivity between the two in the Middle

Tertiary, that no longer applies (except for brief, episodic, potential connectivity through the Torres Strait).

If provincial names are required, the old name Dampierian Province is available for the northwestern province. Further study is needed on a much wider range of invertebrate and fish taxa to determine the degree of dissimilarity between the faunas on the two sides and confirm (or dispute) this provisional conclusion.

9.6 SUBDIVISIONS OF THE DAMPIERIAN PROVINCE

As redefined above, the continental shelf of the Dampierian Province comprises four geomorphic parts:

1. The Rowley Shelf—North West Cape to the Londonderry Rise.
2. The Sahul Shelf—Londonderry Rise to longitude 130°E.
3. The Arafura Shelf—longitude 130°E to the southern coast of New Guinea.
4. The Gulf of Carpentaria.

Each of these sections has distinctive geomorphic characteristics that are outcomes of different geological (tectonic) and climatic histories (eustatic sea-level change). Large areas of the Rowley, Sahul, and Arafura Shelves have been exposed during Quaternary periods of low sea level. The Gulf of Carpentaria has been entirely exposed during those periods or reduced to a shallow lagoon east of the Arafura Sill. The Arafura Shelf represents a remnant of the Middle Tertiary epicontinental sea that once lay across the northern continental margin of Australia.

It is not proposed that these units should be regarded as biogeographic regions although further consideration of their ecological histories and faunistic assemblages might lead to that conclusion.

The IMCRA Meso-scale Bioregions, proposed by the Australian Government (Section 1.3.3 in Chapter 1), are based on a characterization of coastal geomorphology and habitats and an assumption that these may be used for management purposes as surrogates for changes in species and ecosystem distribution around the coast. Chapters 3–7 of this book broadly follow that arrangement in regard to the distribution of major habitat types along the coast of the North West Shelf.

9.7 LAST WORDS

This study of the biogeography of the North West Shelf has not finished in the expected place. It began with a concept of the North West Shelf as an Australian biogeographic region adopted from the geological literature. It has ended with the conclusions that the key to understanding its biogeo-graphic history lies further north in Indonesian waters and in the history of the Late Tertiary epicontinental sea of northern Australia, and that the North West Shelf is a geomorphic unit within a larger biogeographic province.

The initial concept envisaged the North West Shelf as a geomorphic unit with a discrete southern end (Cape Range Peninsula and the narrow Ningaloo Shelf) and a more or less arbitrary northern boundary at longi-tude 130°E at the eastern end of the Sahul Shelf. Yet, the continental mar-gin does not end there. It continues in an arc along the margin of the Arafura Sea to the southern shore of New Guinea. For more than 25 million years in the Late Tertiary, it was also continuous with the vast carbonate platform that lay across the wide continental margin of northern Austra-lia. That northern end of the Australasian continental shelf is where the biogeographic history of the modern marine fauna of the region began. Though now reduced in area, it may continue to have an important influ-ence through a role as reservoir of recruits for recolonization of southern populations in times of stress.

There are two diagrams in this book that together summarize the biogeo-graphic history of the North West Shelf. Figure 8.1 in Chapter 8 illustrates the close proximity of the northern end of the shelf to the EIT at different stages of the Holocene climate cycle. It relates to the matter of origins and the initial establishment of a tropical center of marine biodiversity in the region. Figure 8.9 in Chapter 8 illustrates the primary ocean circulation along the North West Shelf margin as it is today. This relates to the matter of ongoing connectivity and the colonization and recolonization of southern reef and possibly benthic shelf habitats from a northern source.

Figure 8.1A in Chapter 8 shows the northern Australian coastline as it is today, and it is natural that Australian biogeographers should have focused on that. But that alignment of land and sea is a transient thing and has existed for only a few thousand years. It is misleading in terms of historical biogeography (both land and sea).

Figure 8.1B in Chapter 8 provides a much more meaningful image of how things were through most of recent geological time, that is, since the Late Oligocene. It shows the shape of the northwestern continental margin as it was through much of the Holocene. Its curving northern end embraces the modern arc of the eastern Indonesian Moluku and Nusa Tenggaran provinces along the eastern edge of Wallacea, the transition zone between the Australasian shelf and the Sunda Shelf. This is the world's center of maximum marine biodiversity, and the history of exchange of marine species between it and the Australian shelf is the root of historical biogeography of Australia's northern margin.

Figure 8.9 in Chapter 8 illustrates the primary ocean circulation along the North West Shelf margin as it is today, providing, intermittently, the means for the carriage of pelagic larvae of reef and benthic shelf animals and the colonization and recolonization of reef and benthic shelf habitats from a northern source. An assumption may be made that intermittent flow of Pacific water through the eastern Indonesian archipelago and onto the North West Self has operated since contact of the Australasian and Eurasian plates in the Late Oligocene. Conceptually, it provides a model for the colonization of the North West Shelf by marine species from the Central Indo-West Pacific region, their spread southward in times of global warming, and the maintenance of their populations in times of climatic stress.

This illustration emphasizes the mixing of the ITF-Holloway Currents' oceanic, oligotrophic water with the coastal water of the outer shelf. From this, the notion is drawn that this current/dispersal system fosters the episodic delivery of pelagic larvae over geological time and establishment of oceanic reef species on reefs along the shelf margin and the Pilbara (offshore) Bioregion but largely bypasses the Kimberley, Canning, and Eighty Mile Beach Bioregions. Migration of benthic shelf invertebrates, also from a northern source, may have been a slower process with much more localized, self-sustaining connectivity processes. These populations may not depend on recruitment or recolonization from ancestral northern source populations.

There is evidence of *in situ* speciation and radiation of benthic shelf species on the North West Shelf, especially in genera that lack pelagic larvae. Given an origin in the extremely biodiverse EIT, these evolutionary processes have produced a regional marine fauna of high biodiversity with a high degree of endemicity.

Finally, the notion of a Northern Australian Province, as a biogeographic subunit of the Indo-West Pacific Realm, has credibility for the period from the Late Oligocene until the Early Pliocene, but tectonic and sedimentation events in the Pliocene split it apart into northwestern (Dampierian) and northeastern parts that warrant recognition as modern, functionally independent provinces with a common biogeographic origin in the Central Indo-West Pacific Realm.

References

1. Reuter M, Pillar WE, Harzhauser M, Mandic O, Berning B, Rögl F, et al. The Oligo-/Miocene Qom Formation (Iran): evidence for an early Burdigalian restriction of the Tethyan Seaway and closure of its Iranian gateways. *Int J Earth Sci* 2009;**98**(3):627–50.
2. Ekman S. *Zoogeography of the sea.* London: Sidgwick and Jackson; 1953, 417 pp.
3. Hall R. Cenozoic reconstructions of SE Asia and the SW Pacific: changing patterns of land and sea. In: Metcalf I, editor. *Faunal and floral migrations and evolution in SE Asia-Australasia.* Lisse: Swets and Zetlinger; 2001.
4. Hall R. Cenozoic tectonics of SE Asia and Australasia. In: Howes JVC, Noble RA, editors. *Proceedings of the international conference on petroleum systems of SE Asia and Australia.* Jakarta: Indonesian Petroleum Association; 1997. p. 47–62.

5. Hall R. The plate tectonics of Cenozoic SE Asia and the distribution of land and sea. In: Hall R, Holloway D, editors. *Biogeography and geological evolution of SE Asia*. Leiden: Backhuys Publishers; 1998. p. 99–131.

6. Pigram CJ, Davies PJ, Feary DA, Symonds PA. Tectonic controls on carbonate platform evolution in southern Papua New Guinea: passive margin to foreland basin. *Geology* 1989;**17**:199–202.

7. Kennett JP, Keller G, Srinivasan MS. Miocene planktonic foraminiferal biogeography and paleogeographic development of the Indo-West Pacific region. In: Kennett JP, editor. *The Miocene ocean: paleogeography and biogeography. Geological Society of America Memoir*, vol. 163. Boulder, USA: Geological Society of America; 1985. p. 197–236.

8. Darragh T. A revision of the Australian fossil species of *Zoila* (Gastropoda: Cypraeidae). *Mem Museum Vic* 2011;**68**:1–28.

9. Wilson BR, Clarkson P. *Australia's spectacular cowries. A review and field study of two endemic genera, Zoila and Umbilia*. San Diego: Odyssey Publishing; 2004, 396 pp.

10. Meyer CP. Molecular systematics of cowries (Gastropoda: Cypraeidae) and diversification patterns in the tropics. *Biol J Linn Soc* 2003;**79**:401–59.

11. Schilder M, Schilder FA. A catalogue of living and fossil cowries. *Memoires Institute Royal des Sciences Naturellesde BelBelgique (2nd series)* 1971;**85**:1–246.

12. Dharma B. *Recent and fossil Indonesian shells*. Hackenheim, Germany: Conch Books; 2005, 424 pp.

13. Wilson BR. *Australian marine shells.* , vol. 2. Perth: Odyssey Publishing; 1994, 370 pp., 53 col. pls.

Index

Note: Page numbers followed by *f* indicate figures and *t* indicate tables.

Printed and bound by CPI Group (UK) Ltd, Croydon, CR0 4YY

08/05/2025

01864871-0002